T0314034

Annals of Mathematics Studies

Number 45

ANNALS OF MATHEMATICS STUDIES

Edited by Robert C. Gunning, John C. Moore, and Marston Morse

1. Algebraic Theory of Numbers, *by* HERMANN WEYL
3. Consistency of the Continuum Hypothesis, *by* KURT GÖDEL
11. Introduction to Nonlinear Mechanics, *by* N. KRYLOFF *and* N. BOGOLIUBOFF
16. Transcendental Numbers, *by* CARL LUDWIG SIEGEL
17. Problème Général de la Stabilité du Mouvement, *by* M. A. LIAPOUNOFF
19. Fourier Transforms, *by* S. BOCHNER *and* K. CHANDRASEKHARAN
20. Contributions to the Theory of Nonlinear Oscillations, Vol. I, *edited by* S. LEFSCHETZ
21. Functional Operators, Vol. I, *by* JOHN VON NEUMANN
22. Functional Operators, Vol. II, *by* JOHN VON NEUMANN
24. Contributions to the Theory of Games, Vol. I, *edited by* H. W. KUHN *and* A. W. TUCKER
25. Contributions to Fourier Analysis, *edited by* A. ZYGMUND, W. TRANSUE, M. MORSE, A. P. CALDERON, *and* S. BOCHNER
27. Isoperimetric Inequalities in Mathematical Physics, *by* G. POLYA *and* G. SZEGO
28. Contributions to the Theory of Games, Vol. II, *edited by* H. W. KUHN *and* A. W. TUCKER
29. Contributions to the Theory of Nonlinear Oscillations, Vol. II, *edited by* S. LEFSCHETZ
30. Contributions to the Theory of Riemann Surfaces, *edited by* L. AHLFORS *et al.*
31. Order-Preserving Maps and Integration Processes, *by* EDWARD J. MCSHANE
33. Contributions to the Theory of Partial Differential Equations, *edited by* L. BERS, S. BOCHNER, *and* F. JOHN
34. Automata Studies, *edited by* C. E. SHANNON *and* J. MCCARTHY
36. Contributions to the Theory of Nonlinear Oscillations, Vol. III, *edited by* S. LEFSCHETZ
37. Lectures on the Theory of Games, *by* HAROLD W. KUHN. In press
38. Linear Inequalities and Related Systems, *edited by* H. W. KUHN *and* A. W. TUCKER
39. Contributions to the Theory of Games, Vol. III, *edited by* M. DRESHER, A. W. TUCKER *and* P. WOLFE
40. Contributions to the Theory of Games, Vol. IV, *edited by* R. DUNCAN LUCE *and* A. W. TUCKER
41. Contributions to the Theory of Nonlinear Oscillations, Vol. IV, *edited by* S. LEFSCHETZ
42. Lectures on Fourier Integrals, *by* S. BOCHNER. In press
43. Ramification Theoretic Methods in Algebraic Geometry, *by* S. ABHYANKAR
44. Stationary Processes and Prediction Theory, *by* H. FURSTENBERG
45. Contributions to the Theory of Nonlinear Oscillations, Vol. V, CESARI, LASALLE, *and* LEFSCHETZ
46. Seminar on Transformation Groups, *by* A. BOREL *et al.*
47. Theory of Formal Systems, *by* R. SMULLYAN

CONTRIBUTIONS TO THE THEORY OF NONLINEAR OSCILLATIONS

VOLUME V

J. P. LaSALLE
J. JARNIK
J. KURZWEIL
J. CRONIN
J. K. HALE
L. CESARI
A. STOKES
L. MARKUS
G. REEB
P. MENDELSON
J. ANDRÉ
P. SEIBERT
C. COLEMAN
R. P. de FIGUEIREDO

EDITED BY

L. Cesari, J. LaSalle, and S. Lefschetz

PRINCETON, NEW JERSEY

PRINCETON UNIVERSITY PRESS

1960

L. C. Card 60-12225

Printed in the United States of America

PREFACE

This volume is the fifth in a series of contribution to the Nonlinear Oscillations published by the Annals of Mathematics Studies. As usual there are quite a variety of topics covered. The first paper by LaSalle is on a central problem on optimal control — a problem which is agitating mathematicians both in the United States and in the Soviet Union. This is followed by a contribution from Harnik and Kurzweil on a generalized differential equation depending on a parameter. This is a continuation of a series of papers published on the topic by Kurzweil. We have next an elegant paper by Jane Cronin on the perturbation theorem of Poincaré. This is followed by two papers by Hale and one by Cesari. The first paper concerns a discussion of the critical points for a linear system with periodic coefficients. The second paper of Hale deals with the important question of the stability of periodic solutions of periodic and autonomous differential systems. Simple criteria for stability are proved by the use of a convergent process of successive approximations. The paper by Cesari gives existence theorems for periodic solutions of periodic and autonomous differential systems satisfying a Lipschitz condition. These theorems are proved by application of fixed point theorems in functional spaces and from this approach an interpretation is given of the method of successive approximations used in the paper by Hale. Stokes, in his paper, applies a fixed point theorem in linear spaces to the study of certain stability and boundedness properties of the solutions of differential equations. By the use of such methods he is able to reduce the study of systems to that of related first order equations. The next paper by Markus deals with a curious application of abstract algebra to the classification of linear differential systems. The paper by Reeb deals with a property of the totality of bounded solutions of certain dynamical systems. Mendelson's contribution deals with the stable motions in a neighborhood of critical points in topological dynamics. It is connected with a topological method introduced and promoted by Ważewski. The joint paper by André and Seibert deals with piecewise continuous differential equations. It is part of the noteworthy work done by those authors in organizing the earlier work of Mrs. Flugge-Lotz. Coleman's paper is a study of asymptotic stability of a 3-dimensional system where the leading terms are a homogeneous function of degree m. In the final paper of this volume Figueiredo studies the self-sustained oscillations of a second order system.

All but the papers by Jane Cronin, Jarnik and Kurzweil, Reeb, and Figueiredo were carried out at RIAS and supported in part by contract number AF 49(638)-382.

The work by J. Cronin and A. Stokes was supported in part by the Office of Naval Research.

Editors

L. Cesari
J. P. LaSalle
S. Lefschetz

CONTENTS

CONTRIBUTIONS TO THE

THEORY OF

NONLINEAR OSCILLATIONS

VOLUME V

I. THE TIME OPTIMAL CONTROL PROBLEM

J. P. La Salle

§1. INTRODUCTION

It has been an intuitive assumption for some time that if a control system is being operated from a limited source of power and if one wishes to have the system change from one state to another in minimum time, then this can be done by at all times utilizing properly all of the power available. This hypothesis is called "the bang-bang principle." Bushaw accepted this hypothesis and in 1952 in [2] showed for some simple systems with one degree of freedom that of all bang-bang systems (that is, systems which at all times utilize maximum power) there is one that is optimal. In 1953 I made the observation in [7] that the best of all bang-bang systems, if it exists, is then the best of all systems operating from the same power source. More recently, more general results have been obtained by Bellman, Glicksberg, and Gross in [1], and later — but seemingly independently — by Krasovskii in [6] and Gamkrelidze in [4].

We confine our attention to the time optimal problem for control systems which are linear in the sense that the elements being controlled are linear and as a function of time the steering of the system enters linearly. The differential equation for such systems is

$$\dot{x}(t) = A(t)x(t) + B(t)u(t) + f(t) \quad , \tag{1}$$

where x and f are n-dimensional vector functions ($x(t)$ is the state of the system at time t), A is an $(n \times n)$ matrix function, and B is an $(n \times r)$ matrix function. The vector equation (1) represents the system of differential equations

$$\frac{dx_i(t)}{dt} = \sum_{j=1}^{n} a_{ij}(t)x_j(t) + \sum_{k=1}^{r} b_{ik}(t)u_k(t) + f_i(t), \quad i = 1, \ldots, n \ .$$

Our ability to control the system lies in the freedom we have to choose the "steering" function u. We assume that the admissible steering functions are piecewise continuous (or measurable) and have components less than 1 in absolute value $(|u_k(t)| \leq 1)$. The number 1 is selected for convenience. We could just as well assume $-a_k \leq u_k(t) \leq b_k$ where a_k and b_k are any positive numbers. Given an initial state x_0 and a moving particle $z(t)$, the problem of time optimal control is to hit the particle in minimum time. Let $x(t, u)$ be the solution of (1) satisfying $x(0) = x_0$. An admissible steering function u^* is <u>optimal</u> if $x(t^*, u^*) = z(t^*)$ for some $t^* > 0$ and if $x(t, u) \neq z(t)$ for $0 < t < t^*$ and all admissible u.

In [1] Bellman, Glicksberg, and Gross consider the system

$$\dot{x}(t) = Ax(t) + Bu(t) \tag{2}$$

and restrict themselves to the problem of starting at x_0 and reaching the origin in minimum time. The $(n \times n)$ matrix A is constant and its characteristic roots are assumed to have negative real parts. B was assumed to be a constant non-singular $(n \times n)$ matrix, and it is this restriction that is quite unrealistic. Gamkrelidze in [4] considered the same problem, removed the restriction that B be non-singular, and showed for systems which are later in this paper called "normal" the existence and uniqueness of an optimal steering function. The form of the optimal steering function is the same as that given in [1], and for normal systems one can conclude that the optimal steering is bang-bang $(|u_k(t)| = 1)$. Krasovskii, in [6] considers the more general control system (1) and the more general control problem of hitting a moving particle. Using results of Krein on the L-problem in abstract spaces, he proves the existence of an optimal steering function for, what are called in this paper, "proper" control systems. We prove the same existence theorem without restricting ourselves to proper systems. Krasovskii makes the further assertion that the optimal steering function is unique and simple examples show that this is not true even for proper systems.

To date, therefore, the most general bang-bang principle has been proved by Gamkrelidze for normal control systems of the form (2) and for the special problem of reaching the origin in minimum time. We shall show for the general control problem that if an admissible steering function can bring the system from one state to another in time t then there is a bang-bang steering function that can do the same thing in the same time. This extends my result in [7] and at the same time extends the bang-bang principle. This does not mean that all optimal steering functions are bang-bang. Even for proper control systems there are examples where the objective can be reached in minimum time using a steering function which during part of the time has some zero components. As in the special case considered by Gamkrelidze, it is shown that normal systems have unique optimal steering functions and for such systems there is a true bang-bang principle: the only way to reach the objective in minimum time is by bang-bang steering. In Theorem 5 a result is established that should be of some practical importance in the synthesis problem, which is the problem of determining the optimal steering u^* as a function of the state of the system. This result shows that for some control systems optimal steering can be determined by what amounts to running the system backwards. In Section 4 we discuss the controllability properties of proper control systems. The examples have been given at the end of the paper in Section 5, and they serve to illustrate the ideas and the application of the theory.

§2. THE GENERAL PROBLEM

The problem, described in the Introduction, for the system (1) of reaching a moving particle in minimum time will be called the <u>general problem</u>. For the control system (1) the state $x(t, u)$ of the system at time t is given by

$$(3) \qquad x(t, u) = X(t)x_0 + X(t)\int_0^t Y(\tau)u(\tau)d\tau + X(t)\int_0^t X^{-1}(\tau)f(\tau)d\tau \quad .$$

$X(t)$ is the principal matrix solution of $\dot{X}(t) = A(t)X(t)$, and $Y(\tau) = X^{-1}(\tau)B(\tau)$. We want at some time t to have $x(t, u) = z(t)$; that is, to have

$$(4) \qquad w(t) = \int_0^t Y(\tau)u(\tau)d\tau \quad ,$$

where

$$w(t) = X^{-1}(t)z(t) - x_0 - \int_0^t X^{-1}(\tau)f(\tau)d\tau \quad .$$

We assume throughout that $A(t)$, $B(t)$, $f(t)$ and $z(t)$ are continuous for $0 \leq t < \infty$. The proof of the following lemma was pointed out to me by L. Pukanszky.

LEMMA 1. Let M be the set of all real-valued measurable functions $\alpha(t)$ on $[0, 1]$ with

$|\alpha(t)| \leq 1$. Let M^o be the subset of functions in M with $|\alpha(t)| = 1$. Let $y(t)$ be any n-dimensional function in $L^1([0, 1])$. Define

$$K = \left\{ \int_0^1 \alpha(t)y(t)dt; \quad \alpha \in M \right\}$$

and

$$K^o = \left\{ \int_0^1 \alpha^o(t)y(t)dt; \quad \alpha^o \in M^o \right\} \quad .$$

Then K^o is closed and $K = K^o$.

PROOF. For each measurable set E in $[0, 1]$ define

$$\mu(E) = \int_E y(t)dt \quad .$$

Let R_μ denote the range of this vector measure. Let $c_E(t)$ be the characteristic function of E, and let $\alpha^o(t) = 2c_E(t) - 1$. Then $\alpha^o(t) \in M^o$, and each $\alpha^o(t) \in M^o$ can be so represented. Then clearly $K^o = 2R_\mu - \bar{y}$, where

$$\bar{y} = \int_0^1 y(t)dt \quad .$$

By Liapounov's Theorem ([8], [5]) R_μ is closed and convex, and therefore K^o is closed and convex. Let

$$z = \int_0^1 \alpha(t)y(t)dt$$

be any n-vector in K. What we wish to show is that z is a limit of vectors in K^o, and hence is in K^o. It will then follow that $K^o = K$. Let $\beta(t) = \frac{1}{2}(\alpha(t) + 1)$ and $\bar{z} = \frac{1}{2}(z + \bar{y})$. Note that $0 \leq \beta(t) \leq 1$ and

$$\bar{z} = \int_0^1 \beta(t)y(t)dt \quad .$$

Define

$$\bar{z}_m = \sum_{j=1}^{m} \frac{j}{m} \int_{E_j} y(t)dt$$

where

$$E_j = \left\{ t;\ \frac{j-1}{m} < \beta(t) \le \frac{j}{m} \right\}$$

Then

$$|\bar{z} - \bar{z}_m| = \sum_{j=1}^{m} \int_{E_j} \left(\frac{j}{m} - \beta(t)\right) y(t)dt \le \frac{1}{m} \sum_{j=1}^{m} \int_{E_j} |y(t)|dt = \frac{1}{m}\int_0^1 |y(t)|dt \quad .$$

which shows that $\bar{z}_m \longrightarrow \bar{z}$ as $m \longrightarrow \infty$. Letting

$$F_j = \bigcup_{i=j}^{m} E_i \quad ,$$

we obtain

$$\bar{z}_m = \frac{1}{m} \sum_{j=1}^{m} \int_{F_j} y(t)dt \quad .$$

Since R_μ is convex, it follows that $\bar{z}_m \in R_\mu$; therefore $z_m = 2\bar{z}_m - \bar{y} \in K^o$ and $z_m \longrightarrow z$ as $m \longrightarrow \infty$. This completes the proof.

For our purposes here we wish to restate this lemma in the following more general fashion.

LEMMA 2. Let Ω be the set of all r-dimensional vector functions $u(\tau)$ measurable on $[0, t]$ with $|u_i(\tau)| \le 1$. Let Ω^o be the subset of functions $u^o(\tau)$ with $|u_i^o(\tau)| = 1$. Let $Y(\tau)$ be any $(n \times r)$ matrix function in $L^1([0, t])$. Define

$$A(t) = \left\{ \int_0^t Y(\tau)u(\tau)d\tau; \quad u \in \Omega \right\}$$

and

$$A^o(t) = \left\{ \int_0^t Y(\tau)u^o(\tau)d\tau; \quad u^o \in \Omega^o \right\} \quad .$$

Then $A^o(t)$ is closed and $A(t) = A^o(t)$.

PROOF. Let $y^1(\tau), y^2(\tau), \ldots, y^r(\tau)$ be the column vectors in $Y(\tau)$, and define

$$A_j(t) = \left\{ \int_0^t y^j(\tau)u_j(\tau)d\tau; \quad u \in \Omega \right\}$$

and

$$A_j^o(t) = \left\{ \int_0^t y^j(\tau)u_j^o(\tau)d\tau; \quad u^o \in \Omega^o \right\}$$

Since

$$\int_0^t Y(\tau)u(\tau)\, d\tau = \sum_{j=1}^r \int_0^t y^j(\tau)u_j(\tau)d\tau \quad ,$$

we see that

$$A(t) = A_1(t) + A_2(t) + \dots + A_r(t)$$

and

$$A^o(t) = A_1^o(t) + A_2^o(t) + \dots + A_r^o(t) \quad .$$

By Lemma 1 $A_j(t) = A_j^o(t)$ for each $j = 1, \dots, r,$ and hence $A(t) = A^o(t)$. Each $A_j^o(t)$ is bounded, and it follows from Lemma 1 that each $A_j^o(t)$ is compact. Therefore $A^o(t)$ is closed, and this completes the proof. It is convenient to point out also two elementary properties of convex sets that we will use later.

LEMMA 3. Consider sets $A(t)$ of E_n, $t \leq t^*$, with the following properties:

a. Each $A(t)$ is convex.

b. Corresponding to each $\varepsilon > 0$ there is a $\delta > 0$ such that $d(p, A(t)) < \varepsilon$ for each $p \in A(t^*)$ and all $t^* - \delta < t < t^*$. [$d(p, A(t))$ is the distance of p from $A(t)$.]

Then, if q is in the interior of $A(t^*)$, there is a $t_1 < t^*$ such that q is an interior point of $A(t_1)$.

PROOF. Let q be in the interior of $A(t^*)$, and let N be a neighborhood of q of radius $\varepsilon > 0$ inside $A(t^*)$. Suppose for each $t < t^*$ that q is not an interior point of $A(t)$. Then for each $t < t^*$ there is a support plane P_t through q such that there are no points of $A(t)$ on one side of P_t ([3]). Because of the neighborhood N about q that lies in $A(t^*)$ we see that for each $t < t^*$ there is a point p in $A(t^*)$ whose distance from $A(t)$ is at least ε. This contradicts (b).

LEMMA 4. Let M be a convex set in E_n containing the origin with the property: given any number K and any non-zero vector η in E_n there is a vector y in M such that $(\eta, y) > K$. Then $M = E_n$.

PROOF. Suppose that M is not E_n. Then there is some non-zero μ that is not in M, and hence $a\mu$ is a boundary point of M for some $a \geq 0$. Then $a\mu$ has a support plane P ([3]). Let η be a non-zero vector normal to P and directed toward the side of P that contains no points of M. Then (η, y) is bounded above for y in M — a contradiction. Therefore $M = E_n$.

Turning our attention back to the control problem, we see that the set Ω in Lemma 2 is the set of admissible steering functions and Ω^0 is the set of bang-bang steering functions. The set $A(t)$ is related by equation 4 to the set of states that can be reached in time t by the admissible steering functions, and $A^0(t)$ is similarly related to the set of states that can be reached in time t by the bang-bang steering functions. Lemma 2 then states that anything that can be done by an admissible steering function can also be done by a bang-bang function. Suppose that there is a steering function $u \in \Omega$ such that $x(t_1, u) = z(t_1)$; the particle is then hit at time t_1, and

$$w(t_1) = \int_0^{t_1} Y(\tau)u(\tau)d\tau \quad .$$

The lemma then states that there will be a $u^0 \in \Omega^0$ with the property that $x(t_1, u^0) = z(t_1)$; bang-bang steering can accomplish the same thing in the same time. As a direct consequence of this we obtain the following pair of theorems.

THEOREM 1. If of all bang-bang steering functions there is an optimal one relative to Ω^0, then it is optimal (relative to Ω).

THEOREM 2. If there is an optimal steering function, then there is always a bang-bang steering function that is optimal.

The first of these theorems extends the result in [7], and the second is a general bang-bang principle. Although we did not state the intuitive hypothesis this way, the feeling actually is that not only should there always be bang-bang steering that is optimal but no other type of steering should be optimal. If at some time all of the available power is not being used, then it should be possible to improve the performance by using, properly, the additional power that is available. Perhaps this is true in some more general sense, but under our restriction on the amplitude of each component of the steering function this is not true without placing further restrictions on the control system. We return to this question in the next section.

We can now extend the results that have been obtained previously on the existence of and the form of optimal steering. The set $A(t)$ is, as we have said, related by equation (4) to the set of all states of the system that can be reached in time t. The existence is a simple consequence of the fact that $A(t)$ is a closed set, and the form (5) below for optimal steering follows from the convexity of $A(t)$. [For r-dimensional vectors a and b, $a = \operatorname{sgn} b$ means that $a_i = \operatorname{sgn} b_i$, $i = 1, \ldots, r$; $\operatorname{sgn} b_j = 1$ if $b_j > 0$, $\operatorname{sgn} b_j = -1$ if $b_j < 0$, and is considered to be undetermined if $b_j = 0$. In using this notation a is always a column vector and b is a row vector.]

THEOREM 3. If for the general problem there is a steering function u in Ω such that $x(t, u) = z(t)$ for some $t > 0$, then there is an optimal steering function in Ω. Moreover, all optimal steering functions u^* are of the form

$$u^*(t) = \operatorname{sgn} [\eta Y(t)] \tag{5}$$

where η is some non-zero n-dimensional vector.

PROOF. Our assumption that $x(t, u) = z(t)$ for some $t > 0$ and some $u \in \Omega$ is equivalent to assuming that $w(t) \in A(t)$ for some $t > 0$. Let t^* be the greatest lower bound of all positive t's with this property. Let

$$y(t, u) = \int_0^t Y(\tau)u(\tau)d\tau \quad .$$

Then there is a non-increasing sequence t_n converging to t^* and $u_n \in \Omega$ such that $w(t_n) = y(t_n, u_n) \in A(t_n)$. Now for some number K

it is easy to see that

$$|y(t_n, u_n) - y(t^*, u_n)| \leq K|t_n - t^*| \quad .$$

Therefore $w(t^*)$ is the limit of points in $A(t^*)$, and since $A(t^*)$ is closed, $w(t^*)$ is in $A(t^*)$. This proves the existence of a steering function u^* in Ω such that $w(t^*) = y(t^*, u^*)$. By the definition of t^*, u^* is optimal.

We now want to show that u^* is of the form (5). In order to do this we note first that $w(t^*)$ cannot be an interior point of $A(t_1)$ for $0 < t_1 < t^*$. Suppose this were true. Let N be a neighborhood of $w(t^*)$ contained in $A(t_1)$. Then, since $A(t_1) \subset A(t)$ for all $t_1 < t$, N_1 is contained in $A(t)$ for all $t > t_1$. The trajectory $w(t)$ is continuous and this implies $w(t_2) \in A(t_2)$ for some $t_2 < t^*$. This contradicts the definition of t^*, and therefore, $w(t^*)$ cannot be an interior point of $A(t_1)$ for $t_1 < t^*$. It then follows from Lemma 3 that $w(t^*)$ is not an interior point of $A(t^*)$. Therefore $w(t^*)$ is a boundary point of the convex set $A(t^*)$. There is then a support plane P through $w(t^*)$ (see [3]). Let η be a non-zero vector normal to P and directed to the side which contains no points of $A(t^*)$. Therefore for each $y(t^*, u)$ in $A(t^*)$

$$(\eta, y(t^*, u) - w(t^*)) \leq 0 \quad .$$

Now

$$w(t^*) = y(t^*, u^*) = \int_0^{t^*} Y(\tau)u^*(\tau)d\tau \quad ,$$

and this inequality implies

$$\int_0^{t^*} (\eta, Y(\tau)u(\tau))d\tau \leq \int_0^{t^*} (\eta, Y(\tau)u^*(\tau))d\tau$$

for all $u \in \Omega$. Let $v(\tau) = \eta Y(\tau)$. Then

$$\int_0^{t^*} (\eta, Y(\tau)u^*(\tau))d\tau \leq \sum_{j=1}^{r} \int_0^{t^*} |v_j(\tau)|d\tau = \sum_{j=1}^{r} \int_0^{t^*} v_j(\tau) \operatorname{sgn} v_j(\tau)d\tau \quad .$$

Hence we see that on any interval of positive length where $v_j(\tau) \neq 0$ it must be that $u_j^*(\tau) = \operatorname{sgn} v_j(\tau)$. This completes the proof.

The form (5) for an optimal steering function enables us to state a uniqueness criterion. Let $y^j(t)$ be the jth column vector of $Y(t)$. The control system (1) is said to be normal if for each $j = 1, \ldots, r$ the functions $y_1^j(t), \ldots, y_n^j(t)$ are linearly (functionally) independent on each interval of positive length. This is equivalent to saying that no component of $\eta Y(t)$, $\eta \neq 0$, is identically zero on an interval of positive length. Therefore $u^*(t)$ is uniquely determined by (5), and for normal control systems there is at most one optimal steering function, and if there is one, it is bang-bang. Thus, for normal control systems, the only way of reaching the objective in minimum time is by bang-bang steering of the form (5). For other control systems the most that can be said is that if it is possible to reach the objective in finite time then there is a bang-bang steering function that is optimal. The optimal steering will be of the form (5) but there may be an infinity of optimal steering functions, some of which will not be bang-bang steering.

§3. THE SPECIAL PROBLEM

The control problem for the system

$$\dot{x}(t) = A(t)x(t) + B(t)u(t) \tag{6}$$

where the objective is to start at an initial state x_0 at times $t = 0$ and to reach the origin (the equilibrium state) in minimum time will be called the special problem. The objective in the special problem is to have [see equation (4)] for some $t > 0$

$$- x_0 = \int_0^t Y(\tau)u(\tau)d\tau \quad . \tag{7}$$

For this special problem we can show that if there is a steering function of the form (5) that brings the system from the initial state x_0 to the origin in finite time then it is an optimal steering function.

THEOREM 5. If for some $t > 0$ and some n-vector η there is a solution $u = \bar{u}$ of (7) of the form

$$\bar{u}(\tau) = \text{sgn}\,[\eta Y(\tau)] \quad , \tag{5}$$

and if $\eta Y(\tau) \not\equiv 0$ on an interval of positive length, then it is an optimal steering function for the special problem.

PROOF. By Theorem 3 the existence of a $\bar{u}$ and a $\bar{t} > 0$ satisfying

$$- x_o = y(\bar{t}, \bar{u}) = \int_0^{\bar{t}} Y(t)\bar{u}(t)dt$$

implies the existence of an optimal $u*$ and a minimal time $t* \leq \bar{t}$. But then $y(t*, u*) = - x_o$, and

$$(\eta, y(\bar{t}, \bar{u}) - y(t*, u*)) = \int_0^{t*} \eta Y(t)[\bar{u}(t) - u*(t)]dt$$

$$+ \int_{t*}^{\bar{t}} \eta Y(t)\bar{u}(t)dt = 0 \quad .$$

Since $\bar{u}$ is of the form (5), we can conclude that

$$\eta Y(t) = 0 \quad \text{on} \quad [t*, \bar{t}]$$

and that $u_j^*(t) = \bar{u}_j(t)$ on $[0, t*]$ wherever the jth component of $\eta Y(t) \neq 0$. Therefore, by the assumption that $\eta Y(t) \not\equiv 0$ on an interval of positive length, it follows that $t^* = \bar{t}$. Hence $\bar{u}$ is an optimal steering function. This raises in a quite natural way the concept of a "proper" control system which is discussed in the next section.

Although the above result is quite elementary, it is of considerable practical significance and does offer a means of solving the synthesis problem. It has been mathematically convenient up to now to treat the steering function u as though it were a function of time. What one actually wants, is to know the optimal steering function $u*$ as a function of the state x of the system, and this is the "synthesis problem". If the control system is autonomous (equation 2), then we can replace t by $- t$ in equation 2, use a steering function of the form (5), start at the origin, and look at the solution. Theorem 5 states that the particular steering function used is optimal for all states of the system through which the solution passes. The time at which the solution passes through a state of the system is then the minimal time for the system to go from that state to the origin. For normal systems the optimal steering is unique, and this procedure of reversing the system determines the optimal steering as a function of the state of the system. It is even true for some systems which are not normal that the synthesis problem can be solved in this way. This procedure leads to the determination of the switching surfaces, which are surfaces where certain of the components change sign. This method is illustrated in Section 5, and the switching surfaces (curves in the examples considered) are easily obtained by quite elementary reasoning. In Example 3 of Section 5 the synthesis problem is solved for a system that is not normal.

§4. PROPER CONTROL SYSTEMS AND CONTROLLABILITY

In this section we introduce the concept of a proper control system and establish some controllability properties of such systems. We say that a control system is proper if $\eta Y(t) = 0$ on an interval of positive length implies $\eta = 0$. This is equivalent to saying that the row vectors $y_1(t)$, $y_2(t)$, ..., $y_n(t)$ of $Y(t)$ are linearly independent functions on each interval of positive length. It is clear that every normal system is proper. If the steering function has only one component $(r = 1)$, then the concepts of proper and normal are equivalent. However, it is not in general true that every proper system is normal, as is shown by Example 3 in Section 5. It is a direct consequence of Theorem 3 that in proper control systems optimal steering u^* has the property that at any given time some component of u^* assumes an extreme value.

Think now of removing all of the constraints on the admissible control functions, and consider any two states x_1 and x_2 and any two times t_1 and t_2. If for each pair of states and pair of times there is a steering function such that starting at x_1 at time t_1 · the system is brought to the state x_2 at time t_2, then the system is said to be completely controllable.

THEOREM 6. Proper control systems are completely controllable.

PROOF. We may certainly assume that $t_1 = 0$ and $t_2 > 0$. From (4) we see that complete controllability is equivalent to the property that for each $t > 0$ the set M_t of all vectors

$$\int_0^t Y(\tau)u(\tau)d\tau \quad ,$$

using all possible steering functions u, is the whole phase space E_n. Let $v(\tau) = \eta Y(\tau)$, and define

$$y = \int_0^t Y(\tau)\ \mathrm{sgn}\ v(\tau)d\tau \quad .$$

Then

$$(\eta,\ y) = \sum_{j=1}^{r} \int_0^t |v_j(\tau)|d\tau \quad .$$

Since the control system is proper,

$$\sum_{j=1}^{r} \int_0^t |v_j(\tau)|d\tau > 0$$

if $\eta \neq 0$. Therefore corresponding to each direction η there is a vector y in M_t such that $(\eta,\ y) > 0$. M_t is obviously a linear manifold, and therefore $M_t = E_n$. [It is of interest to note that the theorem is also a consequence of the fact that for proper control systems the linear transformation F on E_n defined by

$$F(\eta) = \int_0^t Y(\tau)Y'(\tau)\eta' d\tau$$

is non-singular for each $t > 0$. The prime denotes the transpose. Thus the system is completely controllable with steering functions restricted to those of the form $u(\tau) = Y'(\tau)\eta'$.]

Complete controllability assumes that there is no restriction on the admissible control functions, and if unlimited power is available to a proper control system the above theorem says that it is always possible to move the system from any one state to any other state as quickly as we please. For each r-vector v define $\|v\| = |v_1| + |v_2| + \dots + |v_r|$. If we replace the concept of being proper by the requirement that

$$\int_0^\infty \|\eta Y(\tau)\|d\tau > 0$$

for each $\eta \neq 0$, then it is easily seen that with no constraints on the steering functions there is a time $T > 0$ such that the system can be started at any state x_0 at time $t = 0$ and can be steered to any other state in time T. This fact does not appear to be of any particular importance, and for autonomous systems this condition is equivalent to the condition that the system be proper. We wish now to return to the assumption that the system has limited power and to the special problem. We shall say that a control system is asymptotically proper if

$$\int_0^\infty \|\eta Y(\tau)\| d\tau = \infty$$

for each $\infty \neq 0$. For the special problem we are interested in systems with the property that given any initial state x_0 there is a steering function in Ω that brings the system to the origin in finite time. We shall say that such a system is controllable. Of course, by Theorem 3 we know that if a system is controllable then for each initial state x_0 there is a steering function in Ω that is optimal for the special problem.

THEOREM 7. Asymptotically proper control systems of the form (6) are controllable.

PROOF. For systems of the form (6) we know from (4) that the system is controllable if corresponding to each x_0 in E_n there is a u in Ω with the property that

$$- x_0 = y(t, u) = \int_0^t Y(\tau)u(\tau)d\tau$$

for some $t > 0$. Let $A = \{y(t, u); u \in \Omega, t > 0\}$. Clearly A is a convex set, and what we wish to show is that $A = E_n$.

Taking $u(t) = \mathrm{sgn}[\eta Y(t)]$, we have

$$(\eta, y(t, u)) = \int_0^t \|\eta Y(\tau)\| d\tau \quad .$$

Since the system is asymptotically proper, we know that $(\eta, y(t, u)) \longrightarrow \infty$ as $t \longrightarrow \infty$ for each $\eta \neq 0$. By Lemma 4, $A = E_n$ and the system is controllable.

We now examine some special properties of autonomous control systems (equation 2). Let $\varphi(\lambda) = \det(A + \lambda I) = \lambda^n + c_1\lambda^{n-1} + \ldots + c_n$; $\varphi(- \lambda)$ is the characteristic polynomial of A, and we know that $\varphi(- A) = 0$. Hence $v(t) = \eta Y(t) = \eta e^{-At}B$ satisfies the nth order linear differential equation with constant coefficients

$$v^{(n)}(t) + c_1 v^{(n-1)}(t) + \dots + c_n v(t) = 0 \quad ,$$

where $v^{(k)}(t)$ is the kth derivative. Thus $v(t) \equiv 0$ on an interval of positive length is equivalent to $v(0) = \eta B = 0$, $v'(0) = -\eta AB = 0$, $\dots$, $v^{(n-1)}(0) = (-1)^{n-1}\eta A^{n-1}B = 0$. It then follows that an autonomous control system is proper if and only if one of the following hold ($M = B(E_r)$ is the range of B and $b^1, \dots, b^r$ are the column vectors of B):

1. For each non-zero vector η in E_n there is an integer k, $0 \leq k \leq n - 1$, such that $\eta A^k B \neq 0$.

2. $M + A(M) + \dots + A^{(n-1)}(M) = E_n$.

3. The set of vectors $b^1, \dots, b^r, Ab^1, \dots, Ab^r, \dots, A^{(n-1)}b^1, \dots, A^{(n-1)}b^r$ contains a set of n linearly independent vectors.

The requirement that a system be normal is, in general, much stronger. The system is normal if and only if no component of $v(t) = \eta Y(t)$ is identically zero on an interval of positive length. Therefore an autonomous system is normal if and only if for each integer j, $0 < j \leq r$, the vectors $b^j, Ab^j, \dots, A^{(n-1)}b^j$ are linearly independent.

Let us now assume that an autonomous control system is both proper and stable [the characteristic roots of A have non-positive real parts]. For $\eta \neq 0$ we know that one component of $v(t) = \eta Y(t) = \eta e^{-At}B$ is not identically zero. We may suppose that $v_1(t) \not\equiv 0$. Then

$$v_1(t) = \sum_{j=1}^{m} P_j(t) e^{\alpha_j t} \cos(\beta_j t + \delta_j) \quad ,$$

$\alpha_j \geq 0$ and $P_j(t)$ are polynomials. Hence for t sufficiently large, say $t \geq a$,

$$|v_1(t)| \geq |p(t) + q(t)| \quad ,$$

where $p(t)$ is almost periodic, is not identically zero, and $|q(t)| < \frac{K}{t}$. It then follows that

$$\lim_{T \to \infty} \frac{1}{T} \int_a^T |p(t) + q(t)|^2 dt = \lim_{T \to \infty} \frac{1}{T} \int_a^T |p(t)|^2 dt = 2c > 0 \quad .$$

Consequently, for T sufficiently large,

$$\int_a^T |v_1(t)|dt \geq \int_a^T |p(t) + q(t)|dt \geq T^{-\frac{1}{2}} \int_a^T |p(t) + q(t)|^2 dt \geq cT^{\frac{1}{2}} \quad .$$

Therefore

$$\int_0^\infty |v_1(t)|dt = \infty \quad ,$$

and the system is asymptotically proper. What we have shown is that:

> <u>If an autonomous system is both stable and proper, then it is asymptotically proper and therefore controllable</u>.

Let us now see what can be said about the existence of a solution and the determination of optimal steering for the general problem of starting at x_0 at time $t = 0$ and hitting a moving particle $z(t)$. The control system is

$$\text{(1)} \qquad \dot{x}(t) = A(t)x(t) + B(t)u(t) + f(t) \quad .$$

We wish to know whether or not there is a $u \in \Omega$ and a $t_1 > 0$ such that [equation (4)]

$$w(t_1) = \int_0^{t_1} Y(\tau)u(\tau)d$$

where

$$w(t_1) = X^{-1}(t_1)z(t_1) - x_0 - \int_0^{t_1} X^{-1}(\tau)f(\tau)d\tau \quad ;$$

that is, whether or not $w(t_1) \in A(t_1)$ for some $t_1 > 0$. Think for the moment that t_1 is fixed and ask whether or not, for some $t > 0$, $w(t_1) \in A(t)$. This is then the special problem for the control system

$$\text{(8)} \qquad \dot{x}(t) = A(t)x(t) + B(t)u(t)$$

of starting at $-w(t_1)$ and reaching the origin in finite time. Suppose that the synthesis problem for this special problem can be solved. Then for each point $-w(t_1)$ of the curve $-w(t)$, we would know the optimal steering function $\bar{u}$ and the minimum time $\bar{t}_1$ for starting at $-w(t_1)$ and reaching the origin. If $\bar{t}_1 \leq t_1$, then

$$w(t_1) \in A(\bar{t}_1) \subset A(t_1) \quad ,$$

and we would know by Theorem 3 that the general problem has an optimal steering function u^* and a minimum time t^*; we would know also that $t^* \leq t_1$. Thus, we have the following relation between the general problem and the special problem:

<u>The general problem</u> $x(0, u) = x_0$, $x(t, u) = z(t)$ <u>has a solution if and only if in the special problem it is possible, starting at some point</u> $-w(t_1)$ <u>of the curve</u>

$$-w(t) = -X^{-1}(t)z(t) + x_0 + \int_0^t X^{-1}(\tau)f(\tau)d\tau \quad ,$$

<u>to reach the origin in time</u> $t_2 \leq t_1$.

For proper control systems we can say a bit more. Suppose that the general problem has a solution. Let u^* be the optimal steering function and let t^* be the minimum time. Let $\bar{u}^*$ be the optimal steering for the special problem of starting at $-w(t^*)$ and reaching the origin. Let $\bar{t}^*$ be the minimum time. Then $\bar{t}^* \leq t^*$. Now, using the steering function $\bar{u}^*$ during the time interval $[0, \bar{t}^*]$ and coasting $(u(t) = 0)$ during the interval $[\bar{t}^*, t^*]$, we hit the particle at time t^*. But we know for proper systems that optimal steering is never identically zero on an interval of positive length, and therefore it must be that $\bar{t}^* = t^*$. Thus, we have shown that

<u>For proper systems, if</u> u^* <u>is optimal steering for the general problem and</u> t^* <u>is the minimum time, then</u> u^* <u>is optimal for the special problem of starting at</u> $-w(t^*)$ <u>and reaching the origin;</u> t^* <u>is the minimum time for doing this</u>.

Thus, if the system is proper and if the synthesis problem for the special problem can be solved, one can move along the trajectory $-w(t)$, $t > 0$, use optimal steering from each point, and locate the first point where one can go from $-w(t)$ to the origin in time t. The steering function that does this would then be optimal steering for the general problem. Let $m(x)$ be the minimum time to go from the initial state x to the origin. What we wish to find is the smallest $t > 0$ satisfying $m(-w(t)) = t$.

§5. EXAMPLES

The purpose of these examples is to illustrate the concepts that have been introduced and to indicate how the theory can be applied to solve the synthesis problem. Although similar examples have been given before, the more complete theory simplifies the reasoning and the computation leading to a solution. We had pointed out previously that the restriction $|u_i| \leq 1$ was a matter of convenience and that this can be replaced by the more general condition $-a_i \leq u_i \leq b_i$, $a_i > 0$, $b_i > 0$. This is illustrated in Example 1.

Example 2 was solved in [1], and we consider the same example to show how a more complete theory leads to a simpler solution.

In Example 3 a system which is proper but not normal is considered. The optimal steering is not unique, not necessarily bang-bang, but yet the synthesis problem, which is to define optimal steering uniquely as a function of the state of the system, can be solved.

EXAMPLE 1. The differential equation of a controlled, undamped harmonic oscillator is

$$\ddot{x} + x = u.$$

We consider the constraint $-a \leq u \leq b$, $a > 0$, $b > 0$. An equivalent system is

$$\dot{x} = y$$
$$\dot{y} = -x + u \quad .$$

$$A = \begin{pmatrix} 0 & 1 \\ -1 & 0 \end{pmatrix}, \; B = \begin{pmatrix} 0 \\ 1 \end{pmatrix}, \; X(t) = \begin{pmatrix} \cos t & -\sin t \\ \sin t & \cos t \end{pmatrix}, \; Y(t) = X^{-1}(t)B = \begin{pmatrix} \sin t \\ \cos t \end{pmatrix} ,$$

$$\eta Y(t) = \eta_1 \cos \; t + \eta_2 \sin t = A \cos(t + \delta) \quad .$$

The system is proper, and since $r = 1$, is also normal. It is stable and is therefore controllable. We know therefore that for each initial point (x_0, y_0) there is a unique optimal control function of the form

$$u(t) = \operatorname{sgn}^* [A \cos(t + \delta)] = \begin{cases} b, & \text{if } A \cos(t + \delta) > 0 \\ -a, & \text{if } A \cos(t + \delta) < 0 \end{cases}$$

which brings the system to the origin in minimum time. Conversely, we know that any steering function of this form which brings the system to the origin is optimal. We replace t by $-\tau$, obtain

$$\frac{dx}{d\tau} = -y$$

$$\frac{dy}{d\tau} = x - u \quad ,$$

and consider steering functions of the form $u(\tau) = \mathrm{sgn}^*[A\cos(\delta - \tau)]$. When $u = c$ the trajectory in the phase plane is a circle with center at $(c, 0)$ and as τ increases (t decreases) the direction on the trajectory is counterclockwise. The arrows in the figures are in the direction of increasing t. Thus, the trajectory can leave the origin, as indicated in Figure 1, going counterclockwise on either the circle with center at $(b, 0)$ or the circle with center at $(-a, 0)$. Depending on the choice of η, the steering function can change sign at

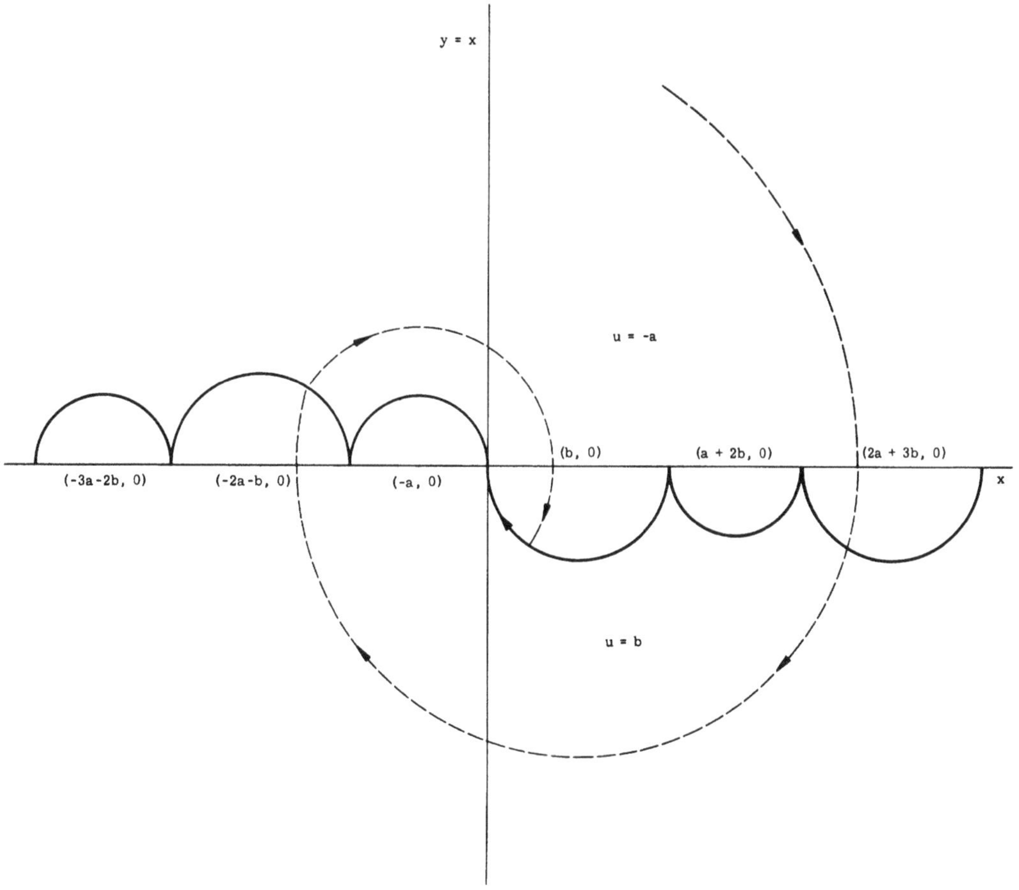

FIGURE 1

any time before a half-revolution and cannot go past a half-revolution without changing sign. The switching curves are therefore the chain of semicircles shown in Figure 1. Above these semicircles $u = -a$ and below them $u = b$. Thus starting in the first quadrant, as illustrated, optimal steering opposes the motion and does not change sign until past the point of maximum displacement except at those exceptional points where the semicircles join. Because of the simple nature of the trajectories we know that there is a unique optimal trajectory through each point of the phase plane. This is also a consequence of the fact that these are the only possible switching curves and that the system is both controllable and normal.

EXAMPLE 2. This example was solved in [1] by reasoning, which we can now see, was more complex than is necessary. In its normalized form the system is

$$\dot{y}_1 = -2y_1 + u_1 + u_2$$

$$\dot{y}_2 = -y_2 + u_1 + 2u_2, \quad |u_1| \leq 1, \quad |u_2| \leq 1 \quad .$$

Here

$$Y(t) = \begin{pmatrix} \eta_1 e^{2t} + \eta_2 e^{t} \\ \\ \eta_1 e^{2t} + 2\eta_2 e^{t} \end{pmatrix}$$

and it is clear that the system is normal and controllable. In order to solve the synthesis problem we consider the solutions, starting at the origin of

$$\frac{dy_1}{d\tau} = 2y_1 - u_1 - u_2$$

$$\frac{dy_2}{d\tau} = y_2 - u_1 - 2u_2$$

using steering functions of the form

$$u_1(\tau) = \operatorname{sgn}(\eta_1 e^{-2\tau} + \eta_2 e^{-\tau})$$

$$u_2(\tau) = \operatorname{sgn}(\eta_1 e^{-2\tau} + 2\eta_2 e^{-\tau}) \quad .$$

Taking $\eta_1 + \eta_2 < 0$, $\eta_1 + 2\eta_2 > 0$, $\eta_2 > 0$, we begin at the origin with $u_1 = -1$, $u_2 = -1$, and move out along the parabola (Figure 2)

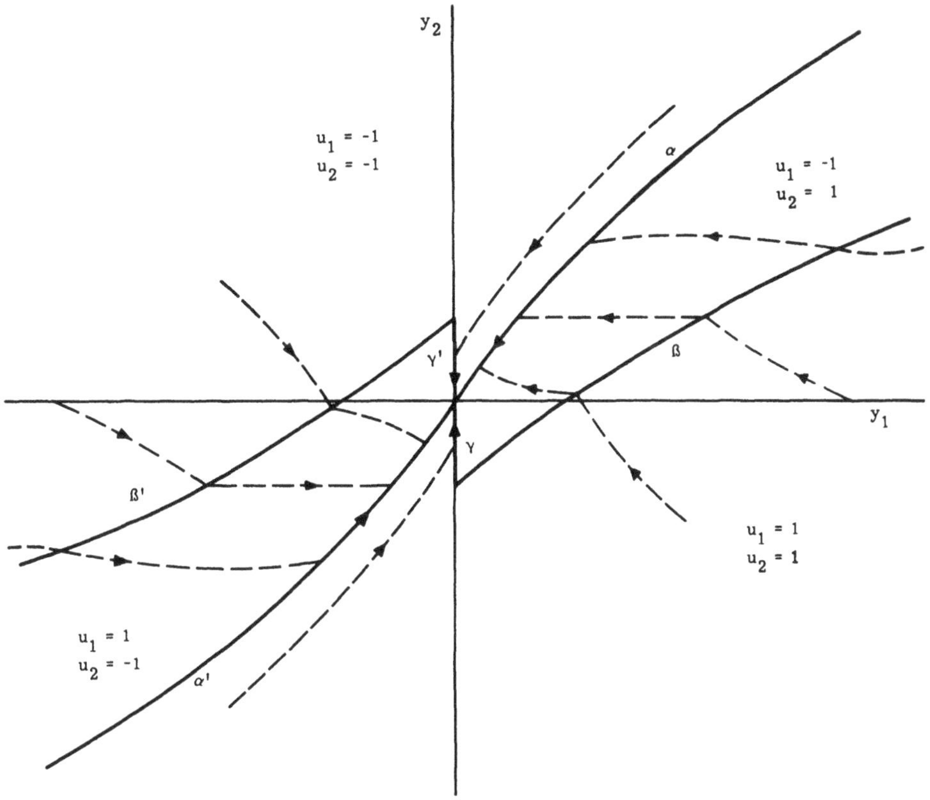

FIGURE 2

$$\alpha : y_1(\tau) = e^{2\tau} - 1, \quad y_2(\tau) = 3(e^{\tau} - 1) \quad .$$

Now $u_2(\tau)$ changes sign at $\tau_1 = \ell n(-\eta_1/2\eta_2)$, and then $u_1(\tau)$ changes sign at $\tau_2 = \ell n(-\eta_1/\eta_2)$. Hence $\tau_2 - \tau_1 = \ell n 2$. Therefore, starting at the point $(y_1(\tau_1), y_2(\tau_1))$ of α with $u_1 = -1$, $u_2 = 1$, we leave α along the parabola

$$y_1(\tau) = y_1(\tau_1)e^{2\tau}$$

$$y_2(\tau) = y_2(\tau_1)e^{\tau} - (e^{\tau} - 1) \quad ,$$

and the shift in the sign of u_1 occurs at

$$y_1(\ell n2) = 4y_1(\tau_1)$$
$$y_2(\ell n2) = 2y_2(\tau_1) - 1 \quad .$$

This linear transformation applied to α gives the switching curve β where u_1 changes sign. Starting at the origin with $\eta_1 + \eta_2 < 0$ and $\eta_1 + 2\eta_2 \geq 0$, we have $u_1 = -1$, $u_2 = 1$, and u_1 must change sign some time not later than $\tau = \ell n2$ and can change at any time along the line from $(0, 0)$ to $(-1, 0)$. This is the switching curve γ. The switching curves α', β', γ' are obtained by symmetry. These are the only possible switching curves, and, since the system is controllable and normal, the steering function is determined uniquely at each point of the phase plane as indicated in Figure 2.

EXAMPLE 3. Here we consider a system which is proper and controllable but not normal. The difficulty is that it is not sufficient to determine all of the possible switching curves, since the optimal steering is not unique. What one wants to do is to determine a set of switching curves for which there is a unique optimal trajectory through each point. The system is

$$\dot{x}_1 = -x_1 + u_1$$

$$\dot{x}_2 = -2x_2 + u_1 + u_2, \qquad |u_1| \leq 1, \qquad |u_2| \leq 1 ,$$

where the matrix A is in diagonal form.

$$X(t) = \begin{pmatrix} e^{-t} & 0 \\ 0 & e^{-2t} \end{pmatrix}, \qquad B = \begin{pmatrix} 1 & 0 \\ 1 & 1 \end{pmatrix},$$

$$Y(t) = \begin{pmatrix} e^{t} & 0 \\ e^{2t} & e^{2t} \end{pmatrix}, \qquad \eta Y(t) = \begin{pmatrix} \eta_1 e^{t} + \eta_2 e^{2t} \\ \eta_2 e^{2t} \end{pmatrix}.$$

The matrix B is non-singular, and the system is proper and controllable but not normal. We shall obtain a solution in which one component of the steering is zero in two regions of the phase plane. Letting $t = -\tau$, we examine the solutions of

$$\frac{dx_1}{d\tau} = x_1 - u_1$$

$$\frac{dx_2}{d\tau} = 2x_2 - u_1 - u_2$$

leaving the origin with steering

$$u_1(\tau) = \operatorname{sgn}(\eta_1 e^{-\tau} + \eta_2 e^{-2\tau}), \qquad u_2(\tau) = \operatorname{sgn}(\eta_2 e^{-2\tau}) \quad .$$

Taking $\eta_2 = 0$, we can choose u_2 to be any value between -1 and 1 and switch values any time we please. We restrict ourselves to values -1, 0, or 1.

The curve α in Figure 3 corresponds to $\eta_1 < 0$, $\eta_2 = 0$, and $u_1 = -1$, $u_2 = 1$. If we switch to $u_2 = 0$, the trajectories leaving α are parabolas with vertices at $(-1, -\frac{1}{2})$. The curve β corresponds to switching at time $\tau = 0$, and every point between α and β has an optimal trajectory passing through it. The curve β corresponds to leaving the origin with $u_1 = -1$, $u_2 = 0$. Switching to $u_2 = -1$, we leave β at any point we please along parabolas with vertices at $(-1, 1)$. The curve γ corresponds to switching to $u_2 = -1$ at time $t = 0$; that is, to leaving the origin with $u_1 = -1$, $u_2 = -1$. All points between β and γ can be reached in this way. Picking $\eta_1 > 0$ and $\eta_1 + \eta_2 < 0$, we see that we can switch to $u_1 = 1$ at any point along γ we please. The trajectories leaving γ are parabolas with vertices at $(1, 0)$, and each point between γ and α can be reached in this way. The switching curves α', β', and γ' are obtained by symmetry.

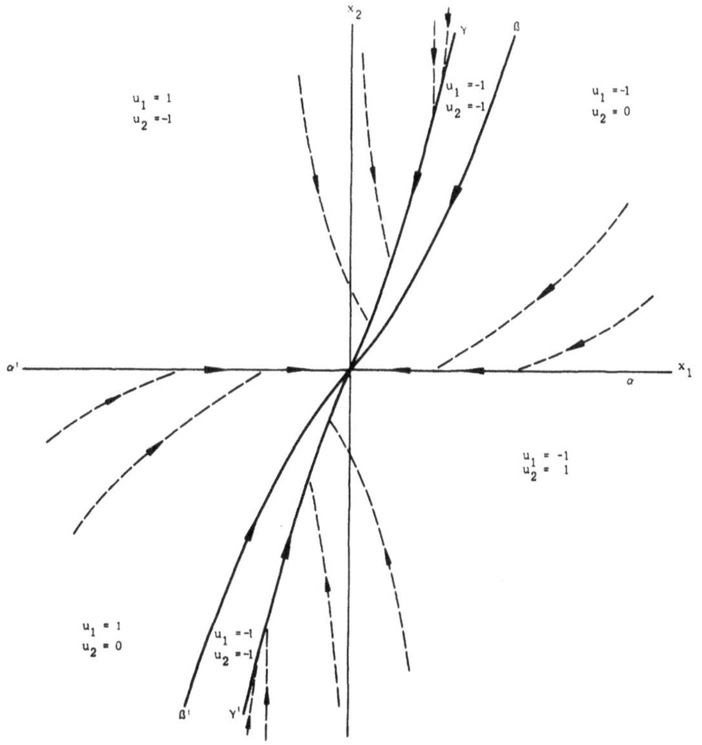

FIGURE 3

REFERENCES

[1] BELLMAN, R., GLICKSBERG, I., and GROSS, O., On the "bang-bang" control problem, Quart. Appl. Math., 14 (1956), 11-18.

[2] BUSHAW, D. W., Ph. D. Thesis, Department of Mathematics, Princeton University, 1952; Differential equations with a discontinuous forcing term, Experimental Towing Tank, Stevens Institute of Technology, Report No. 469 (January 1953); Optimal discontinuous forcing terms, Contributions to the Theory of Nonlinear Oscillations, IV, Princeton, 1958.

[3] EGGLESTON, H. G., Convexity, Cambridge Tracts in Math. and Math. Phys., No. 47, Cambridge, 1958.

[4] GAMKRELIDZE, R. V., Theory of time-optimal processes for linear systems, Izv. Akad. Nauk SSSR. Ser. Mat., 22 (1958), 449-474 (Russian).

[5] HALMOS, P. R., The range of a vector measure, Bull. Amer. Math. Soc., 54 (1948), 416-421.

[6] KRASOVSKII, N. N., Concerning the theory of optimal control; Avtomat. i Telemeh., 18 (1957), 960-970 (Russian).

[7] LASALLE, J. P., Abstract 247t, Bull. Amer. Math. Soc., 60 (1954), 154; Study of the basic principle underlying the "bang-bang" servo, Goodyear Aircraft Corp. Report GER-5518 (July 1953).

[8] LIAPUNOV, A., Sur les fonctions-vecteurs completement additives, Bull. Acad. Sci. URSS (Izv. Akad. Nauk SSSR. Ser. Mat.) 4 (1940), 465-478.

II. CONTINUOUS DEPENDENCE ON A PARAMETER

J. Jarník and J. Kurzweil

Gichman [1] was the first to show that the averaging principle is a consequence of a modified theorem on continuous dependence on a parameter (cf. also [2] and [3]). Let the solutions of

$$\frac{dx}{dt} = f(x, t) \tag{1}$$

be unique. In [4] it was proved that the solutions of

$$\frac{dx}{dt} = f_k(x, t), \qquad k = 1, 2, 3, \ldots \tag{2}$$

tend uniformly to the solutions of (1) with $k \longrightarrow \infty$ if

$$\int_0^t f_k(x, \tau)d\tau \longrightarrow \int_0^t f(x, \tau)d\tau \tag{3}$$

uniformly and if the following Condition A is fulfilled. (x, f, f_k are n-vectors, f, f_k are continuous for $x \in G$, $t \in \langle 0, T\rangle$, G open.)[1]

Let $\omega_1(\eta)$, $\omega_2(\eta)$ be non-decreasing for $0 \leq \eta \leq \sigma$, $\psi(\eta) = \omega_1(\eta)\omega_2(\eta)$, $\eta^{-1}\psi(\eta)$ non-decreasing.[2]

CONDITION A:

$$\left| \int_{t_1}^{t_2} f_k(x, t)\right| dt \leq \omega_1(|t_2 - t\,|) \tag{4}$$

[1] Cf. [4], Theorems 4, 2, 1, which is formulated for generalized differential equations. In the special case of equations (1), (2) with f, f_k continuous, every solution is necessarily regular (cf. Definitions 4, 2, 1).

[2] It is sufficient that $\psi(\eta)$ be non-decreasing. This will be proved in a paper by J. Kurzweil, which will be published in Czechoslovak Math. Journal.

for $|t_2 - t_1| \leq \sigma$, $k = 1, 2, 3, \ldots$

(5) $$\left| \int_{t_1}^{t_2} (f_k(x_2, t) - f_k(x_1, t))dt \right| \leq |x_2 - x_1| \omega_2(|t_2 - t_1|)$$

for $|t_2 - t_1| \leq \sigma$, $|x_2 - x_1| \leq 2\omega_1(\sigma)$, $k = 1, 2, 3, \ldots$

(6) $$\sum_{i=1}^{\infty} 2^i \psi\left(\frac{\sigma}{2^i} \right)$$

Consider the special sequence

(2') $$\frac{dx}{dt} = xk^{1-\alpha}\cos kt + k^{1-\beta}\sin kt$$
$$0 < \alpha,\ \beta \leq 1,\ k = 1, 2, 3, \ldots$$

and the equation

(1') $$\frac{dx}{dt} = 0$$
$$(n = 1,\ G = E(x;\ |x| < 2).$$

Then (3) is obviously fulfilled, (4) and (5) hold with $\omega_1(\eta) = K_1\eta^{\gamma}$, $\omega_2(\eta) = K_2\eta^{\alpha}$, $\gamma = \min(\alpha, \beta)$ (K_1, K_2 independent of k) and (6) holds for $\alpha + \gamma > 1$. Consequently the solutions of (2') tend to the solutions of (1') with $k \longrightarrow \infty$ if $\alpha + \gamma > 1$, i.e.,

(7) $$\alpha + \beta > 1, \quad \alpha > \frac{1}{2} .$$

By direct calculation it is easy to show that the solutions of (2') tend to the solutions of (1') precisely if $\alpha + \beta > 1$ (cf. [4], p. 419).

Our aim is to prove that condition (6) cannot be weakened even if we restrict ourselves to linear equations. In the case of linear equations we shall always suppose that

$$\omega_1(\eta) \geq c\omega_2(\eta),\ c > 0 .$$

As $\omega_1(\eta)$, $\omega_2(\eta)$ are moduli of continuity (cf. (4) and (5)), we shall suppose that

$$(8) \qquad \omega_j(\eta_1) + \omega_j(\eta_2) \geq \omega_j(\eta_1 + \eta_2), \qquad j = 1, 2 \quad .$$

THEOREM. Let $\eta^{-\lambda}\omega_1(\eta)$, $\eta^{-\lambda}\omega_2(\eta)$, $(0 \leq \eta \leq 1)$ be non-decreasing for a positive λ, $\psi(\eta) = \omega_1(\eta)\omega_2(\eta)$,

$$(9) \qquad \omega_1(\eta) \geq c\omega_2(\eta), \qquad c > 0 \quad ,$$

$$(10) \qquad \sum_{i=1}^{\infty} 2^i \psi\left(\frac{i}{2^i}\right) .$$

Then there exists a sequence of linear differential equations

$$(11) \qquad \frac{dx}{dt} = a_k(t)x + b_k(t) \quad ,$$

$(n = 1,\ G = E(x;\ |x| < 2))$, such that (4) and (5) are fulfilled,

$$\int_0^t a_k(\tau)d\tau \longrightarrow 0, \qquad \int_0^t b_k(\tau)d\tau \longrightarrow 0$$

uniformly, while the sequence of solutions $x_k(t)$ of (11) (with $x_k(0) = 0$) does not converge.

Let us explain the rôle of the inequality $\alpha > \frac{1}{2}$ in (7). Consider the sequence

$$(2'') \qquad \frac{dx}{dt} = a_k(t)x + b_k(t) \quad ,$$

where x, b_k are n-vectors, a_k are $n \times n$-matrices. Suppose that $a_k(t_1)a_k(t_2) = a_k(t_2)a_k(t_1)$ for $t_1, t_2 \in \langle 0, T \rangle$, $k = 1, 2, 3, \ldots$. Then the solutions of (2") tend to the solutions of

$$(1'') \qquad \frac{dx}{dt} = a(t)x + b(t)$$

if

$$A_k(t) = \int_0^t a_k(\tau)d\tau \longrightarrow A(t) = \int_0^t a(\tau)d\tau$$

uniformly,

$$B_k(t) = \int_0^t b_k(\tau)d\tau \longrightarrow B(t) = \int_0^t b(\tau)d\tau$$

uniformly,

$$|B_k(t_2) - B_k(t_1)| \leq \omega_1(|t_2 - t_1|),\ |A_k(t_2) - A_k(t_1)| \leq \omega_2(|t_2 - t_1|),$$ [3],

if (6) holds (with $\psi = \omega_1\omega_2$) and if $\psi(\eta)$ is non-decreasing. This follows from the representation

$$x_k(t) = \exp\{A_k(t)\}\left[2 + \int_0^t \exp\{-A_k(\tau)\}dB_k(\tau)\right] \quad ,$$

where

$$\int_0^t \exp\{-A_k(\tau)\}dB_k(\tau)$$

tends to

$$\int_0^t \exp\{-A(\tau)\}dB(\tau)$$

if $\eta^{-1}\psi(\eta)$ is non-decreasing according to [4], Theorem 3,2. (In the case that merely $\psi(\eta)$ is non-decreasing, cf. footnote 2.) Let $b_k(t) = b(t) = 0$. Then the solutions of (2") converge to the solutions of (1") if (and only if) $A_k(t) \longrightarrow A(t)$. This situation specially occurs for $n = 1$.

In the non-commutative case it follows from the Theorem that there exists such a sequence of linear homogeneous equations that (3), (4) and (5) are fulfilled (with $\omega_1(\eta) = \omega_2(\eta)$) and the sequence of solutions of these equations does not converge, if

$$\sum_{i=1}^{\infty} 2^i \psi_2\left(\frac{1}{2^i}\right) = \infty \quad ,$$

$\psi_2(\eta) = \omega_2^2(\eta)$, $\eta^{-\lambda}\psi_2(\eta)$ non-decreasing, $\lambda > 0$. (It suffices to take $n = 2$ and

---

[3] If Z is a matrix and Z_{ij} are its elements, we put, for example, $|Z| = (\Sigma Z_{ij}^2)^{\frac{1}{2}}$.

$$\frac{dx_1}{dt} = a_k(t)x_1 + b_k(t)x_2 ,$$

$$\frac{dx_2}{dt} = 0$$

$$x_{1k}(0) = 0, \quad x_{2k}(0) = 1, \quad k = 1, 2, 3, \dots$$

with $a_k(t)$, $b_k(t)$ from the Theorem). This explains the appearance of the inequality $\alpha > \frac{1}{2}$ in (7), as a general result on the continuous dependence on a parameter applied to the sequence (2').

PROOF. Let

$$\exp\{-A_k(t)\} = C_1 + \sum_{i=k}^{m_k} \ell_i \omega_2\left(\frac{1}{2^i}\right) \sin 2^i t ,$$

$$B_k(t) = C_2 \sum_{i=k}^{m_k} \ell_i \omega_1\left(\frac{1}{2^i}\right) \cos 2^i t ,$$

$$a_k(t) = \frac{d}{dt} A_k(t), \ b_k(t) = \frac{d}{dt} B_k(t), \ k = 1, 2, 3, \dots$$

where the constants $C_1, C_2, \ell_1, \ell_2, \ell_3, \dots, m_1, m_2, m_3, \dots, m_k > k$, will be chosen later.

We shall prove

(1) $$A_k(t) - A_k(0) = \int_0^t a_k(\tau)d\tau \longrightarrow 0 \text{ uniformly,}$$

$$B_k(t) - B_k(0) = \int_0^t b_k(\tau)d\tau \longrightarrow 0 \text{ uniformly with } k \longrightarrow \infty.$$

(2) $\vartheta\omega_2(\eta)$ is a modulus of continuity of $A_k(t) = \vartheta = \min\left(\frac{c}{4}, \frac{1}{4}\right)$ (cf. (9)) $\frac{1}{2}\omega_1(\eta)$ is a modulus of continuity of $B_k(t)$, $k = 1, 2, 3, \dots$.

(3) Consequently (4) and (5) are fulfilled. $(G = E(x; |x| < 2))$ the sequence $x_k^{(t)}$ diverges for $t \neq 0$.

Let us choose the sequence ℓ_i. Let

$$L_s = E\left[i;\ i > s;\ \omega_j\left(\frac{1}{2^i}\right) \leq 2^{s-i}\omega_j\left(\frac{1}{2^s}\right)\frac{1}{C}\left(1 + \frac{1}{C}\right)^{i-s-1} \text{ for } j = 1 \text{ or } j = 2\right],$$

$$s = 1, 2, 3, \ldots \qquad\qquad i = 2, 3, 4, \ldots ,$$

$$L = \bigcup_{s=1}^{\infty} L_s, \qquad M = Q - L ,$$

where Q is the set of all natural numbers and $C > 1$ will be chosen later. We put

$$\ell_i = \begin{cases} 1 & \text{for } i \in M , \\ 0 & \text{for } i \in L . \end{cases}$$

As $\eta^{-\lambda}\omega_j(\eta)$ is non-decreasing,

$$(12) \qquad \omega_j\left(\frac{1}{2^{i+1}}\right) \leq 2^{-\lambda}\omega_j\left(\frac{1}{2^i}\right), \quad j = 1, 2, \quad i = 1, 2, 3, \ldots .$$

Let us estimate

$$\sum_{i\in L_s} 2^i\psi\left(\frac{1}{2^i}\right) \leq \sum_{i>s} 2^i\left[\omega_1\left(\frac{1}{2^i}\right)2^{s-i}\omega_2\left(\frac{1}{2^s}\right) + \right.$$

$$\left. + \omega_2\left(\frac{1}{2^i}\right)2^{s-i}\omega_1\left(\frac{1}{2^s}\right)\right]\frac{1}{C}\left[1 + \frac{1}{C}\right]^{i-s-1} \leq$$

$$\leq 2^s\psi\left(\frac{1}{2^s}\right)\frac{2}{C}\sum_{i>s}\left[2^{-\lambda}\left(1 + \frac{1}{C}\right)\right]^{i-s-1} .$$

Let C be so large that

$$4\sum_{i=0}^{\infty}\left[2^{-\lambda}\left(1 + \frac{1}{C}\right)\right]^i < C ;$$

it follows that

$$\sum_{i \in L_s} 2^i \psi\left(\frac{1}{2^i} \right) < \frac{1}{2}\, 2^s \psi\left(\frac{1}{2^s} \right) .$$

Consequently

$$\sum_{i=1}^{k} \ell_i 2^i \psi\left(\frac{1}{2^i} \right) \geq \sum_{i=1}^{k} 2^i \psi\left(\frac{1}{2^i} \right) - \sum_{s=1}^{k} \sum_{i \in L_s} 2^i \psi\left(\frac{1}{2^i} \right) \geq$$

$$\geq \frac{1}{2} \sum_{i=1}^{k} 2^i \psi\left(\frac{1}{2^i} \right)$$

and

$$\sum_{i=1}^{\infty} \ell_i 2^i \psi\left(\frac{1}{2^i} \right) = \infty \tag{13}$$

(cf. (10)).

We shall further prove that

$$\sum_{i=1}^{r-1} \ell_i 2^{i-r} \omega_j\left(\frac{1}{2^i} \right) \leq C(C + 2)\omega_j\left(\frac{1}{2^r} \right) \tag{14}$$

$$j = 1, 2, \quad r = 2, 3, 4, \ldots .$$

If $r \in M$, then

$$\omega_j\left(\frac{1}{2^r} \right) \geq 2^{i-r} \omega_j\left(\frac{1}{2^i} \right) \frac{1}{C} \left[1 + \frac{1}{C} \right]^{r-i-1} , \quad i = 1, 2, \ldots, r - 1 ,$$

$$\sum_{i=1}^{r=1} \ell_i \omega_j\left(\frac{1}{2^i} \right) 2^{i-r} \leq C\omega_j\left(\frac{1}{2^r} \right) \sum_{i=1}^{r-1} \left(1 + \frac{1}{C} \right)^{i+1-r} \leq C(C + 1)\omega_j\left(\frac{1}{2^r} \right) .$$

If $r \in L$ (obviously $1 \in M$), then there exists such an $r_1 < r$, $r_1 \in M$ that $i \in L$ for $r_1 < i \leq r$. Then

$$\sum_{i=1}^{r-1} \ell_i \omega_j\left(\frac{1}{2^i}\right) 2^{i-r} = 2^{r_1-r} \sum_{i=1}^{r_1-1} \ell_i \omega_j\left(\frac{1}{2^i}\right) 2^{i-r_1} + 2^{r_1-r} \omega_j\left(\frac{1}{2^{r_1}}\right) \leq$$

$$\leq C(C+1)\omega_j\left(\frac{1}{2^{r_1}}\right) 2^{r_1-r} + 2^{r_1-r}\omega_j\left(\frac{1}{2^{r_1}}\right) \leq$$

$$\leq C(C+2)2^{r_1-r}\omega_j\left(\frac{1}{2^{r_1}}\right) \leq C(C+2)\omega_j\left(\frac{1}{2^r}\right)$$

$(C > 1)$ (cf. (8)) and (14) holds for $r = 2, 3, 4, \ldots$.

Let us prove 1), 2), 3).

1) According to (12),

$$\sum_{i=1}^{\infty} \omega_j\left(\frac{1}{2^i}\right) < \infty .$$

If

$$C_1 > \sum_{i=1}^{\infty} \omega_2\left(\frac{1}{2^i}\right) ,$$

then $A_k(t) - A_k(0) \longrightarrow 0$, $B_k(t) - B_k(0) \longrightarrow 0$ uniformly.

2) Let

$$S_k(t) = \sum_{i=1}^{\infty} d_i \omega_1\left(\frac{1}{2^i}\right) \cos 2^i t ,$$

$$T_k(t) = \sum_{i=1}^{\infty} d_i \omega_2\left(\frac{1}{2^i}\right) \sin 2^i t ,$$

$$d_i = \ell_i \quad \text{for} \quad i = k, k+1, \ldots, m_k ,$$

$$d_i = 0 \quad \text{for} \quad i < k \text{ or } i > m_k ,$$

$$\frac{1}{2^{r-1}} \geq |t_2 - t_1| > \frac{1}{2^r} .$$

Then

$$|S_k(t_2) - S_k(t_1)| \leq \sum_{i=1}^{r-1} d_i\omega_1\left(\frac{1}{2^i}\right)\left|\sin 2^i \frac{t_2 - t_1}{2}\right| + \sum_{i=r}^{\infty} \omega_1\left(\frac{1}{2^i}\right) \leq$$

$$\leq \sum_{i=1}^{r-1} \ell_i\omega_1\left(\frac{1}{2^i}\right)2^{i-r} + \omega_1\left(\frac{1}{2^r}\right)\sum_{i=o}^{\infty} 2^{-\lambda i} \leq K\omega_1\left(\frac{1}{2^r}\right) \leq$$

$$\leq K\omega_1(|t_2 - t_1|) \quad ,$$

$$K = C(C + 2) + (1 - 2^{-\lambda})^{-1}$$

(cf. (12) and (14)). Similarly

$$|T_k(t_2) - T_k(t_1)| \leq K\omega_2(|t_2 - t_1|) \quad .$$

If C_1, C_2 are large enough, then 2) is fulfilled.

3) Obviously

$$b_k(t) = - C_2 \sum_{i=k}^{m_k} \ell_i 2^i \omega_1\left(\frac{1}{2^i}\right) \sin 2^i t \quad ,$$

$$x_k(t) = \exp\{A_k(t)\} \int_0^t \exp\{- A_k(\tau)\} b_k(\tau) d\tau =$$

$$= - \left[C_1 + \sum_{i=k}^{m_k} \ell_i\omega_2\left(\frac{1}{2^i}\right) \sin 2^i t\right]^{-1} \times$$

$$\int_0^t \left[C_1 + \sum_{i=k}^{m_k} \ell_i\omega_2\left(\frac{1}{2^i}\right) \sin 2^i\tau\right] C_2 \sum_{i=k}^{m_k} 2^i \ell_i\omega_1\left(\frac{1}{2^i}\right) \sin 2^i\tau \, d\tau \quad .$$

As

$$C_1 > \sum_{i=1}^{\infty} \omega_2\left(\frac{1}{2^i}\right) \quad ,$$

it follows that

$$|x_k(t)| \geq C_3 \Big[\sum_{i=k}^{m_k} 2^i \ell_i \psi\Big(\frac{1}{2^i}\Big) \int_0^t \sin^2 2^i\tau \, d\tau$$

$$- C_1 \sum_{i=k}^{m_k} 2^i \ell_i \omega_1\Big(\frac{1}{2^i}\Big) \Big| \int_0^t \sin 2^i\tau \, d\tau \Big| -$$

$$- \sum_{i,j=k}^{m_k} 2^i \ell_i \ell_j \omega_1\Big(\frac{1}{2^i}\Big) \omega_2\Big(\frac{1}{2^j}\Big) \Big| \int_0^t \sin 2^i\tau \sin 2^j\tau \, d\tau \Big| \Big],$$

$$C_3 > 0 \ .$$

If $j \neq i$, then

$$2^i \Big| \int_0^t \sin 2^i\tau \sin 2^j\tau \, d\tau \Big| \leq \frac{2^i}{|2^i - 2^j|} + \frac{2^i}{2^i + 2^j} \leq 3$$

$$\sum_{\substack{i,j=k \\ j\neq i}}^{m_k} 2^i \ell_i \ell_j \omega_1\Big(\frac{1}{2^i}\Big) \omega_2\Big(\frac{1}{2^j}\Big) \Big| \int_0^t \sin 2^i\tau \sin 2^j\tau \, d\tau \Big| \leq$$

$$\leq 3 \Big[\sum_{i=k}^{m_k} \omega_1\Big(\frac{1}{2^i}\Big) \Big] \Big[\sum_{i=k}^{m_k} \omega_2\Big(\frac{1}{2^i}\Big) \Big] \ .$$

Consequently

$$|x_k(t)| \geq C_3 \Big\{ \frac{t}{2} \sum_{i=k}^{m_k} 2^i \ell_i \psi\Big(\frac{1}{2^i}\Big) - \frac{1}{4} \sum_{i=k}^{m_k} \ell_i \psi\Big(\frac{1}{2^i}\Big) \sin 2^{i+1} t$$

$$- 2C_1 \sum_{i=k}^{m_k} \omega_1\Big(\frac{1}{2^i}\Big) - 3 \Big[\sum_{i=k}^{m_k} \omega_1\Big(\frac{1}{2^i}\Big) \Big] \Big[\sum_{i=k}^{m_k} \omega_2\Big(\frac{1}{2^i}\Big) \Big] \Big\} \ .$$

Let us choose m_k in such a manner that

$$\sum_{i=k}^{m_k} 2^i \ell_i \psi\left(\frac{1}{2^i}\right) \longrightarrow \infty$$

with $k \longrightarrow \infty$ (cf. (13)). Then $|x_k(t)| \longrightarrow \infty$ with $k \longrightarrow \infty$ for $t \neq 0$ (cf. (12)).

REFERENCES

[1] GICHMAN, I. I.,Concerning a theorem of N. N. Bogoljubov, Ukrain. Math. Journal, 4, 2, 215-219 (in Russian).

[2] KRASNOSELSKIJ, M. A. and KREJN, S. G., Averaging principle in non-linear mechanics, Uspechi mat. nauk., 10, 3 (1955), 147-152 (in Russian).

[3] KURZWEIL, J. and VOREL, Z., Continuous dependence on a parameter of solutions of differential equations, Czechoslovak math. journal, 7, (82), 4 (1957), 568-583 (in Russian).

[4] KURZWEIL, J., Generalized ordinary differential equations and continuous dependence on a parameter, Czech. math. journal, 7 (82), 1957, 3, 418-449.

III. POINCARÉ'S PERTURBATION METHOD AND TOPOLOGICAL DEGREE

Jane Cronin

§1. INTRODUCTION

We study the periodic solutions of a system of differential equations

$$\frac{dx}{dt} = F(x, t, \mu) \tag{1.1}$$

where x, F are n-vectors, appropriate differentiability and periodicity conditions are imposed on the components of F, and μ is a real parameter. Poincaré's perturbation method is used and the results extended by using topological degree. The technique developed gives a general approach to the existence problem for the degenerate case (the case for which the variational equation has periodic solutions) and new existence theorems are obtained if the degree of degeneracy exceeds one. (In problems of mechanical or electrical oscillations, the degree of degeneracy is frequently greater than one.) The technique consists in using topological degree to make an 'in the large' study of the bifurcation system (Verzweigungsgleichungen) without imposing any local uniqueness conditions on the solutions of the bifurcation system. The results obtained are generalizations and refinements of the results in a previous paper [7], (Numbers in brackets refer to the bibliography at the end of this paper.) although the present paper is independent of [7]. Existence theorems are obtained for a wider class of equations; autonomous systems are studied; and it is shown that for the totally degenerate case (degree of degeneracy equals dimension of the system), the topological degree is a kind of lower bound for the number of distinct periodic solutions.

The literature on this subject is extensive. General treatments are described briefly in [7]. Bass [2] has given a large bibliography including not only general treatments but papers on applications. The work of Friedrichs [8, 9, 10], Coddington and Levinson [3, 4] and

Lefschetz [11, 12] are most closely related to our approach. We base our treatment on the results of Coddington and Levinson concerning properties of the bifurcation system. As in Lefschetz's work, we need not impose local uniqueness conditions on the solutions of the bifurcation system, and our results may be regarded as an extension of the criterion for the existence of real periodic solutions given by Lefschetz.

In Section 2, we describe how the bifurcation system is set up. Periodic solutions of the original differential equation correspond to solutions of the bifurcation system. We study solutions of the bifurcation system by determining the topological degree of the mapping defined by the bifurcation system. By using the properties of the bifurcation system obtained by Coddington and Levinson [3, 4], we show in Section 2 that the problem of determining the topological degree can be reduced to that of determining the topological degree of a mapping in Euclidean q-space where q is the degree of degeneracy.

In Section 3, we apply the method to obtain some existence theorems for periodic solutions of non-autonomous differential equations. In particular we show that if the function $F(x, t, \mu)$ in (1.1) is such that $F(x, t, \mu) = Ax + \mu f(x, t, \mu)$ where A is a constant matrix and $f(x, t, \mu) = f_1(x, \mu) + f_2(t, \mu)$ where the components of $f_1(x, 0)$ behave like polynomials for large x, then, except for cases for which the topological degree is not defined, equation (1.1) has, for sufficiently small μ, at least one periodic solution near the initial solution.

In Section 4, the exceptional case in which the topological degree is not defined is discussed. In Section 5, the case in which the topological degree is even is briefly considered. In Section 6, we prove that the topological degree is, for the totally degenerate case, a kind of lower bound for the number of distinct periodic solutions. Finally in Section 7, application of topological degree to an autonomous differential equation is described.

I am indebted to Professor S. Lefschetz for a number of extremely helpful discussions of this material.

Most of the work on this paper was done under the sponsorship of the Office of Naval Research contract Nonr-1858(04).

§2. THE BIFURCATION SYSTEM

In [8], Friedrichs treated the equation

$$\dot{x} = F(x, t, \mu) \qquad (\cdot = d/dt) \tag{2.1}$$

where x, F are n-vectors; the components of F have continuous derivatives with respect to x, t, and μ; these derivatives have continuous derivatives with respect to x; $F(x, t, \mu)$ is periodic in t with period $T(\mu)$ where $d^2T/d\mu^2$ is continuous; and (2.1) has for $\mu = 0$ a solution $x_0(t)$ of period $T(0)$. The problem is to determine if for sufficiently small μ, equation (2.1) has a solution $x(t, \mu)$ periodic of period $T(\mu)$. The classical result of Poincaré ([13], Vol. 1, Chapter IV) states that if the variational equation $\dot{x} = A(t)x$, where

$$A(t) = \frac{\partial F}{\partial x}\Big]_{x=x_0(t),\mu=0} ,$$

has no non-zero solutions of period $T(0)$, then for sufficiently small μ, equation (2.1) has a unique solution $x(t, \mu)$, periodic of period $T(\mu)$, such that

$$\lim_{\mu \to 0} x(t, \mu) = x_0(t) .$$

Here we treat the degenerate case in which the variational equation has q linearly independent solutions, periodic of period $T(0)$, where $1 \leq q \leq n$. The number q is called the degree of degeneracy of the problem.

Friedrichs shows that if

$$\frac{\partial F}{\partial \mu}\Big]_{x=x_0(t),\mu=0} = 0 ,$$

then the problem can be reduced to the study of the equation

$$\dot{x} = A(t)x + \mu f(x, t, \mu) \tag{2.2}$$

and then derives a bifurcation system for (2.2). Coddington and Levinson ([3], [4]), by imposing the further condition that $A(t)$ is a constant matrix and then putting this constant matrix in a canonical form, derive a more explicit form for the bifurcation system. (Coddington and Levinson also assume that $T(\mu)$ is a constant function. This is not a restriction in the generality because Friedrichs shows in [8] that the general case may be reduced to the case $T(\mu)$ a constant.)

Now we describe the bifurcation system derived by Coddington and Levinson. The system is obtained by using the variation of constants formula to describe the solution and then imposing the condition that the solution be periodic of period 2π (i.e., assume $T(0) = 2\pi$.) The system is:

$$(2.3)\qquad (e^{2\pi A} - E)c + \mu \int_0^{2\pi} \{e^{(2\pi - s)A} f[x(s, \mu, c), s, \mu]\} ds = 0 \quad ,$$

(this is equation (1.12), p. 23, of [3] or equation (3.20), p. 360 of [4]) where E is the identity matrix, c is an n-vector such that $x(t, \mu, c)$ is a solution of (2.2) with initial value c, i.e., $x(0, \mu, c)\ c$, and $x(t, \mu, c)$ has period 2π. Thus (2.3) is a system of n equations in the n unknowns which are the components $c_1, \ldots, c_n$ of c. The problem of finding periodic solutions for (2.2) is reduced to that of solving (2.3) for $c_1, \ldots, c_n$.

In order to study the properties of (2.3), Coddington and Levinson assumed that matrix A already has the following real canonical form

$$A = \begin{pmatrix} A_1 & & & & & & & \\ & \ddots & & & & & & \\ & & A_k & & & & & \\ & & & B_1 & & & & \\ & & & & \ddots & & & \\ & & & & & B_m & \\ & & & & & & C \end{pmatrix}$$

where the elements not shown are zeros. Each A_j, $j = 1, \ldots, k$, is a matrix of α_j (α_j even) rows and columns of the form

$$A_j = \begin{pmatrix} S_j & & & \\ E_2 & S_j & & \\ & \ddots & \ddots & \\ & & E_2 & S_j \end{pmatrix}$$

where all elements are zero except S_j and E_2, and

$$S_j = \begin{pmatrix} 0 & -N_j \\ N_j & 0 \end{pmatrix} , \qquad E_2 = \begin{pmatrix} 1 & 0 \\ 0 & 1 \end{pmatrix}$$

where N_j is a positive integer. A matrix A_j may have only two rows and columns in which case $A_j = S_j$. Each matrix B_j has β_j rows and columns, $j = 1, \ldots, m$, and is of the form

$$B_j = \begin{pmatrix} 0 & 0 & \cdot & \cdot & \cdot & \cdot & 0 \\ 1 & 0 & & & & & \cdot \\ 0 & 0 & \ddots & & & & \cdot \\ \vdots & \ddots & \ddots & \ddots & & & \cdot \\ \cdot & & \ddots & \ddots & \ddots & & \cdot \\ \cdot & & & \ddots & \ddots & \ddots & \cdot \\ 0 & \cdots & & 0 & & 1 & 0 \end{pmatrix}$$

where B_j may have only one row and column in which case B_j consists of the single element 0. The matrix C has $(n - \Sigma_{j=1}^{k}\alpha_j - \Sigma_{j=1}^{m}\beta_j)$ rows and columns and has no characteristic roots of the form iN for any integer N including $N = 0$. Matrix C need not be in canonical form.

If $(c_1, \ldots, c_i, \ldots, c_n)$ is an n-vector, the indices i corresponding to the last two rows of any A_j or to the last row of any B_j are called exceptional indices. They are the indices with the following form:

$$i = \alpha_1 + \alpha_2 + \ldots + (\alpha_j - 1)$$
$$i = \alpha_1 + \alpha_2 + \ldots + \alpha_j$$

where $j = 1, \ldots, k$ and

$$i = \alpha_1 + \ldots + \alpha_k + \beta_1 + \ldots + \beta_j$$

where $j = 1, \ldots, m$. The indices i corresponding to the first two rows of any A_j or to the first row of any B_j are called singular indices. They are the indices with the following form:

$$\begin{aligned} i = 1, 2, \alpha_1 + 1, \alpha_1 + 2, \ldots, \alpha_1 + \alpha_2 + \ldots + \alpha_{k-1} + 1, \\ \alpha_1 + \alpha_2 + \ldots + \alpha_{k-1} + 2, \alpha_1 + \ldots + \alpha_k + 1 \quad , \\ \alpha_1 + \ldots + \alpha_k + \beta_1 + 1, \ldots, \alpha_1 + \ldots + \alpha_k + \beta_1 + \ldots + \beta_{m-1} + 1 \; . \end{aligned}$$

There are $(2k + m)$ exceptional indices and $(2k + m)$ singular indices. The number $q = 2k + m$ is the degree of degeneracy of the problem. Throughout our study, we assume that $q > 0$, i.e., that there is at least one A_j or one B_j in the canonical form of matrix A.

Now let $(c_1', \ldots, c_{n-q}')$ be the (n-q)-vector whose components are the components (in the same order) of $(c_1, \ldots, c_n)$ which have non-exceptional indices and let $(c''_1, \ldots, c''_q)$ be the q-vector whose components are the components of $(c_1, \ldots, c_n)$ which have exceptional indices. Let subscript j denote a singular index and j' a non-singular index. It is shown in [3] and [4] that system (2.3) can be replaced by the system

$$N(c'_1, \ldots, c'_{n-q}) + \mu \left\{ \int_0^{2\pi} e^{(2\pi - s)A} f[x(s, \mu, c), s, \mu] ds \right\}_{(j')} = 0 \tag{2.4}$$

$$\left\{ \int_0^{2\pi} e^{(2\pi - s)A} f[x(s, \mu, c), s, \mu] ds \right\}_{(j)} = 0$$

where N is a non-singular $(n-q) \times (n-q)$ matrix acting on vector $(c'_1, \ldots, c'_{n-q})$ and

$$\left\{ \int_0^{2\pi} e^{(2\pi - s)A} f[x(s, \mu, c), s, \mu] ds \right\}_{(j')}$$

denotes the vector composed of the $(n-q)$ components of

$$\left\{ \int_0^{2\pi} e^{(2\pi - s)A} f[x(s, \mu, c), s, \mu] ds \right\}$$

which have non-singular indices. Similarly

$$\left\{ \int_0^{2\pi} e^{(2\pi - s)A} f[x(s, \mu, c), s, \mu] ds \right\}_{(j)}$$

denotes the vector composed of the q components which have singular indices.

The left side of system (2.4) describes a continuous mapping (call it $\mathcal{M}$) of real Euclidean n-space into itself. Let c'' denote the vector c in which the $(n - q)$ components with non-exceptional indices have been set equal to zero. Then

$$\left\{ \int_0^{2\pi} e^{(2\pi - s)A} f[e^{sA} c'', s, 0] ds \right\}_{(j)}$$

defines a continuous mapping of real q-space into itself which we denote by $\mathcal{M}_0$. We call the system

$$\left\{ \int_0^{2\pi} e^{(2\pi - s)A} f[e^{sA} c'', s, 0] ds \right\}_{(j)} = 0$$

the reduced bifurcation system.

Friedrichs [8] and Coddington and Levinson, [3] and [4] assume the existence of a solution c'' of the reduced bifurcation system such that the Jacobian of the reduced bifurcation system is non-zero at c''. From this, the existence of a solution of (2.3) and hence of a periodic solution of (2.2) is obtained. We use instead the notion of topological degree.

LEMMA 2.1. For given r, there is an $\varepsilon > 0$ such that if $|\mu| < \varepsilon$, the topological degree of $\mathcal{M}$ at the origin and relative to a solid (n-1)-sphere with radius r and center at the origin is equal (except possibly for sign) to the topological degree of $\mathcal{M}_0$ at the origin (in q-space) and relative to a solid (q-1)-sphere with radius r and center at the origin.

PROOF. The proof follows from the definition of topological degree and the invariance under homotopy of the topological degree. (see [1, Deformationssatz, p. 424].)

Lemma 2.1 shows that in order to demonstrate the existence of periodic solutions $x(t, \mu, c)$ of (2.2) it is sufficient to show that the topological degree of $\mathcal{M}_0$ is non-zero. For then the degree of $\mathcal{M}$ is non-zero; hence for given small μ, system (2.3) has at least one solution $c_0 = (c_1^0, \ldots, c_n^0)$. This implies that (2.2) has a solution $x(t, \mu, c_0)$ of period 2π such that $x(0, \mu, c_0) = (c_1^0, \ldots, c_n^0)$. From the fundamental properties of topological degree, the initial values $(c_1, \ldots, c_n)$ are continuous in μ in the following sense: if for given μ_0, $(c_1^0, \ldots, c_n^0)$ is an isolated solution of (2.3) which has non-zero topological index, then given $\varepsilon > 0$, there is a $\delta > 0$ such that if $|\mu_1 - \mu_0| < \delta$, there is at least one periodic solution $x(t, \mu, c_1)$ with initial value $(c_1', \ldots, c_n')$ such that $|c_i' - c_i^0| < \varepsilon$ for $i = 1, \ldots, n$; if $(c_1^0, \ldots, c_n^0)$ is not an isolated solution of (2.3), i.e., $(c_1^0, \ldots, c_n^0)$ is a limit point of a set S of solutions $\{(c_1^\nu, \ldots, c_n^\nu)\}$ of (2.3) for $\mu = \mu_0$, then suppose that the topological degree of an open set N containing S but no other solutions of (2.3) is different from zero. Then given $\varepsilon > 0$, there is a $\delta > 0$ such that if $|\mu - \mu_0| < \delta$, then there is at least one solution $x(t, \mu, c)$ with initial value $(c_1, \ldots, c_n)$ such that for some $(c_1^\nu, \ldots, c_n^\nu) \in S$, we have $|c_i^\nu - c_i| < \varepsilon$ for $i = 1, \ldots, n$. From the usual existence theorems for differential equations, it follows that the solutions $x(t, \mu, c)$ are themselves continuous in μ in the same sense.

§3. EXISTENCE THEOREMS FOR THE NON-AUTONOMOUS CASE

We obtain existence theorems for periodic solutions of the equation

$$\dot{x} = Ax + \mu f(x, t, \mu) \tag{3.1}$$

where A is a constant matrix; μ is a real parameter; and f, $\partial f/\partial x_i$ are continuous in (x, t, μ) for small $|\mu|$ and all (x, t); and f has period 2π in t. The periodic solutions of the variational equation $\dot{x} = Ax$ correspond to the characteristic roots of A that are of the form iN where N is an integer.

ASSUMPTION 1. Suppose that A has just one pair of characteristic roots of the form iN, $-iN$ where we assume the integer N is non-zero. Then A may be put in the standard form

$$A = \left(\begin{array}{cc|c} 0 & -N & 0 \\ N & 0 & \\ \hline \multicolumn{2}{c|}{0} & A_0 \end{array}\right)$$

where A_0 is not necessarily canonical but is real and has no characteristic roots of the form iN. We assume that A is in this standard form.

ASSUMPTION 2. The components $f_1(x, t, \mu)$ and $f_2(x, t, \mu)$ of $f(x, t, \mu)$ have the form

$$f_i(x, t, \mu) = H_i(x_1, \ldots, x_n, \mu) + K_i(t, \mu) \qquad (i = 1, 2)$$

where $x_1, \ldots, x_n$ are the components of x, and H_i, K_i are functions of the indicated variables which have continuous second derivatives in these variables. The functions $K_1(t, \mu)$ and $K_2(t, \mu)$ have period 2π in t. Also there exist polynomials $P_1(x_1, x_2)$ and $P_2(x_1, x_2)$ such that

$$\lim_{r \to \infty} \frac{H_i(x_1, x_2, 0, \ldots, 0, 0)}{P_i(x_1, x_2)} = 1 \qquad (i = 1, 2,)$$

where $r^2 = x_1^2 + x_2^2$.

Two real independent periodic solutions of the variational equation are:

$$x^{(1)}(t) = (\cos Nt, \sin Nt, 0, \ldots, 0)$$
$$x^{(2)}(t) = (-\sin Nt, \cos Nt, 0, \ldots, 0) \;.$$

Following Coddington and Levinson, we may write the reduced bifurcation system as:

$$(3.2)\qquad \int_0^{2\pi} \Big\{\cos(Ns)H_1\big[c_1x^{(1)}(-s) + c_2x^{(2)}(-s),\ 0\big] - \sin(Ns)H_2(c_1x^{(1)} + c_2x^{(2)},\ 0)\Big\}\,ds + \int_0^{2\pi} \Big\{\cos(Ns)K_1(-s,\ 0) - \sin(Ns)K_2(-s,\ 0)\Big\}\,ds = 0 \ .$$

$$(3.3)\qquad \int_0^{2\pi} \Big\{\sin(Ns)H_1\ c_1x^{(1)}(-s) + c_2x^{(2)}(-s),\ 0 + \cos(Ns)H_2(c_1x^{(1)} + c_2x^{(2)},\ 0)\Big\}\,ds + \int_0^{2\pi} \Big\{\sin(Ns)K_1(-s,\ 0) + \cos(Ns)K_2(-s,\ 0)\Big\}\,ds = 0 \ .$$

(Here for convenience we have not indicated the components of $c_1x^{(1)} + c_2x^{(2)}$ separately although H_1 and H_2 are actually functions of these components. We use this convention with P_1 and P_2 also.) Let k_1 and k_2 denote the constants which are the second integrals on the left sides of equations (3.2) and (3.3). Let $\mathscr{H}_1(c_1,\ c_2)$ and $\mathscr{H}_2(c_1,\ c_2)$ denote the first integrals on the left sides of (3.2) and (3.3), and let $\mathscr{P}_1(c_1,\ c_2)$ and $\mathscr{P}_2(c_1,\ c_2)$ denote the polynomials obtained if in $\mathscr{H}_1$ and $\mathscr{H}_2$, the terms $H_1(c_1x^{(1)} + c_2x^{(2)},\ 0)$ and $H_2(c_1x^{(1)} + c_2x^{(2)},\ 0)$ are replaced by $P_1(c_1x^{(1)} + c_2x^{(2)})$ and $P_2(c_1x^{(1)} + c_2x^{(2)})$.

ASSUMPTION 3. If $\mathscr{P}_1$ and $\mathscr{P}_2$ are of degrees n_1 and n_2 respectively in c_1 and c_2, let Q_1 and Q_2 be the homogeneous polynomials of degrees n_1 and n_2 in $\mathscr{P}_1$ and $\mathscr{P}_2$ respectively. Let $M : (c_1,\ c_2) \longrightarrow (c_1',\ c_2')$ be the mapping defined by

$$Q_1(c_1,\ c_2) = c_1'$$
$$Q_2(c_1,\ c_2) = c_2' \ .$$

We assume that the topological index at $(0,\ 0)$ of M is defined.

From standard arguments, we have:

LEMMA 3.1. Let $M_1 : (c_1, c_2) \longrightarrow (c_1', c_2')$ be defined by:

$$\mathcal{P}_1(c_1, c_2) + k_1 = c_1'$$

$$\mathcal{P}_2(c_1, c_2) + k_2 = c_2' \quad .$$

The topological degree of mapping M_1 at $(0, 0)$ and relative to any sufficiently large circle with center $(0, 0)$ is equal to the topological index at $(0, 0)$ of M.

THEOREM 3.1. Let Assumptions 1, 2, 3 be satisfied. Then for all sufficiently small μ, Equation (3.1) has at least one solution of period 2π.

PROOF. Let $\mathcal{M}_0 : (c_1, c_2) \longrightarrow (c_1', c_2')$ be defined by:

$$\mathcal{H}_1(c_1, c_2) + k_1 = c_1'$$

$$\mathcal{H}_2(c_1, c_2) + k_2 = c_2' \quad .$$

It is sufficient by Lemma 2.1 to show that the topological degree of $\mathcal{M}_0$ at $(0, 0)$ and relative to a circle with center at $(0, 0)$ is different from zero.

We show first that relative to any sufficiently large circle, the topological degrees of M_1 and $\mathcal{M}_0$ at $(0, 0)$ are the same. From Assumption 3 and the fact that $\mathcal{P}_1$ and $\mathcal{P}_2$ are polynomials, it follows that given $m > 0$, there exists $r_1 > 0$ such that if $c_1^2 + c_2^2 > r_1^2$, then

$$[\mathcal{P}_1(c_1, c_2) + k_1]^2 + [\mathcal{P}_2(c_1, c_2) + k_2]^2 > m \quad .$$

Further there exists $\varepsilon > 0$ and $r_2 > 0$ such that if $|\eta_1| < \varepsilon$, $|\eta_2| < \varepsilon$, and $c_1^2 + c_2^2 > r_2^2$, then

$$\Big[\int_0^{2\pi} \Big\{ \cos(Ns)[P_1(c_1 x^{(1)} + c_2 x^{(2)})](1 + \eta_1)$$

$$- \sin(Ns)[P_2(c_1 x^{(1)} + c_2 x^{(2)})](1 + \eta_2) \Big\} \, ds + k_1 \Big]^2 \quad +$$

$$+ \left[\int_0^{2\pi} \left\{ \sin(Ns)[P_1(c_1x^{(1)} + c_2x^{(2)})](1 + \eta_1) \right.\right.$$

$$\left.\left. + \cos(Ns)[P_2(c_1x^{(1)} + c_2x^{(2)})](1 + \eta_2) \right\} ds + k_2 \right]^2 > \frac{1}{2} m \ .$$

By Assumption 2, there exists $r_3 > 0$ such that if $c_1^2 + c_2^2 > r_3^2$, then

$$H_i(c_1x^{(1)} + c_2x^{(2)}, 0) = (1 + \eta_i)P_i(c_1x^{(1)} + c_2x^{(2)})$$

where $|\eta_i| < \varepsilon$ for $i = 1, 2$. Hence the homotopy $h: (c_1, c_2, t) \longrightarrow (c_1'(t), c_2'(t))$ defined by:

$$c_1'(t) = \int_0^{2\pi} \left\{ \cos(Ns)[P_1(c_1x^{(1)} + c_2x^{(2)})](1 + t\eta_1) \right.$$

$$\left. - \sin(Ns)[P_2(c_1x^{(1)} + c_2x^{(2)})](1 + t\eta_2) \right\} ds + k_1$$

$$c_2'(t) = \int_0^{2\pi} \left\{ \sin(Ns)[P_1(c_1x^{(1)} + c_2x^{(2)})](1 + t\eta_1) \right.$$

$$\left. - \cos(Ns)[P_2(c_1x^{(1)} + c_2x^{(2)})](1 + t\eta_2) \right\} ds + k_2$$

shows that the topological degrees of M_1 and $\mathcal{M}_0$ at $(0, 0)$ and relative to any sufficiently large circle with center $(0, 0)$ are the same.

A trivial computation based on the fact that

$$\int_0^{2\pi} \sin^m(x) \cos^n(x)\, dx \neq 0$$

only if m and n are even shows that $\mathcal{P}_1$ and $\mathcal{P}_2$ are polynomials of odd degree in c_1 and c_2. Then the topological index at $(0, 0)$ of M is odd [5] and hence the topological degrees of M_1 and $\mathcal{M}_0$ are different from zero. This completes the proof of Theorem 3.1.

As a second example, we consider Equation (3.1) under the following assumptions.

ASSUMPTION 4. Matrix A has just one pair of characteristic roots iN, $-iN$ where N is a non-zero integer and A has 0 as a characteristic root. Also A has the form

$$A = \begin{pmatrix} \begin{matrix} 0 & -N & 0 \\ N & 0 & 0 \\ 0 & 0 & 0 \end{matrix} & 0 \\ 0 & A_0 \end{pmatrix}$$

where A_0 is not necessarily canonical but is real and has no characteristic roots of the form iN.

ASSUMPTION 5. The components $f_1(x, t, \mu)$, $f_2(x, t, \mu)$ and $f_3(x, t, \mu)$ of $f(x, t, \mu)$ have the form:

$$f_i(x, t, \mu) = H_i(x_1, \ldots, x_n, \mu) + K_i(t, \mu) \qquad (i = 1, 2, 3)$$

where $x_1, \ldots, x_n$ are the components of x; and H_i and K_i are functions of the indicated variables and have continuous second derivatives in these variables; and K_1, K_2, K_3 have period 2π in t. There exist polynomials $P_1(x_1, x_2, x_3)$, $P_2(x_1, x_2, x_3)$, $P_3(x_1, x_2, x_3)$ such that

$$\lim_{r \to \infty} \frac{H_i(x_1, x_2, x_3, 0, \ldots, 0)}{P_i(x_1, x_2, x_3)} = 1 \qquad (i = 1, 2, 3) \quad .$$

Also the degree of P_3 is k, odd, and there is a term ax_3^k (with $a \neq 0$) in P_3.

Three real independent periodic solutions of the variational equation are:

$$x^{(1)} = (\cos(Nt), \sin(Nt), 0, \ldots, 0)$$
$$x^{(2)} = (-\sin(Nt), \cos(Nt), 0, \ldots, 0)$$
$$x^{(3)} = (0, 0, 1, \ldots, 0).$$

The reduced bifurcation system is:

$$\int_0^{2\pi} [\cos(Ns)H_1(c_1x^{(1)} + c_2x^{(2)} + c_3x^{(3)}, 0)$$
$$- \sin(Ns)H_2(c_1x^{(1)} + c_2x^{(2)} + c_3x^{(3)}, 0)] \, ds$$
$$+ \int_0^{2\pi} [\cos(Ns)K_1(-s, 0) - \sin(Ns)K_2(-s, 0)] \, ds = 0$$

$$\int_0^{2\pi} [\sin(Ns)H_1 + \cos(Ns)H_2]\, ds + \int_0^{2\pi} [\sin(Ns)K_1 + \cos(Ns)K_2]\, ds = 0$$

$$\int_0^{2\pi} H_3(c_1 x^{(1)} + c_2 x^{(2)} + c_3 x^{(3)}, 0)\, ds = 0 \quad .$$

Let k_i, $\mathcal{H}_i$, $\mathcal{P}_i$, and Q_i $(i = 1, 2, 3)$ be defined analogously to their counterparts in the preceding example.

ASSUMPTION 6. Let $M : (c_1, c_2, c_3) \longrightarrow c_1', c_2', c_3'$ be the mapping defined by $c_i' = Q_i(c_1, c_2, c_3)$ for $i = 1, 2, 3$. We assume that the topological index at $(0, 0)$ of M is defined.

By the same type of argument as for the preceding example, we obtain:

THEOREM 3.2. Let Assumptions 4, 5, 6 be satisfied. Then for all sufficiently small μ, Equation (3.1) has at least one solution of period 2π.

More complicated examples may be treated in a similar manner.

§4. EXCEPTIONAL CASES

By Assumption 3, the topological index at $(0, 0)$ of mapping M is defined. Now suppose that this assumption is not satisfied, i.e., suppose the topological index is not defined. This means that Q_1, Q_2 have a common real factor. Hence by varying one coefficient in Q_1 (i.e., varying H_1) arbitrarily slightly, we obtain a mapping for which the topological index is defined. When such a change is made, two different cases arise. First suppose that after Q_1 is changed slightly in some definite manner, the topological index j of the resulting mapping is defined and suppose that j is odd or that $|j| > 2$. Then by using the analysis of [5, Section 3] it can be seen that regardless of how Q_1 or Q_2 is changed (provided the magnitude of the change is sufficiently small) the topological index will be non-zero and there will be at least one periodic solution.

Now suppose that after Q_1 is changed slightly in some definite way, the index j is zero. Then no conclusion can be drawn about whether there is a periodic solution. Further suppose that $j = 2$ or -2. Then, again by using the analysis of [5, Section 3] it can be shown that a different small change in Q_1 will yield a mapping of index zero. Again no conclusion about the existence of periodic solutions can be drawn.

Another exceptional case which may occur is this. In the first example of Section 3, functions $\mathscr{H}_1(c_1, c_2)$ or $\mathscr{H}_2(c_1, c_2)$ may be identically zero. If $\mathscr{H}_1(c_1, c_2) \equiv 0$ and $k_1 \neq 0$, the bifurcation system is inconsistent and there are no periodic solutions with initial values continuous in μ for small μ, If $H_1(c_1, c_2) \equiv 0$, and $k_1 = 0$, no immediate conclusion can be drawn. It is worth noting that if we consider a problem in which

$$\left.\frac{dT}{d\mu}\right]_{\mu=0} \neq 0 \quad ,$$

then as is shown in [8], neither $\mathscr{H}_1$ or $\mathscr{H}_2$ is identically zero.

§5. IF THE TOPOLOGICAL DEGREE IS EVEN

The condition (in Assumptions 2 and 5) that the $f_1(x, t, \mu)$ can be written as the sum of a term that depends only on x and μ and a term that depends only on t and μ implies that the topological degree is odd and therefore non-zero. If we do not impose this condition, the topological degree may be odd or even. By the results of [5] and Lemma 3.1, if there are two equations in the reduced bifurcation system, the topological degree can be computed in all cases and the computation consists simply in approximating the real roots of certain polynomials. (See [7], Theorems 1 and 2.) If there are three or more equations in the reduced bifurcation system, the degree can be computed in many special cases.

If the topological degree is zero, Equation (3.1) may or may not have periodic solutions as the example in [7, Section 5] shows.

§6. THE NUMBER OF DISTINCT PERIODIC SOLUTIONS

For the totally degenerate case $(q = n)$, the topological degree also yields an estimate of the number of distinct periodic solutions. We illustrate this by considering the first example of Section 3 with the additional hypothesis that $n = q$. As before, the reduced bifurcation system is:

$$\mathscr{H}_1(c_1, c_2) + k_1 = 0$$

$$\mathscr{H}_2(c_1, c_2) + k_2 = 0 \quad .$$

LEMMA 6.1. Let M be a continuous mapping defined on the closure $\bar{O}$ of an open set $O \subset R^n$, real Euclidean n-space, and suppose M is differentiable in O. Suppose the topological degree of M at point p_0 and relative to O is $d \neq 0$. Then there is a neighborhood U of p_0 and a set E of n-dimensional measure zero, $E \subset U$, such that if $p \in U - E$, then $M^{-1}(p)$ is a finite set consisting of at least $|d|$ points.

This lemma is proved in [6, p. 213].

Now suppose that the topological degree at $(0, 0)$ and relative to a solid circle S with center $(0, 0)$ and radius r of the mapping defined by the reduced bifurcation system is d for all sufficiently large r. Let mapping M_3 be defined by $c_1' = \mathscr{H}_1(c_1, c_2)$, $c_2' = \mathscr{H}_2(c_1, c_2)$. Then if r is sufficiently large, the topological degree of M_3 at $(-k_1, -k_2)$ and relative to S is also equal to d. Applying this fact and Lemma 6.1 to the bifurcation system, we see that if k_1, k_2 are changed arbitrarily slightly, there will be at least $|d|$ distinct solutions of the bifurcation system. Since

$$k_1 = \int_0^{2\pi} \left\{\cos(Ns)K_1(-s, 0) - \sin(Ns)K_2(s, 0)\right\} ds$$

and similarly for k_2, we obtain:

THEOREM 6.1. Suppose K_1 and K_2 are independent of μ and suppose the topological degree at $(0, 0)$ of the mapping defined by the left side of the reduced bifurcation system is equal to d if the topological degree is taken relative to any sufficiently large solid circle with center at $(0, 0)$. Let μ have a fixed value, say μ_0. Then given $\eta > 0$, there exist functions $K_{11}(t)$, $K_{12}(t)$, both of period 2π, with continuous second derivatives, such that

$$\max_{0 \leq t \leq 2\pi} |K_{1i}(t) - K_i(t)| < \eta \qquad (i = 1, 2)$$

and such that if $\mu = \mu_0$ in Equation (3.1) and $K_1(t)$, $K_2(t)$ are replaced by $K_{11}(t)$, $K_{12}(t)$, then the resulting equation has at least $|d|$ distinct solutions of period 2π.

§7. THE AUTONOMOUS CASE

Topological degree can also be used to study autonomous systems, but because the bifurcation system is quite different in the autonomous case, a different approach must be used. We describe the application to an example.

Consider the equation:

$$\dot{x} = Ax + \mu f(x, \mu) \quad . \tag{7.1}$$

Assume A has just one pair of characteristic roots of the form iN, $-iN$, where N is a non-zero integer, that A has zero as a characteristic root, and that A has the standard form described in Assumption 4. Assume further that $n = q$. The bifurcation system may be written:

$$- \nu c_2 + \int_0^{2\pi} \Big\{ \cos(Ns) f_1(c_1 x^{(1)} + c_2 x^{(2)} + c_3 x^{(3)}, 0)$$

$$- \sin(Ns) f_2(c_1 x^{(1)} + c_2 x^{(2)} + c_3 x^{(3)}, 0) \Big\} ds = 0$$

$$\nu c_1 + \int_0^{2\pi} \Big\{ - \sin(Ns) f_1 + \cos(Ns) f_2 \Big\} ds = 0$$

$$\nu c_3 + \int_0^{2\pi} \Big\{ f_3(c_1 x^{(1)} + c_2 x^{(2)} + c_3 x^{(3)}, 0) \Big\} ds = 0 \quad ,$$

where $f_1(x, \mu)$, f_2, f_3 are the components of f, and $x^{(1)}$, $x^{(2)}$, and $x^{(3)}$ have the same meaning as before and

$$\nu = \lim_{\mu \to 0} \frac{\tau(\mu)}{\mu}$$

where $\tau(\mu)$ is the period of the solutions of (7.1). The physical problem represented by the autonomous case is such that the corresponding mathematical problem can be posed as: set one $c_i = 0$ and solve for the remaining c_i's and ν. (See [3, pp. 30-31] and [4, pp. 364-366].)

Let us assume that $f_1(x, 0)$, $f_2(x, 0)$, $f_3(x, 0)$ are polynomials in the components of x. Then the bifurcation system may be written:

$$-\nu c_2 + P_1(c_1, c_2, c_3) = 0 \tag{7.2}$$

$$\nu c_1 + P_2(c_1, c_2, c_3) = 0 \tag{7.3}$$

$$\nu c_3 + P_3(c_1, c_2, c_3) = 0 \tag{7.4}$$

where P_1, P_2, P_3 are polynomials in c_1, c_2, c_3. We set $c_3 = 0$ and solve (7.2), (7.3), (7.4) for $c_1, c_2,$ and ν. Multiplying (7.2) by c_1 and (7.3) by c_2, and adding, we obtain the system:

$$c_1P_1(c_1, c_2, 0) + c_2P_2(c_1, c_2, 0) = 0 \tag{7.5}$$

$$P_3(c_1, c_2, 0) = 0 \quad . \tag{7.6}$$

If μ is varied slightly, Equations (7.5) and (7.6) become

$$c_1R_1(c_1, c_2, \mu) + c_2R_2(c_1, c_2, \mu) = 0 \tag{7.5)'$$

$$R_3(c_1, c_2, \mu) = 0 \tag{7.6)'$$

where R_1, R_2, R_3 are continuous functions of $c_1, c_2,$ and μ. Suppose the topological degree of the mapping defined by the left sides of (7.5) and (7.6) is non-zero. Then for a sufficiently small μ, (7.5)' and (7.6)' have a solution (c_1^0, c_2^0). When μ is varied slightly in (7.2), (7.3), and (7.4) (but keeping $c_3 = 0$) we obtain:

$$-\nu c_2 + R_1(c_1, c_2, \mu) = 0 \tag{7.2)'$$

$$\nu c_1 + R_2(c_1, c_2, \mu) = 0 \tag{7.3)'$$

$$R_3(c_1, c_2, \mu) = 0 \quad . \tag{7.4)'$$

Now suppose one of the c_1^0, c_2^0, say c_1^0, is different from zero. (This will occur if, for example, $P_3(c_1, c_2, 0)$ has a constant term.) We set

$$\nu_0 = \frac{R_2(c_1^0, c_2^0, \mu)}{c_1^0} \quad .$$

Then c_1, c_2, ν_0 are a solution of (7.2)', (7.3)', and (7.4)'.

REFERENCES

[1] ALEXANDROFF, P., and HOPF, H., Topologie 1, Berlin, 1935.

[2] BASS, R. W., Extension of frequency method of analyzing relay-operated servomechanisms, Section III, Final report, Contract DA-36-034-ORD-1273 RD, Johns Hopkins Institute for Cooperative Research (1955).

[3] CODDINGTON, E. A., and LEVINSON, N., "Perturbations of linear systems with constant coefficients possessing periodic solutions", Contributions to the theory of nonlinear oscillations, Vol. II, Annals of Mathematics Studies No. 29, Princeton, 1952.

[4] CODDINGTON, E. A., and LEVINSON, N., Theory of ordinary differential equations, New York, 1953.

[5] CRONIN, J., "Topological degree of some mappings", Proc. Amer. Math. Soc., 5 (1954), pp. 175-178.

[6] CRONIN, J., "Branch points of solutions of equations of Banach Space, II, Trans. Amer. Math. Soc., 76 (1954), pp. 207-222.

[7] CRONIN, J., "Note to Poincaré's perturbation method", Duke Mathematical Journal, 26 (1959), pp. 251-262.

[8] FRIEDRICHS, K. O., Advanced ordinary differential equations (mimeographed) New York University, 1949.

[9] FRIEDRICHS, K. O., "Fundamentals of Poincaré's theory", Proceedings of the symposium on nonlinear circuit analysis, New York, 1953, pp. 56-67.

[10] FRIEDRICHS, K. O., Special topics in analysis (mimeographed), New York University, 1954.

[11] LEFSCHETZ, S., "Complete families of periodic solutions of differential equations", Comment. Math. Helv., 28 (1954), pp. 341-345.

[12] LEFSCHETZ, S., Differential equations: geometric theory, New York, 1957.

[13] POINCARÉ, H., Les méthodes nouvelles de la mécanique céleste, Vol. I, II, III, (1892-1899). Reprinted by Dover Publications, New York, 1957.

Polytechnic Institute of Brooklyn

IV. ON THE BEHAVIOR OF THE SOLUTIONS OF LINEAR PERIODIC DIFFERENTIAL SYSTEMS NEAR RESONANCE POINTS

Jack K. Hale

§1. INTRODUCTION

In the last few years, systems of differential equations of the form

$$(1.1) \qquad y_j'' + \sigma_j^2 y_j = \varepsilon \sum_{k=1}^{n} \varphi_{jk} y_k, \qquad j = 1, \ldots, n ,$$

where ε is a real small parameter, the constants σ_j are positive, and the real functions $\varphi_{jk}(t)$ are periodic of period $T = 2\pi/\omega$ and L-integrable in $[0, T]$, have been investigated by many authors [1, 3a, 4a, 5, 7, 8, 9, 10b, 10e, 12, 13]. (See also the book [3b]). For $n = 1$, classical results of O. Haupt [9] assure that all solutions of (1.1) are bounded for $|\varepsilon|$ sufficiently small and an arbitrary periodic function φ_{11} provided $2\sigma_1 \not\equiv 0 \pmod{\omega}$.

For the study of the system (1.1) for $|\varepsilon|$ small and $n \geq 1$, L. Cesari [3a] in 1940 considered a method of successive approximations, which was successively developed by L. Cesari, J. K. Hale and R. A. Gambill [6, 4b, 10c, 10e] and applied to questions of existence and stability of periodic solutions of weakly nonlinear differential systems (cf. the book [3b] and [3c]). By using this method it was first proved [3a, 10b] that all solutions of (1.1) are bounded provided $2\sigma_j = 0$, $\sigma_j \pm \sigma_k \not\equiv 0 \pmod{\omega}$, $j \neq k$, $j, k = 1, 2, \ldots, n$, and the matrix $\Phi(t) = [\varphi_{jk}(t)]$ is either symmetric or even in t. Under the same restrictions on the numbers σ_j more general boundedness theorems have then been proved [5a, 10e] by the same method, but they all involve some type of "symmetry" conditions on the matrix $\Phi(t)$. Again using the same method it has been proved [3a, 5b] that some "symmetry" condition is necessary to assure that all solutions of (1.1) are bounded for ε

sufficiently small. For systems more general than (1.1), see [1, 10e].

Analogous boundedness theorems involving "symmetry" conditions have been proved by M. Golomb [7] by a different method. Finally, J. Moser[13], J. M. Gel'fand and V. B. Lidskii [14] have recently shown that it is sufficient to require $2\sigma_j \not\equiv 0$, $\sigma_j + \sigma_k \not\equiv 0 \pmod{\omega}$, $j \neq k$, $j, k = 1, 2, \ldots, n$, when Φ is a symmetric matrix.

In the present paper a procedure based on the same method developed by Cesari, Hale and Gambill is given for the study of the behavior of the solutions of (1.1) (and more general systems) near the "resonance points" $2\sigma_j = s\omega$, $\sigma_j \pm \sigma_k = s\omega$, for some integer s. The ultimate goal, of course, is to try to classify systems of type (1.1) according to the "resonance points" for which some solutions are unbounded for every $\varepsilon \neq 0$. This general problem is not solved in this paper. However, we do give some general theorems concerning the behavior of the solutions of a class of differential equations (1.1) (and even a more general type) around the resonance points (A)$\sigma_1 - \sigma_2 = s\omega$, s an integer or zero, $2\sigma_j \not\equiv 0$, $\sigma_j \pm \sigma_k \not\equiv 0 \pmod{\omega}$, $j \neq k$, $j, k = 2, 3, \ldots, n$ [Theorems (5.2) and (5.3)] and (B) $\sigma_1^2 = \varepsilon\, \sigma^2$, $\sigma_j \not\equiv 0$, $2\sigma_j \not\equiv 0$, $\sigma_j \pm \sigma_k \not\equiv 0 \pmod{\omega}$, $j \neq k$, $j, k = 2, 3, \ldots, n$ [Theorem (5.4)].

More specifically, if the σ_j satisfy (A) and Theorem (5.2) is applied to system (1.1) with (C) $\Phi = (\Phi_{jk})$, where each $\Phi_{jk}(-t) = (-1)^{k+j}\, \Phi_{jk}(t)$ is a matrix, then the AC (absolutely continuous) solutions of (1.1) are bounded for ε sufficiently small if a certain function $G(s, \sigma)$ of the Fourier coefficients of the functions $\varphi_{jk}(t)$ is is positive and some AC solutions are unbounded for every $\varepsilon \neq 0$ if this function is negative. If Φ is symmetric and also satisfies (C), then $G(s, \sigma) > 0$ (Remark 5.3), which agrees with the conclusions of J. Moser [13]. Some examples are given in §5 to show that $G(s, \sigma)$ may be < 0 for some matrices Φ if Φ satisfies (C) and is not symmetric. If Theorem (5.4) is applied to system (1.1) satisfying (B), (C), then all of the AC solutions of (1.1) are bounded for ε sufficiently small.

In §6, the above method is applied to systems of Mathieu type equations of the form (1.1) where each $\varphi_{jk}(t) = d_{jk} \cos 2t$, each d_{jk} is a constant and $2\sigma_j \not\equiv 0$, $\sigma_j \pm \sigma_k \not\equiv 0 \pmod{1}$, $j \neq k$, $j, k = 2, 3, \ldots, n$, $\sigma_1(0) = m$, where m is a positive integer. Sufficient conditions are given (Theorem 6.1) to insure that there are unbounded absolutely continuous solutions of (1.1) in every neighborhood of the point $(m, 0)$ in the (σ_1, ε)-plane, where m is a positive integer. A corollary of Theorem (6.1) is the well known fact [11] that there are unbounded solutions of the Mathieu equation $x'' + \sigma^2 x + \varepsilon(\cos 2t)x = 0$, in every neighborhood of the point $(m, 0)$ of the (σ, ε)-plane for every positive integer m.

Using determinants, the behavior of the solutions of (1.1), $n > 1$, near the "resonance points" has also been discussed by E. Mettler [12] and W. Haacke [3] for the case where $\Phi(t)$ is even in t. General theorems of the above type seem to be easier to obtain using the method of this paper since it does not involve determinants.

Another application of the method discussed here concerns the stability of periodic solutions of weakly nonlinear differential systems. The linear variational equations associated with such a periodic solution is a linear differential equation with periodic coefficients of the form (1.1) and, in many cases, some of the basic frequencies σ_j are "in resonance" with ω. The asymptotic stability of periodic solutions of weakly nonlinear periodic differential systems has been discussed recently by H. R. Bailey and R. A. Gambill [2] using essentially the same method. For a more complete discussion of the application of this method, in particular, Theorems (2.1) and (3.1), to the stability of periodic solutions of both weakly nonlinear autonomous and weakly nonlinear periodic differential systems, see [10g].

§2. DESCRIPTION OF THE METHOD

Let C_ω denote the family of all functions which are finite sums of functions of the form $f(t) = e^{\alpha t}\varphi(t)$, $-\infty < t < +\infty$, where α is any complex number and $\varphi(t)$ is any complex-valued function of the real variable t, periodic of period $T = 2\pi/\omega$, L-integrable in $[0, T]$. Following L. Cesari [3a], if $f(t) = e^{\alpha t}\varphi(t) \in C_\omega$ and $\varphi(t)$ has the Fourier series

$$\varphi(t) \sim \sum_{k=-\infty}^{+\infty} c_n e^{in\omega t} ,$$

$i = \sqrt{-1}$, then we denote by mean value, $M[f]$, the number $M[f] = 0$ if $in\omega + \alpha \neq 0$ for all n; $M[f] = c_n$ if $in\omega + \alpha = 0$ for some n. It is clear that $M[f]$ is uniquely defined. Finally define $M[f]$ in the class C_ω as an additive functional. Obviously $M[f]$ reduces to the usual mean value for periodic functions $f(t)$ of period T. The following lemma is needed in the sequel.

LEMMA (2.1) If $f(t) \in C_\omega$ and $M[f] = 0$, then there is a unique primitive of $f(t)$, say $F(t) = \int f(t)dt$, which belongs to C_ω and $M[f] = 0$. (L. Cesari [3a], J. K. Hale [10a]).

Consider a linear system of differential equations of the form

$$z' = Az + \varepsilon\, C(t)z \tag{2.1}$$

where ε is a parameter, A, C are $N \times N$ matrices, $C = (c_{jk}(t))$, each $c_{jk}(t)$ is periodic in t of period $T = 2\pi/\omega$, L-integrable in $[0, T]$, $A = \operatorname{diag}(\rho_1, \ldots, \rho_N)$, where the ρ_j are complex numbers such that

$$\begin{aligned} &\rho_j - \rho_k = m_{jk}\omega i,\ m_{jk} \text{ an integer or zero}, \quad j, k = 1, 2, \ldots, \nu, \\ &\rho_j \not\equiv \rho_k \pmod{\omega i},\ j = 1, 2, \ldots, \nu; \quad k = \nu + 1, \ldots, N . \end{aligned} \tag{2.2}$$

Notice that all of the ρ_j, $j = 1, 2, \ldots, \nu$, could be equal and also the remaining $N - \nu$ of the ρ_k, $k = \nu + 1, \ldots, N$, can be two by two congruent mod ωi. This becomes important in discussing the stability of periodic solutions of nonlinear differential systems [10g].

In this section, a method of successive approximations is given for the determination of the characteristic exponents as well as the solutions of (2.1). This method coincides with the one given by L. Cesari [3a] for the case where $\nu = 1$ in (2.2). More specifically, we try to determine solutions of (2.1) of the form

$$z(t) = e^{\tau t}p(t),\ p(t + T) = p(t) , \tag{2.3}$$

where τ is a complex number to be determined. The equation for $p(t)$ is then

$$p' = (A - \tau I)p + \varepsilon\, C(t)p . \tag{2.4}$$

In the following, the characteristic exponent τ which is "close to" ρ_1 is to be determined, i.e., τ as a function of ε is such that $\tau(0) = \rho_1$. From (2.2) it follows that $\rho_k - \tau$ will be "close to" a multiple of ωi for $k = 1, 2, \ldots, \nu$. Rather than solve (2.4) directly, consider an auxiliary equation

$$p' = Bp + \varepsilon\, C(t)p , \tag{2.5}$$

where

$$\begin{aligned} &B = \operatorname{diag}(\rho_1^*, \ldots, \rho_N^*) , \\ &\rho_1^* = 0,\ \rho_k^* = m_{k1}\omega i,\ k = 2, 3, \ldots, \nu,\ \rho_k^* = \rho_k - \tau,\ k = \nu + 1, \ldots, N. \end{aligned} \tag{2.6}$$

and the m_{jk} are defined in (2.2). Suppose that S_1 is a circle in the complex plane with the center at ρ_1 such that

(2.7) $$\rho_j^* \not\equiv 0 \pmod{\omega i},\ j = \nu + 1, \ldots, N ,$$

for all $\tau \in S_1$.

If the mth approximation to $p(t)$ is denoted by

(2.8) $$p^{(m)} = x^{(0)} + \varepsilon x^{(1)} + \ldots + \varepsilon^m x^{(m)} ,$$

where each $x^{(r)}$ is periodic in t of period $T = 2\pi/\omega$, let

(2.9) $$s^{(r)}(t) = e^{-Bt}C(t)x^{(r-1)}(t) ,$$
$$s^{(r)} = (s_1^{(r)}, \ldots, s_N^{(r)}) ,$$

and define

(2.10) $$S^{(r)} = \mathrm{diag}(S_1^{(r)}, \ldots, S_\nu^{(r)}, 0, \ldots, 0)$$
$$S_j^{(r)} = 0 \quad \text{if} \quad M[s_j^{(r)}] = 0 ,$$
$$d_j S_j^{(r)} = M[s_j^{(r)}] \quad \text{if} \quad M[s_j^{(r)}] \neq 0 \quad j = 1, 2, \ldots, \nu ,$$

where $M[s_j^{(r)}]$ denotes the mean value for functions of the class C_ω and $d_1, \ldots, d_\nu$ are complex numbers. If $M[s_j^{(r)}] \neq 0$ for any r, it is understood in definition (2.10) that the corresponding d_j is chosen $\neq 0$ so that $S_j^{(r)}$ will be well defined in all cases. In fact, if d_j is a function of ε, then d_j is $\neq 0$ for $\varepsilon = 0$.

With these notations, the successive approximations are defined by

(2.11) $$x^{(0)} = (d_1 e^{\rho_1^* t}, \ldots, d_\nu e^{\rho_\nu^* t}, 0, \ldots, 0)$$
$$x^{(r)} = e^{Bt} \int e^{-Bt}[C(t)x^{(r-1)} - (S^{(r)}x^{(0)} + \ldots$$
$$S^{(1)}x^{(r-1)})]dt, \ r = 1, 2, 3, \ldots .$$

It is clear that condition (2.7) assures that each of the integrands in (2.11) belongs to the class C_ω and has mean value zero provided that the integrations are always performed so as to obtain the unique primitive of mean value zero. It is assumed that the integrations are performed in this manner. Furthermore, the $x^{(r)}(t)$ defined by (2.11) are periodic of period T.

The method of successive approximations (2.11) is precisely the one given in [6] for obtaining periodic solutions of nonlinear systems of differential equations. Exactly as in [6], it follows that $p^{(m)}$ defined by (2.8), (2.11) converges uniformly in t for $|\varepsilon|$ sufficiently small to a solution of the equation

$$p' = [B - \varepsilon H(\tau, d, \rho, \varepsilon)]p + \varepsilon C(t)p \quad , \tag{2.12}$$

where $d = (d_1, \ldots, d_\nu)$, $\rho = (\rho_1, \ldots, \rho_N)$ and $H = \mathrm{diag}(h_1, \ldots, h_\nu, 0, \ldots, 0)$ and each $h_j(\tau, d, \rho, \varepsilon)$ is an analytic function of ε at $\varepsilon = 0$ and its power series expansion is given by

$$h_j = S_j^{(1)} + \varepsilon S_j^{(2)} + \varepsilon^2 S_j^{(3)} + \ldots, \; j = 1, 2, \ldots, \nu \quad . \tag{2.13}$$

Therefore, if the numbers $\tau, d_1, \ldots, d_\nu, \rho_1, \ldots, \rho_\nu$ are chosen so that the matrix equation,

$$B - \varepsilon H(\tau, d, \rho, \varepsilon) = A - \tau I \tag{2.14}$$

or, equivalently, the equations,

$$\begin{gathered} m_{k1}\omega i - \varepsilon h_k(\tau, d, \rho, \varepsilon) = \rho_k - \tau, \\ k = 1, 2, \ldots, \nu, \; m_{11} = 0 \quad , \end{gathered} \tag{2.14'}$$

are satisfied, then the periodic solution $p(t)$ of (2.12) will be a periodic solution of (2.4) and $z(t)$ defined by (2.3) will be a solution of (2.1). We shall refer to the equations (2.14), (2.14') as the <u>determining equations</u> for $\tau, d_1, \ldots, d_\nu, \rho_1, \ldots, \rho_\nu$. These represent ν equations for the $2\nu + 1$ parameters $\tau, d_1, \ldots, d_\nu, \rho_1, \ldots, \rho_\nu$, and the particular problem of interest will, in general, determine which of these parameters are to be determined and which are fixed.

REMARK 2.1. In the above discussion, it was assumed that the numbers ρ_j, $j = 1, 2, \ldots, N$, and the matrix $C(t)$ were independent of ε. This is not necessary and systems of the form

$$z' = A(\varepsilon)z + \varepsilon C(t, \varepsilon)z \tag{2.15}$$

could have been considered, where $A(\varepsilon) = \mathrm{diag}(\rho_1(\varepsilon), \ldots, \rho_N(\varepsilon))$, each $\rho_j(\varepsilon)$ is an analytic function of ε at $\varepsilon = 0$ with

$$\rho_j(0) - \rho_k(0) = m_{jk}\omega i, \ m_{jk} \text{ an integer or zero}, \ j, k = 1, 2, \ldots, \nu,$$

(2.16)

$$\rho_j(0) \not\equiv \rho_k(0), \ (\text{mod } \omega i), \ j = 1, 2, \ldots, \nu; \ k = \nu + 1, \ldots, N;$$

$C(t, \varepsilon) = (c_{jk}(t, \varepsilon))$, $j, k = 1, 2, \ldots, N$, where each $c_{jk}(t, \varepsilon)$ is analytic in ε for $0 \leqq |\varepsilon| \leqq \varepsilon_0$, $\varepsilon_0 > 0$, $|c_{jk}(t, \varepsilon)| < \eta(t)$, $0 \leqq |\varepsilon| \leqq \varepsilon_0$, $j, k = 1, 2, \ldots, N$, and $\eta(t)$ is L-integrable in $[0, T]$. The matrix B in (2.6) is exactly the same as before and the method of successive approximations is applied in the same manner by replacing everywhere ρ_j by $\rho_j(\varepsilon)$, $j = 1, 2, \ldots, N$, and the matrix $C(t)$ by $C(t, \varepsilon)$. The determining equations are

$$(2.17) \quad m_{k1}\omega i - \varepsilon h_k(\tau, d, \rho(\varepsilon), \varepsilon) = \rho_k(\varepsilon) - \tau, \ k = 1, 2, \ldots, \nu,$$

where $m_{11} = 0$ and the functions h_k are given by (2.13) with the $S_j^{(m)}$ analytic functions of ε for $0 \leqq |\varepsilon| \leqq \varepsilon_0$. The remainder of the discussion will deal with this more general system. It is also true that the above analyticity at $\varepsilon = 0$ can be replaced by continuity in ε at $\varepsilon = 0$ if $\rho_j(\varepsilon) - \rho_j(0) = O(\varepsilon)$, but to make the presentation as simple as possible we will not consider this generalization.

A solution $\tau, d_1, \ldots, d_\nu, \rho_1, \ldots, \rho_\nu$, of (2.17) must be such that $d_j \neq 0$ for $\varepsilon = 0$ if $M[s_j^{(r)}] \neq 0$ for any $r = 1, 2, \ldots$. In case some d_j is $O(\varepsilon)$, then the method of successive approximations must be modified slightly taking the zeroth approximation to have this $d_j = O(\varepsilon)$. We will not give the details of the method for such a case, since it is clear how one would proceed.

In case the numbers $\rho_1, \ldots, \rho_\nu$ are fixed, then the determining equations represent ν equations for the $\nu + 1$ parameters $\tau, d_1, \ldots, d_\nu$. Since (2.1) is a linear system, it is clear that one of the numbers d_j may be chosen equal to one, but, in general, this decision will depend upon the form of equations (2.17). Two solutions $(\tau, d_1, \ldots, d_\nu)$, $(\tau^*, d_1^*, \ldots, d_\nu^*)$ of (2.17) will be called <u>distinct</u> if they correspond to two linearly independent solutions of (2.1) of the form (2.3). A similar definition applies to k solutions of (2.17). If $\rho_j = \rho_k$, $j, k = 1, 2, \ldots, \nu$, one would hope that there would be ν distinct solutions of (2.17). However, this is not always the case since some of the characteristic exponents τ may be equal.

From (2.10) the functions $h_j(\tau, d, \rho(\varepsilon), \varepsilon)$ in (2.17) satisfy the property

$$(2.18)\quad d_j h_j(\tau,\ d,\ \rho(0),\ 0) = \frac{1}{T}\int_0^T \sum_{k=1}^{\nu} c_{jk}(t,0)d_k e^{i(m_{k1}-m_{j1})\omega t}dt,$$

$$j = 1,\ 2,\ \dots,\ \nu\ ,$$

which is a linear function of $d_1, \dots, d_\nu$ and is independent of τ. If $\varepsilon\beta = \tau - \rho_1(0)$, then for $\varepsilon \neq 0$ equations (2.17) are equivalent to the system of equations

$$(2.19)\qquad (G - \beta I)d + \varepsilon\, F(d,\ \tau,\ \rho(\varepsilon),\ \varepsilon) = 0$$

where $G = (g_{jk})$, $j, k = 1, 2, \dots, \nu$, is independent of ε and

$$g_{jj} = \frac{1}{T}\int_0^T c_{jj}(t,\ 0)dt + \lim_{\varepsilon \to 0} \frac{\rho_j(\varepsilon)-\rho_j(0)}{\varepsilon}$$

$$(2.20)\quad g_{jk} = \frac{1}{T}\int_0^T e^{i(m_{k1}-m_{j1})\omega t} c_{jk}(t,\ 0)dt,\ m_{11} = 0\ ,$$

$$j \neq k,\quad j,\ k = 1,\ 2,\ \dots,\ \nu\ ,$$

and $F(d,\ \tau,\ \rho(\varepsilon),\ \varepsilon)$ is an analytic vector function of ε for $0 \leqq |\varepsilon| \leqq \varepsilon_0$. These results are summarized in the following theorem.

THEOREM (2.1). Consider the system of differential equations (2.15), (2.16), and suppose the matrix G is defined by (2.20). If β_0 is a simple characteristic root of the matrix G and the corresponding eigenvector, d_0, has no zero components, then there is a characteristic exponent $\tau(\varepsilon)$ of (2.15) which is analytic in ε for $0 \leqq |\varepsilon| \leqq \varepsilon_0$, $\varepsilon_0 > 0$, and

$$(2.21)\qquad \tau(\varepsilon) = \rho_1(\varepsilon) + \varepsilon\,\beta_0 + O(\varepsilon^2)\ .$$

The proof of this theorem follows immediately by applying the implicit function theorem to (2.19). If the matrix d_0 has some components which are zero, then the above method of successive approximations must be modified slightly taking the first approximation (2.11) to have some $d_j = O(\varepsilon)$. In case the matrix G in Theorem (2.1) is the zero matrix, then the theorem says nothing about the characteristic exponents. However, by obtaining more of the functions $S_j^{(m)}$ in (2.17), one can determine the characteristic exponents. This theorem and the following corollary are not used in this paper but are important for the study of stability of periodic solutions of weakly nonlinear systems of differential equations (see [2], [10g]).

COROLLARY (2.1). Consider the system of differential equations (2.15) with

$$\rho_j(0) \not\equiv \rho_k(0) \pmod{\omega i}, \; k \neq j, \; k = 1, 2, \dots, N \quad , \tag{2.22}$$

where j is a fixed integer, $1 \leqq j \leqq N$. Then there is a characteristic exponent of (2.15), $\tau_j = \tau_j(\varepsilon)$, analytic in ε for $0 \leqq |\varepsilon| \leqq \varepsilon_0$, and

$$\tau_j(\varepsilon) = \rho_j(\varepsilon) + \frac{\varepsilon}{T}\int_0^T c_{jj}(t, 0)dt + O(\varepsilon^2) \quad , \tag{2.23}$$

where the terms of higher order in ε may be calculated by using the above method of successive approximations.

REMARK 2.2. L. Cesari [3a] had previously observed that one could calculate the characteristic exponents one at a time to obtain relation (2.23) but under the more restrictive requirement that $\rho_j \not\equiv \rho_k \pmod{\omega i}$ $j \neq k$, $j, k = 1, 2, \dots, N$. It is also clear from his work that the above corollary holds for a particular characteristic exponent τ_j, with the weaker restriction (2.22).

For specific examples, the method of successive approximations of this section is probably as efficient as any, but to obtain general boundedness theorems and the form of the functions h_j in (2.13) for classes of differential equations, the following formulation is somewhat more convenient when the ρ_k satisfy certain additional properties.

§3. AN ALTERNATIVE FORMULATION OF THE METHOD

Suppose the numbers ρ_k in (2-1) satisfy the property*

$$\begin{gathered} \rho_{2j-1} = \bar{\rho}_{2j}, \; j = 1, 2, \dots, \nu \quad , \\ \rho_{2j-1} - \rho_{2k-1} = m_{jk}\omega i, \; m_{jk} \text{ an integer or zero, } j, k = 1, 2, \dots, \nu, \\ \rho_j \not\equiv \rho_k \pmod{\omega i}, \; j = 1, 2, \dots, 2\nu, \; k = 2\nu + 1, \dots, N \quad . \end{gathered} \tag{3.1}$$

---

* Throughout this paper, $\bar{a}$ will always denote the complex conjugate of a.

Notice that ρ_{2j-1} may or may not be congruent to ρ_{2j} modulo ωi, $j = 1, 2, \ldots, \nu$. Both situations are important in the applications. Rather than make the substitution (2.3), consider the auxiliary equation

$$z' = Bz + \varepsilon C(t)z \tag{3.2}$$

where $B = \operatorname{diag}(\rho_1^*, \ldots, \rho_N^*)$,

$$\begin{aligned} \rho_1^* &= \tau = \bar{\rho}_2^*, \quad \rho_{2k-1}^* = \tau - m_{k1}\omega_i = \bar{\rho}_{2k}^*, \quad k = 1, 2, \ldots, \nu \ , \\ \rho_k^* &= \rho_k, \quad k = 2\nu + 1, \ldots, N \ , \end{aligned} \tag{3.3}$$

where the numbers m_{k1} are defined by (3.1), and τ is an undetermined complex number such that

$$\rho_j^* \not\equiv \rho_k^*, \quad j = 1, 2, \ldots, 2\nu; \quad k = 2\nu + 1, \ldots, N \ . \tag{3.4}$$

If the mth approximation is defined by

$$z^{(m)} = x^{(0)} + \varepsilon x^{(1)} + \ldots + \varepsilon^m x^{(m)} \ , \tag{3.5}$$

$s^{(r)}$ is defined as in (2.9) with B as in (3.3) and

$$\begin{aligned} S^{(r)} &= \operatorname{diag}(S_1^{(r)}, \ldots, S_{2\nu}^{(r)}, 0, \ldots, 0) \\ S_j^{(r)} &= 0 \quad \text{if} \quad M[s_j^{(r)}] = 0 \\ d_j S_j^{(r)} &= M[s_j^{(r)}] \quad \text{if} \quad M[s_j^{(r)}] \neq 0, \ j = 1, 2, \ldots, 2\nu \ , \end{aligned} \tag{3.6}$$

where $M[s_j^{(r)}]$ again denotes the mean value for functions of the class C_ω and $d_1, \ldots, d_{2\nu}$ are complex numbers. If $M[s_j^{(r)}] \neq 0$ for any r, it is understood in definition (3.6) that the corresponding d_j is chosen $\neq 0$ so that $S_j^{(r)}$ will be well defined in all cases. In fact, if d_j is a function of ε, then d_j is $\neq 0$ for $\varepsilon = 0$. The successive approximations are now defined by

$$\begin{aligned} x^{(0)} &= (d_1 e^{\rho_1^* t}, \ldots, d_{2\nu} e^{\rho_{2\nu}^* t}, 0, \ldots, 0) \ , \\ x^{(r)} &= e^{Bt} \int e^{-Bt} [C(t)x^{(r-1)} - (S^{(r)}x^{(0)} + \ldots + S^{(1)}x^{(r-1)})]dt, \\ & \qquad r = 1, 2, 3, \ldots, \end{aligned} \tag{3.7}$$

where the integrations are performed so as to obtain the unique primitive of mean value zero. The proof of convergence of the functions $z^{(m)}$ defined by (3.5), (3.7) to a solution of the equation

$$z' = [B - \varepsilon V(\tau, d, \varepsilon)]z + \varepsilon C(t)z \quad , \tag{3.8}$$

where $V = \operatorname{diag}(V_1, V_2, \ldots, V_{2\nu}, 0, \ldots, 0)$, $(d = (d_1, \ldots, d_{2\nu})$,

$$V_j = S_j^{(1)} + \varepsilon S_j^{(2)} + \ldots \tag{3.9}$$

is exactly the same as before. The determining equations for τ, d_1, $\ldots$, $d_{2\nu}$, ρ_1, $\ldots$, $\rho_{2\nu}$, are

$$\tau - \varepsilon V_j(\tau, d, \varepsilon) = \rho_j, \; j = 1, 2, \ldots, 2\nu \quad . \tag{3.10}$$

Remark (2.1) at the end of §2 also applies to the method defined in this manner. In the following, our choice between these two methods will depend upon the application and the convenience which it affords.

In case system (2.1) arises from a real system of first and second order differential equations, some of the determining equations (3.10) are redudant as we show in the next few lines. Consider the system of real equations

$$\begin{aligned} y_j'' + \alpha_j y_j' + \sigma_j^2 y_j &= \varepsilon \sum_{k=1}^{n} \varphi_{jk} y_k + \varepsilon \sum_{k=1}^{\mu} \psi_{jk} y_k', \; j = 1, 2, \ldots, \mu \quad , \\ y_j' + \beta_j y_j &= \varepsilon \sum_{k=1}^{n} \varphi_{jk} y_k + \varepsilon \sum_{k=1}^{\mu} \psi_{jk} y_k', \; j = \mu + 1, \ldots, n \quad , \end{aligned} \tag{3.11}$$

where α_j, β_j, σ_j, ε are real numbers, $\gamma_j \equiv (4\sigma_j^2 - \alpha_j^2)^{\frac{1}{2}} > 0$, $j = 1, 2, \ldots, \mu$, and the functions $\varphi_{jk}(t)$, $\psi_{jk}(t)$ are real, periodic in t of period $T = 2\pi/\omega$, L-integrable in $[0, T]$. Suppose the roots of the equation $z^2 + \alpha_j z + \sigma_j^2 = 0$ are λ_{2j-1}, $\lambda_{2j} = \bar{\lambda}_{2j-1}$, $j = 1, 2, \ldots, \mu$, and let $\lambda_{\mu+j} = -\beta_j$, $j = \mu + 1, \ldots, n$. The transformation

$$\begin{aligned} y_j = (2i\gamma_j)^{-1}(z_{2j-1} + z_{2j}), \; y_j' &= (2i\gamma_j)^{-1}(\lambda_{2j-1}z_{2j-1} + \lambda_{2j}z_{2j}) \; , \\ & j = 1, 2, \ldots, \mu, \\ y_j = z_{\mu+j}, \; j &= \mu + 1, \ldots, n \quad , \end{aligned} \tag{3.12}$$

yields the equivalent system of first order equations,

$$z' = Az + \varepsilon\, C(t)z \tag{3.13}$$

where $A = \operatorname{diag}(\lambda_1, \ldots, \lambda_{n+\mu})$ and C is an $(n+\mu) \times (n+\mu)$ matrix whose elements $c_{jk}(t)$ are given by

$$c_{2j-1,2k-1} = (2i\gamma_k)^{-1}(\varphi_{jk} + \rho_{2k-1}\psi_{jk});\ c_{2j-1,2k} = (2i\gamma_k)^{-1}(\varphi_{jk} + \rho_{2k}\psi_{jk}),$$

$$k = 1, 2, \ldots, \mu,$$

$$c_{2j-1,\mu+h} = \varphi_{jh},\ h = \mu+1, \ldots, n;\ c_{2j-1,\ell} = -c_{2j,\ell},\ \ell = 1,2,\ldots,n+\mu;$$

$$j = 1, 2, \ldots, \mu; \tag{3.14}$$

$$c_{\mu+h,2k-1} = (2i\gamma_k)^{-1}(\varphi_{hk} + \rho_{2k-1}\psi_{hk});\ c_{\mu+h,2k} = (2i\gamma_k)^{-1}(\varphi_{hk} + \rho_{2k}\psi_{hk}),$$

$$k = 1, 2, \ldots, \mu;\ c_{\mu+h,\mu+\ell} = \varphi_{h\ell},\ \ell = \mu+1, \ldots, n;\ h = \mu+1, \ldots, n\,.$$

We also suppose that the equations in (3.13) are reordered so that

$$A = \operatorname{diag}(\rho_1, \ldots, \rho_{n+\mu}) \tag{3.15}$$

where each ρ_j is one of the numbers λ and the ρ_j satisfy condition (3.1). It is very easy to prove the following lemma.

LEMMA (3.1). If the alternative method of successive approximations is applied to the auxiliary equation of (3.13), (3.15) with the d_j in (3.7) such that $d_{2j-1} = -\bar{d}_{2j}$, $j = 1, 2, \ldots, \nu$, then

$$V_{2j-1}(\tau, d, \varepsilon) = \bar{V}_{2j}(\tau, d\, \varepsilon),\quad j = 1, 2, \ldots, \nu\,, \tag{3.16}$$

for every τ, d where V_j is defined by (3.9). Therefore, the determining equations (3.10) become

$$\begin{aligned} &\tau - \varepsilon\, V_{2j-1}(\tau, d\, \varepsilon) = \rho_1, && j = 1, 2, \ldots, \nu\,, \\ &d = (d_1, \ldots, d_{2\nu}),\ d_{2j} = -\bar{d}_{2j-1}, && j = 1, 2, \ldots, \nu\,. \end{aligned} \tag{3.17}$$

For applications to the stability of periodic solutions of nonlinear differential equations (see [10g]), it is convenient to have the following theorem.

THEOREM (3.1). Consider the system of differential equations (3.11) with

$$\lambda_{2j-1} = \bar{\lambda}_{2j},\ \lambda_{2j-1} \not\equiv \lambda_{2j} \pmod{\omega i},$$

(3.18) $$\lambda_{2j-1} - \lambda_{2k-1} = m_{jk}\omega i,\ m_{jk} \text{ an integer or zero, } j,\ k = 1,\ 2,\ \dots,\ \nu,$$

$$\lambda_j \not\equiv \lambda_k \pmod{\omega i},\ j = 1,\ 2,\ \dots,\ 2\nu;\ k = 2\nu + 1,\ \dots,\ N,$$

where $\lambda_{2j-1} = \frac{1}{2}(-\alpha_j + i\gamma_j)$, $\gamma_j = (4\sigma_j^2 - \alpha_j^2)^{\frac{1}{2}} > 0$, $j = 1, 2, \dots, \mu$, $\lambda_{\mu+j} = -\beta_j$, $j = \mu + 1, \dots, n$, and let $H = (h_{jk})$, $j, k = 1, 2, \dots, \nu$, be the $\nu \times \nu$ matrix defined by

(3.19) $$h_{jk} = \frac{1}{T}\int_0^T c_{2j-1,2k-1}(t)e^{i(m_{k1}-m_{j1})\omega t}\,dt, \qquad j,\ k = 1,\ 2,\ \dots,\ \nu,$$

where the functions $c_{2j-1,2k-1}$ are defined by (3.14). If β_0 is a simple root of the equation $|H - \beta I| = 0$ and the corresponding eigenvector has no zero components, then there are two characteristic exponents τ, $\bar{\tau}$ of (3.11) given by

(3.20) $$\tau = \rho_1 + \epsilon\,\beta_0 + O(\epsilon^2)$$

for ε sufficiently small.

PROOF. For $\varepsilon \neq 0$ the determining equations (3.17) can be written in the form

$$(H - \beta I)d = O(\varepsilon),$$

where $\varepsilon\,\beta = \tau - \rho_1$, $d = (d_1, d_3, \dots, d_{2\nu-1})$ and $H = (h_{jk})$ is the $\nu \times \nu$ matrix given by (3.19). It then follows immediately from the implicit function theorem that Theorem (3.1) is true for ε sufficiently small.

§4. SOME BASIC LEMMAS

Consider the system of linear differential equations,

$$(4.1)\quad \begin{aligned} y_j'' + \sigma_j^2 y_j &= \varepsilon \sum_{k=1}^{\mu} (\varphi_{jk}y_k + \psi_{jk}y_k') + \varepsilon \sum_{k=\mu+1}^{n} \psi_{jk}y_k, \quad j = 1, 2, \ldots, \mu, \\ y_j' &= \varepsilon \sum_{k=1}^{\mu} (\varphi_{jk}y_k + \psi_{jk}y_k') + \varepsilon \sum_{k=\mu+1}^{n} \psi_{jk}y_k, \quad j = \mu + 1, \ldots, n, \end{aligned}$$

where ε is a real parameter, each σ_j is a positive real number (and if $\sigma_j = \sigma_j(\varepsilon)$ is an analytic function of ε at $\varepsilon = 0$, then $\sigma_j > 0$ for $\varepsilon \neq 0$), each $\varphi_{jk}(t)$, $\psi_{jk}(t)$ is a real function, periodic in t of period $T = 2\pi/\omega$, L-integrable in $[0, T]$. Suppose the matrices $\Phi = (\varphi_{jk})$, $j = 1, 2, \ldots, n$, $k = 1, 2, \ldots, \mu$; $\Psi = (\psi_{jk})$, $j, k = 1, 2, \ldots, n$, are partitioned so that $\Phi = (\Phi_{jk})$, $j = 1, 2, 3$, $k = 1, 2$, $\Psi = (\Psi_{jk})$, $j, k = 1, 2, 3$, where Φ_{11}, Ψ_{11} are $p \times p$ matrices, Φ_{22}, Ψ_{22} are $(\mu - p) \times (\mu - p)$ matrices and Ψ_{33} is an $(n - \mu) \times (n - \mu)$ matrix. Throughout the present section, we always assume that

$$(4.2)\qquad \Phi_{jk}(t) = (-1)^{k+j}\,\Phi_{jk}(-t), \quad \Psi_{jk}(t) = (-1)^{k+j+1}\,\Psi_{jk}(-t)$$

for all j, k.

LEMMA (4.1). If $\sigma_j = \sigma_j(\varepsilon)$ is an analytic function of ε at $\varepsilon = 0$, $\sigma_j(0) \not\equiv 0 \pmod{\omega i}$, $j = 1, 2, \ldots, \mu$, and the matrices Φ, Ψ satisfy (4.2), then there exists an $\varepsilon_0 > 0$ such that there are $n - \mu$ linearly independent periodic solutions of (4.1) for $|\varepsilon| < \varepsilon_0$.

PROOF. The transformation $y_{\mu+j} = z_j$, $j = 1, 2, \ldots, n - \mu$, $y_j = (2i\sigma_j)^{-1}(z_{n-\mu+j} + z_{n+j})$, $y_j' = 2^{-1}(z_{n-\mu+j} - z_{n+j})$, $j = 1, 2, \ldots, \mu$, leads to an equivalent system of first order equations

$$(4.3)\qquad z' = Az + \varepsilon\, C(t)z \;,$$

where $A = \mathrm{diag}(0, \ldots, 0, i\sigma_1, \ldots, i\sigma_\mu, -i\sigma_1, \ldots, -i\sigma_\mu)$, $C = (C_{jk})$, $j = 1, 2, \ldots, 5$;

$$C_{jk} = -C_{j+2,k},\ j = 2,3,\ k = 1,2,\dots,5;\ C_{j,k} = -\bar{C}_{j,k+2},\ k = 2,3,\ j = 1,2,\dots,5;$$

$$C_{11} = \Psi_{33};\ C_{21} = \Psi_{13};\ C_{31} = \Psi_{23};\ 2C_{1,k+1} = \Phi_{3k}A_k^{-1} + \Psi_{3k},\ k = 1,2;\ 2C_{j+1,k+1} =$$

$$= \Phi_{j,k}A_k^{-1} + \Psi_{jk},\ j,\ k = 1,2;\ A_1 = \mathrm{diag}(i\sigma_1,\dots,i\sigma_p);\ A_2 = \mathrm{diag}(i\sigma_{p+1},\dots,i\sigma_\mu).$$

With this partitioning of the matrix C it is convenient to also partition the vector $z = (z_{(1)}, \dots, z_{(5)})$, where $z_{(j)}$ has a dimension compatible to the matrix multiplication. We now apply the above method of successive approximations of §2 directly to (4.3) with $x^{(0)} = (d, 0, 0, 0, 0)$, $d = (d_1, \dots, d_{n-\mu})$, where $d_1, \dots, d_{n-\mu}$ are arbitrary real numbers. The solution obtained in this manner will obviously be periodic. If the successive approximations $x^{(r)}$ are partitioned the same as z, i.e., $x^{(r)} = (x^{(r)}_{(1)}, \dots, x^{(r)}_{(5)})$, then we first show by induction that

$$H^{(r)} = \mathrm{diag}(S_1^{(r)}, \dots, S_{n-\mu}^{(r)}) = 0 \quad ,$$

$$x^{(r)}_{(1)}(-t) = x^{(r)}_{(1)}(t),\ x^{(r)}_{(2)}(-t) = -x^{(r)}_{(4)}(t),\ x^{(r)}_{(3)}(-t) = x^{(r)}_{(5)}(t)\ , \tag{4.4}$$

$$r = 0,\ 1,\ 2,\ \dots \ .$$

The induction is on the functions $x^{(r)}_{(j)}$. The assertion is clearly true for $r = 0$. Assume the assertion true for $r = 0, 1, 2, \dots, v-1$. Then the matrix $C(t)x^{(r)}(t)$ satisfies the property

$$C(-t)x^{(r)}(-t) = E\,C(t)x^{(r)}(t),\ E = \mathrm{diag}(-1, -1, 1, -1, 1) \ .$$
$$r = 0,\ 1,\ 2,\ \dots,\ v-1 \ .$$

Since the first vector component of this matrix product is an odd function of t it follows immediately from (2.10) that $H^{(r)} = 0$, $r = 1, 2, \dots, v$, for every $d = (d_1, \dots, d_{n-\mu})$. To complete the induction, we have from (2.11), since $B = A$,

$$x^{(v)}(-t) = e^{-At}\int e^{-A\alpha}C(\alpha)x^{(v-1)}(\alpha)d\alpha =$$

$$= -e^{-At}\int e^{A\alpha}C(-\alpha)x^{(v-1)}(-\alpha)d\alpha =$$

$$= -e^{-At}\int e^{A\alpha}EC(\alpha)x^{(v-1)}(\alpha)d\alpha \ .$$

This is clearly assertion (4.4) for $r = v$, since $C_{j,k} = - C_{j+2,k}$, $j = 2, 3$, $k = 1, 2, \ldots, 5$; therefore, (4.4) is true for all r. Since the d_j, $j = 1, 2, \ldots, n - \mu$, are arbitrary, one can find $n - \mu$ such linearly independent solutions and the lemma is proved.

Suppose the $\sigma_j = \sigma_j(\varepsilon)$ of (4.1) are analytic functions of ε at $\varepsilon = 0$, $\sigma_j(0) > 0$, $j = 1, 2, \ldots, \mu$, $\sigma_j(0) - \sigma_k(0) = m_{jk}\omega$, m_{jk} an integer or zero, $j, k = 1, 2, \ldots, q$, and the transformation (3.12) is applied to (4.1), to obtain the equivalent first order system

$$z' = Az + \varepsilon C(t)z \tag{4.5}$$

where $A = \mathrm{diag}(i\sigma_1, - i\sigma_1, \ldots, i\sigma_\mu, - i\sigma_\mu, 0, \ldots, 0)$ and $C = (\frac{1}{2} C_{jk})$, $j, k = 1, 2, \ldots, n + \mu$, where

$$\begin{aligned} &C_{2j-1,2k-1} = \frac{\varphi_{jk}}{i\sigma_k} + \psi_{jk}, \; C_{2j-1,2k} = \frac{\varphi_{jk}}{i\sigma_k} - \psi_{jk}, \; k = 1, 2, \ldots, \mu; \\ &C_{2j-1,h} = \psi_{j,h}, \; h = \mu + 1, \ldots, n; \; C_{2j,\ell} = - C_{2j-1,\ell}, \\ &\qquad \ell = 1, 2, \ldots, n; \; j = 1, 2, \ldots, \mu; \\ &C_{h,2k-1} = \frac{\varphi_{hk}}{i\sigma_k} + \psi_{hk}, \; C_{h,2k} = \frac{\varphi_{hk}}{i\sigma_k} - \psi_{hk}, \; k = 1, 2, \ldots, \mu; \\ &C_{h,m} = \psi_{hm}, \; m = \mu + 1, \ldots, n; \; h = \mu + 1, \ldots, n \; . \end{aligned} \tag{4.6}$$

The auxiliary equation of (4.5) is chosen as

$$z' = Bz + \varepsilon C(t)z \tag{4.7}$$

where $B = \mathrm{diag}(i\tau_1, i\tau_2, \ldots, i\tau_{2q}, i\sigma_{q+1}, - i\sigma_{q+1}, \ldots, i\sigma_\mu, - i\sigma_\mu, 0, \ldots, 0)$, and the τ's are real numbers such that

$$\begin{aligned} &\tau_{2j-1} = - \tau_{2j}, \; \tau_{2j-1} - \tau_{2k-1} = m_{jk}\omega, \; j, k = 1, 2, \ldots, q \; , \\ &\tau_{2j-1} \not\equiv 0, \; \tau_{2j-1} \pm \sigma_k(0) \not\equiv 0 \pmod{\omega}, \; j = 1, 2, \ldots, q, \\ &\qquad k = q + 1, \ldots, n \; ; \end{aligned} \tag{4.8}$$

and the m_{jk} are defined above by the numbers σ_j at $\varepsilon = 0$. Notice that $2\tau_j$ may or may not be a multiple of ω.

LEMMA (4.2). If the matrices Φ, Ψ associated with (4.1) satisfy (4.2) and if the alternative method of successive approximations is applied to (4.7), (4.8) with the zeroth approximation given by

$$x^{(0)} = (d_1 e^{i\tau_1 t}, \ldots, d_{2q} e^{i\tau_{2q} t}, 0, \ldots, 0),$$

with $d_j = b_j$, $j = 1, 2, \ldots, 2 \min(p, q)$; $d_j = ib_j$, $j = 2p + 1, \ldots, 2q$; where each b_j is a non zero real number, then the numbers V_j, $j = 1, 2, \ldots, 2q$, defined by (3.9) are purely imaginary.

PROOF. For simplicity in notation we only prove the lemma for $p \geq q$, but it will be clear how to discuss the other situation. Applying the method (3.7), it is clear that $x^{(r)}(t)$ has the form

$$x^{(r)}(t) = \sum_{j=1}^{2q} e^{i\tau_j t} \sum_{k=-\infty}^{+\infty} B_{jk}^{(r)} e^{ik\omega t} , \tag{4.9}$$

where $B_{jk}^{(r)}$ is an $n + \mu$ dimensional column vector. We show by induction that

$$B_{jk}^{(r)} = \operatorname{col}(B_{jk1}^{(r)}, i\, B_{jk2}^{(r)}, B_{jk3}^{(r)}),\ r = 0, 1, 2, \ldots, \tag{4.10}$$

and all j, k, where $B_{jk\ell}^{(r)}$, $\ell = 1, 2, 3$, are real column vectors of dimension $2p$, $2\mu - 2p$, $n - \mu$, respectively. The assertion is clearly true for $r = 0$. Assume that (4.10) is true for $r = 0, 1, \ldots, v - 1$ and observe that the Fourier series for

$$C(t) \sim \sum_{k=\infty}^{+\infty} C^{(k)} e^{ik\omega t}$$

in (4.5) is such that $C^{(k)} = (\alpha_{h+j} C_{h,j}^{(k)})$, $h, j = 1, 2, 3$, where $\alpha_{h+j} = 1$ if $h + j$ is odd; $= i$, if $h + j$ is even, and each $C_{h,j}^{(k)}$ is a real matrix with $C_{jj}^{(k)}$, $j = 1, 2, 3$, of dimensions $2p \times 2p$, $(2\mu - 2p) \times (2\mu - 2p)$ and $(n - \mu) \times (n - \mu)$, respectively. Therefore,

$$C(t)x^{(r)}(t) = \sum_{j=1}^{2q} e^{i\tau_j t} \sum_{k=-\infty}^{+\infty} D_{jk}^{(r)} e^{ik\omega t}, r = 1, 2, \ldots, v - 1,$$

where $D_{jk}^{(r)} = \text{col}(iD_{jk1}^{(r)}, D_{jk2}^{(r)}, D_{jk3}^{(r)})$ where each of the real vectors $D_{jk\ell}^{(r)}, \ell = 1, 2, 3,$ has the same dimension as $B_{jk\ell}^{(r)}$. From (3.6) and the assumption on the numbers d_j in the statement of the lemma,

$$\text{(4.11)} \qquad S_j^{(r)} \text{ is purely imaginary}, \quad j = 1, 2, \ldots, 2q;\ r = 1, 2, \ldots, \nu\ .$$

It then follows immediately from (3.7) and the above results that (4.10) is true for $r = \nu$ and, therefore, (4.10) is true for all r. Finally, (4.11) also holds for all r and the statement of the lemma is true.

REMARK (4.1). Lemma (4.2) holds also in a slightly more general situation, namely when q of the σ's differ by a multiple of ω at $\varepsilon = 0$, but the ones which satisfy this property are not necessarily the first q. The notation for the statement of this result as in Lemma (4.2) is extremely complicated, but it should be clear that the same conclusion is true, the only thing necessary is the proper choice of the numbers d_j.

REMARK (4.2). When $\mu = n$ in (4.1), it is obvious that the condition $\tau_{2j-1} \not\equiv 0 \pmod{\omega}$ in (4.8) may be eliminated.

REMARK (4.3). The determining equations for the $2q + 1$ real numbers $b_1, \ldots, b_{2q}$ and τ_1 [only τ_1, since the τ_j are related by (4.8)] in Lemma (4.2) are the $2q$ real equations

$$\text{(4.12)} \qquad \tau_j - \varepsilon\, I\, [V_j(\tau, b, \varepsilon)] = (-1)^{j+1}\, \sigma_j,\ j = 1, 2, \ldots, 2q\ ,$$

$b = (b_1, \ldots, b_{2q})$ and $I[w]$ is the imaginary part of a complex number w. If, in addition to condition (4.8), we assume that $2\tau_j \not\equiv 0 \pmod{\omega}$, $j = 1, 2, \ldots, 2q$; and choose $b_{2j-1} = -b_{2j}$, $j = 1, 2, \ldots, p$, $b_{2j-1} = b_{2j}$, $j = p + 1, \ldots, q$ in Lemma (4.2), then we know from Lemma (3.1) that $V_{2j-1} = \bar{V}_{2j}$, $j = 1, 2, \ldots, q$, and the determining equations (4.12) are equivalent to the equations

$$\text{(4.13)} \qquad \tau_{2j-1} - \varepsilon I[V_{2j-1}(\tau, b, \varepsilon)] = \sigma_j,\ j = 1, 2, \ldots, q\ ,$$

for the $q + 1$ real numbers $b_1, b_3, \ldots, b_{2q-1}$ and τ_1. This remark will be important for the applications.

REMARK (4.4). In the previous remark, equations (4.12), (4.13) were considered as determining equations for b and τ_1, but we could just as well have considered τ_1 as fixed and determined the numbers σ, b so as to obtain a solution of (4.5). This will be important in determining the stable and unstable regions around "resonance points".

§5. SOME BOUNDEDNESS THEOREMS

As an application of the lemmas in §4, we state a few theorems concerning the behavior of the characteristic exponents of (4.1) and also some theorems concerning the boundedness or unboundedness of the AC solutions of (4.1) in $(-\infty, +\infty)$.

THEOREM (5.1). If the matrices Φ, Ψ associated with system (4.1) satisfy (4.2) and the numbers $\sigma_j = \sigma_j(\varepsilon)$ are real analytic functions of ε at $\varepsilon = 0$, with

$$\sigma_j(0) \not\equiv 0 \pmod{\omega}, \quad j = 1, 2, \ldots, \mu, \tag{5.1}$$

then there exists an $\varepsilon_0 > 0$ such that there are always $n - \mu$ linearly independent periodic solutions of (4.1) for $|\varepsilon| < \varepsilon_0$. If one of the numbers σ_j, say σ_1, is such that

$$2\sigma_1(0) \not\equiv 0, \quad \sigma_1(0) \not\equiv \pm \sigma_k(0) \pmod{\omega}, \quad k = 2, 3, \ldots, \mu \ , \tag{5.2}$$

then there are always two characteristic exponents $i\tau_1$, $i\tau_2$ of (4.1) which are purely imaginary for $|\varepsilon| < \varepsilon_0$, $\varepsilon_0 > 0$, are analytic in ε at $\varepsilon = 0$ and $\tau_1 = -\tau_2$, $\tau_1(0) = \sigma_1$.

PROOF. The first part of the theorem is a restatement of Lemma (4.1). The second part of the theorem is an immediate consequence of Lemma (4.2) and Remark (4.3). For since $q = 1$, the implicit function theorem assures us that Equation (4.13) has a real solution τ_1 analytic in ε at $\varepsilon = 0$ with $\tau_1(0) = \sigma_1$.

COROLLARY (5.1). If the conditions of Theorem (5.1) are satisfied and

$$2\sigma_j(0) \not\equiv 0, \quad \sigma_j(0) \not\equiv \pm \sigma_k(0) \pmod{\omega}, \quad j \neq k, \ j, k = 1, 2, \ldots, \mu; \tag{5.3}$$

then all of the AC solutions of (4.1) are bounded for $|\varepsilon| < \varepsilon_0$, $\varepsilon_0 > 0$.

PROOF. This result follows immediately by applying Theorem (5.1) to each of the numbers, σ_j, $j = 1, 2, \ldots, \mu$.

Corollary (5.1) coincides with a previous result obtained by the author [10e]. For $\mu = n - 1$, this corollary has also been obtained by the author [10f] without using successive approximations. For $\mu = n$, this result has also been obtained by M. Golomb [7].

Consider the system of differential equations (4.1) satisfying (4.2) and suppose that $\sigma_1(0) - \sigma_2(0) = s\omega$ where s is an integer or zero and each σ_j is a real positive analytic function of ε at $\varepsilon = 0$ with

$$(5.4)\quad 2\sigma_j(0) \not\equiv 0,\ \sigma_j(0) \pm \sigma_k(0) \not\equiv 0 \pmod{\omega}\ j \neq k,\ j, k = 2, 3, \ldots, \mu.$$

If j, k in (5.4) are allowed to take on all integer values from 1 to μ, it follows from Corollary (5.1) that all of the AC solutions of (4.1) are bounded for $|\varepsilon|$ sufficiently small. We now wish to determine additional conditions on the functions φ_{jk}, ψ_{jk} in (4.1) so that all of the AC solutions are still bounded for $|\varepsilon|$ sufficiently small.

It follows from Theorem (5.1) that there are $n + \mu - 4$ linearly independent bounded AC solutions of (4.1) for $|\varepsilon|$ sufficiently small. Furthermore, the characteristic exponents corresponding to these solutions are either zero of close to one of the numbers $i\sigma_j$, $-i\sigma_j$, $j = 3, 4, \ldots, \mu$. It remains only to determine the four characteristic exponents close to the numbers $i\sigma_j$, $-i\sigma_j$, $j = 1, 2$. Consider the equivalent first order system (4.5), (4.6) and apply the alternative method of successive approximations (3.7) to the auxiliary equation

$$(5.5)\qquad z' = Bz + \varepsilon\, C(t)z \quad ,$$

where

$$(5.6)\quad B = \operatorname{diag}(i\tau, -i\bar{\tau}, i(\tau - s\omega), -i(\bar{\tau} - s\omega), i\sigma_3, -i\sigma_3, \ldots, i\sigma_\mu, -i\sigma_\mu, 0, \ldots, 0),$$

and the zeroth approximation is

$$(5.7)\quad x^{(0)} = (ibe^{i\tau t}, i\bar{b}e^{-i\bar{\tau}t}, iae^{i(\tau - s\omega)t}, i\bar{a}e^{-i(\bar{\tau} - s\omega)t}, 0, \ldots, 0) \quad ,$$

where τ, a, b are undetermined complex numbers, and if τ is a function of ε, then τ is close enough to σ_1 so that (5.4) is satisfied with $\sigma_2(0)$ replaced by $\tau(0) - s\omega$.

From Lemma (3.1), $S_1^{(r)} = \bar{S}_2^{(r)}$, $S_3^{(r)} = \bar{S}_4^{(r)}$ for every $r = 1, 2, \ldots,$ and all complex numbers a, b. Therefore, the determining equations

$$\begin{aligned} i\tau - \varepsilon V_1(\tau, a, b, \varepsilon) &= i\sigma_1 \\ i\tau - \varepsilon V_3(\tau, a, b, \varepsilon) &= i(\sigma_2 + s\omega) \end{aligned} \tag{5.8}$$

represent two equations for the undetermined parameters τ, a. We now calculate the first terms of V_1 and V_3 to obtain the characteristic exponent τ.

Suppose the Fourier series of $\varphi_{jk}(t)$, $\psi_{jk}(t)$ are given by

$$\varphi_{jk}(t) \sim \sum_{\ell=-\infty}^{+\infty} b_{jk\ell} e^{i\ell\omega t}, \quad \psi_{jk}(t) \sim \sum_{\ell=-\infty}^{+\infty} d_{jk\ell} e^{i\ell\omega t}, \quad j, k = 1, 2, \ldots, n . \tag{5.9}$$

From (4.2) it follows that each $b_{jk\ell}$, $d_{jk\ell}$ must be either purely imaginary or real and $b_{jk,\ell} = \bar{b}_{jk,-\ell}$. If

$$b_{jjo} = d_{jjo} = b_{jks} = d_{jks} = 0, \quad j \neq k, \quad j, k = 1, 2 \quad , \tag{5.10}$$

then $S_1^{(1)} = S_3^{(1)} = 0$. Calculating the functions $x_j^{(1)}$, $j = 1, 2, \ldots, n + \mu$, from (3.7) and (4.6), it is then very easy to show that

$$\begin{aligned} S_1^{(2)} &= -\frac{1}{4i} [b\alpha_{11}^{(s)}(\tau) + a\alpha_{12}^{(s)}(\tau)] \quad , \\ S_3^{(r)} &= -\frac{1}{4ia} [b\alpha_{21}^{(s)}(\tau) + a\alpha_{22}^{(s)}(\tau)] \quad , \end{aligned} \tag{5.11}$$

where

$$\begin{aligned} &\alpha_{11}^{(s)}(\tau) = \beta_{11}(0,0), \; \alpha_{12}^{(s)}(\tau) = \beta_{21}(s,0), \; \alpha_{21}^{(s)}(\tau) = \beta_{12}(0,s), \\ &\alpha_{22}^{(s)}(\tau) = \beta_{22}(s, s) \end{aligned} \tag{5.12}$$

$$\beta_{uv}(\ell_1, \ell_2) = \sum_{k=1}^{\mu} \sum_{\ell=-\infty}^{+\infty} \gamma_{ku,\ell+\ell_1}^{(+)} [\gamma_{vk,-(\ell+\ell_2)}^{(+)} \delta_\ell(\nu_k) - \gamma_{vk,-(\ell+\ell_2)}^{(-)} \delta_\ell(-\bar{\nu}_k)] + \sum_{k=\mu+1}^{n} \sum_{\ell=-\infty}^{+\infty} \gamma_{ku,\ell+\ell_1}^{(+)} \cdot b_{vk_1,-(\ell+\ell_2)} \cdot \delta_\ell(0)$$

$$\gamma_{kj\ell}^{(+)} = \left(\frac{b_{kj\ell}}{\sigma_j} + id_{kj\ell}\right),\ \gamma_{kj\ell}^{(-)} = \left(\frac{b_{kj\ell}}{\sigma_j} - id_{kj\ell}\right),\ \delta_\ell(\nu_k) = \frac{1}{\ell\omega + \tau - \nu_k},$$

$$\nu_1 = \tau,\ \nu_2 = \tau - s\omega,\ \nu_j = \sigma_j,\ j = 3, 4, \ldots, \mu,$$

and the index $\ell + \ell_j$ in (5.12) does not take on the values in (5.10).

In order to make the statement of the following theorem meaningful, first observe that the conditions (4.2) and Lemma (4.2) imply that each of the numbers $\alpha_{jk}^{(s)}$ is real if τ is real and either $p \geqq 2$ or $p = 0$, and $\alpha_{11}^{(s)}, \alpha_{22}^{(s)}$ are real, $\alpha_{12}^{(s)}, \alpha_{21}^{(s)}$ are purely imaginary if τ is real and $p = 1$, since the $\alpha_{jk}^{(s)}$ are independent of the numbers d_j in Lemma (4.2).

THEOREM (5.2). Consider the system of differential equations (4.1) with the functions $\varphi_{jk}(t)$, $\psi_{jk}(t)$ satisfying the condition (4.2) and having the Fourier series

$$\varphi_{jk}(t) \sim \sum_{\ell=-\infty}^{+\infty} b_{jk\ell}e^{i\ell\omega t},\quad \psi_{jk}(t) \sim \sum_{\ell=-\infty}^{+\infty} d_{jk\ell}e^{i\ell\omega t},$$

$b_{jjo} = d_{jjo} = b_{jks} = d_{jks} = 0$, $j \neq k$, $j, k = 1, 2$. If $\sigma_1(0) - \sigma_2(0) = s\omega$, where s is an integer or zero and each σ_j is a real positive analytic function of ε at $\varepsilon = 0$ with $2\sigma_j(0) \not\equiv 0$, $\sigma_j(0) \pm \sigma_k(0) \not\equiv 0$ (mod ω), $j \neq k$, $j, k = 2, 3, \ldots, \mu$; if the numbers $\alpha_{uv}^{(s)}(\tau)$ are defined by (5.12) and $G(s, \sigma_1, \sigma) = [\alpha_{22}^{(s)}(\sigma_1) - \alpha_{11}^{(s)}(\sigma_1) + 4\sigma]$, $H(s, \sigma_1) = 4\alpha_{21}^{(s)}(\sigma_1)\alpha_{12}^{(s)}(\sigma_1)$ where $\sigma = \lim_{\varepsilon\to 0}\varepsilon^{-2}[\sigma_2(\varepsilon) - \sigma_1(\varepsilon) + s\omega]$ is finite, then the following conclusions hold:

(i) If $H(s, \sigma_1) \neq 0$, $[G(s, \sigma_1, \sigma)]^2 + H(s, \sigma_1) > 0$ for $\varepsilon = 0$, then there exists an $\varepsilon_0 > 0$ such that all of the AC solutions of (4.1) with the σ_j as above are bounded for $|\varepsilon| < \varepsilon_0$;

(ii) if $H(s, \sigma_1) \neq 0$, $[G(s, \sigma_1, \sigma)]^2 + H(s, \sigma_1) < 0$ for $\varepsilon = 0$, then some of the AC solutions of (4.1) with the σ_j as above are unbounded for every $\varepsilon \neq 0$.

PROOF. From the remarks preceding the statement of the theorem it is sufficient to discuss the four characteristic exponents which are close to the numbers $i\sigma_j$, $-i\sigma_j$. For $\varepsilon \neq 0$, the determining equations for these characteristic exponents are given by

$$(5.13)\quad \begin{aligned} iF_1(\tau,a,b,\epsilon) &\equiv i\tau + \frac{\epsilon^2}{4i}[\alpha_{11}^{(s)}(\tau) + \frac{a}{b}\alpha_{12}^{(s)}(\tau)] - i\sigma_1 + O(\epsilon^3) = 0 \\ F_2(\tau,a,b,\epsilon) &\equiv \frac{a}{b}\alpha_{12}^{(s)}(\tau) - G(s, \sigma_1, \sigma) - \frac{b}{a}\alpha_{21}^{(s)}(\tau) + O(\epsilon) = 0 \end{aligned}$$

where the $\alpha_{jk}^{(s)}$ are given by (5.12). Any solution τ, a, b of Equation (5.13) yields two characteristic exponents $i\tau$, $-i\bar{\tau}$.

Suppose that the conditions of (i) are satisfied. If the integer p associated with system (4.1), (4.2) is ≥ 2, then, from the remark preceding the statement of the theorem, all of the numbers $\alpha_{jk}^{(s)}(\sigma_1)$ are real. Furthermore, from Lemma (4.2), the functions F_1, F_2 in (5.13) are real if τ, a are real and $b = 1$. The numbers $\tau_0 = \sigma_1(0)$, $a_0 = \{G(s, \sigma_1, \sigma) \pm [G^2(s, \sigma_1, \sigma) + H(s, \sigma_1)]^{\frac{1}{2}}\} \cdot [2\alpha_{12}^{(s)}(\sigma_1)]^{-1}$ are such that $F_1(\tau, a_0, 0) = F_2(\tau_0, a_0, 0) = 0$ and the Jacobian of these two functions with respect to τ, a at $\tau = \tau_0$, $a = a_0$, $\varepsilon = 0$ is different from zero. Consequently, Equations (5.13) have two real distinct solutions $\tau = \tau(\varepsilon)$ $a = a(\varepsilon)$ analytic in ε at $\varepsilon = 0$ and

$$\tau = \sigma_1 + \frac{\varepsilon^2}{4}[\alpha_{11}^{(s)}(\sigma_1) + a_0\,\alpha_{12}^{(s)}(\sigma_1)] + O(\varepsilon^3), \quad a = a_0 + O(\varepsilon) .$$

For $p = 1$, take a purely imaginary, $b = 1$ and for $p = 0$, take a purely imaginary, $b = -i$, and apply the same reasoning.

The reasoning for the proof of (ii) of the theorem is exactly the same as above except that we do not have to apply Lemma (4.2) but apply the implicit function theorem directly to the functions F_1, F_2 in Equations (5.13). The two τ's obtained are obviously such that one $i\tau$ has a real part positive and the other has a real part negative. Therefore, the theorem is proved.

REMARK (5.1). In the statement of Theorem (5.2) some cases have been excluded, in particular, the case where $H(s, \sigma_1) \neq 0$ and the other expression in Theorem (5.2) is equal to zero. Reasoning as in the proof of (i) Of Theorem (5.2), one sees that two of the remaining characteristic roots are still purely imaginary, but one cannot decide using the above method whether the solutions are bounded or unbounded. The other cases may be treated by going to higher approximations in the method.

Theorem (5.2) gives some insight into the behavior of the solutions of (4.1) near the resonance point $\sigma_1(0) - \sigma_2(0) = s\omega$. More specifically, if $\sigma_1, \sigma_3, \dots, \sigma_n$ are fixed and say independent of ε, if the conditions of Theorem (5.2) are satisfied and $H(s, \sigma_1) \neq 0$, then all of the AC solutions of (4.1) are bounded for ε sufficiently small along the curve $\sigma_2 = \sigma_1 - s\omega + \varepsilon^2\sigma$ in the (σ_2, ε) plane if the discriminant $D(\sigma) = [G(s, \sigma_1, \sigma)]^2 + H(s, \sigma_1) > 0$ and some solutions are unbounded along this curve for every $\varepsilon \neq 0$ if $D(\sigma) < 0$. Furthermore, if $H(s, \sigma_1) > 0$, then $D(\sigma) > 0$ for every σ and the AC solutions of (4.1) are bounded along every curve $\sigma_2 = \sigma_1 - s\omega + \varepsilon^2\sigma$ in the (σ_2, ε) plane for $\varepsilon = \varepsilon(\sigma)$ sufficiently small. On the other hand if $H(s, \sigma_1) < 0$, then there exists two σ's such that $G(\sigma) = 0$. If these two σ's are distinct then there is a region in the neighborhood of the point $(\sigma_1 - s\omega, 0)$ of the (σ_2, ε) plane where some solutions are unbounded no matter how small $\varepsilon \neq 0$. These results are summarized in the following theorem.

THEOREM (5.3). If the conditions of Theorem (5.2) are satisfied and $\sigma_1, \sigma_j, j = 3, 4, \dots, n$, are fixed numbers independent of ε, then

(i) $H(s, \sigma_1) > 0$ for $\varepsilon = 0$ implies that the AC solutions of (4.1) are bounded in a sufficiently small neighborhood of the point $(\sigma_1 - s\omega, 0)$ in the (σ_2, ε) plane:

(ii) $H(s, \sigma_1) < 0$ for $\varepsilon = 0$ implies that some of the AC solutions of (4.1) are unbounded in every neighborhood of the point $(\sigma_1 - s\omega, 0)$ of the (σ_2, ε) plane.

REMARK (5.2). Theorems (5.2) and (5.3) deal only with the case $\sigma_1(0) - \sigma_2(0) = s\omega$. The same type of analysis could be used to discuss the resonance points $\sigma_1(0) + \sigma_2(0) = s\omega$. The essential element of the argument was (5.11) and Lemma (4.2). It should be clear that a statement similar to Lemma (4.2) is true for the more general situation where $\sigma_j(0) + \sigma_k(0) = n_{jk}\omega$, n_{jk} an integer or zero, $j, k \in T_1$, $\sigma_j(0) \tau \sigma_k(0) = m_{jk}\omega$, m_{jk} an integer or zero, $j, k \in T_2$, where T_1, T_2 are subsets of the set of integers $(1, 2, \dots, \mu)$, if the matrix B in (4.7) is chosen properly and the numbers τ satisfy a condition similar to (4.7). That the functions S_1, S_3 have the form (5.11) when $\sigma_1(0) + \sigma_2(0) = s\omega$, $2\sigma_j(0) \neq 0$, $\sigma_j(0) \pm \sigma_k(0) \neq 0 \pmod{\omega}$, $j \neq k$, $j, k = 2, 3, \dots, \mu$, follows immediately from the method of successive approximations. With the generalization of Lemma (4.2) mentioned above, one can discuss the region of stability and instability of systems (4.1) when any of the numbers σ_j are "in resonance" with ω. However, the results in general cannot be stated explicitly as in Theorems (5.2) and (5.3) since the determining equations are extremely complicated.

REMARK (5.3). As remarked in the Introduction, if $\mu = n$ in (4.1), $\Psi = 0$, and Φ is symmetric, J. M. Gel'fand and V. B. Lidskii [14] and J. Moser [13] have shown that only the points $\sigma_j + \sigma_k = m\omega$, $\sigma_j > 0$ may lead to unbounded solutions. On the other hand, for systems (4.1), (4.2), Theorem (5.2) implies the behavior of the solutions at the resonance points $\sigma_1 - \sigma_2 = s\omega$ depends upon the value of a certain discriminant. Therefore if $\mu = n$, $\Psi = 0$, Φ is symmetric and also satisfies (4.2), it must be true that the function $H(s, \sigma_1)$ in Theorem (5.3) is > 0 for $\varepsilon = 0$. This is indeed the case; for the functions $\alpha_{jk}^{(s)}$ in (5.12) are given by

$$\alpha_{11}^{(s)}(\tau) = \sum_{k=1}^{\mu} \sum_{\ell=-\infty}^{+\infty} \frac{|b_{k1,\ell}|^2}{\sigma_1 \sigma_k} A_{\ell k}; \quad \alpha_{12}^{(s)}(\tau) = \sum_{k=1}^{\mu} \sum_{\ell=-\infty}^{+\infty} \frac{b_{k2,\ell+s}\bar{b}_{k1,\ell}}{\sigma_2 \sigma_k} A_{\ell k} \; ;$$

(5.14)

$$\alpha_{21}^{(s)}(\tau) = \sum_{k=1}^{\mu} \sum_{\ell=-\infty}^{+\infty} \frac{b_{k1\ell}\bar{b}_{k2,\ell+s}}{\sigma_1 \sigma_k} A_{\ell k}; \quad \alpha_{22}^{(s)}(\tau) = \sum_{k=1}^{\mu} \sum_{\ell=-\infty}^{+\infty} \frac{|b_{k2,\ell+s}|^2}{\sigma_2 \sigma_k} A_{\ell k}$$

where $A_{\ell k} = \delta_\ell(\nu_k) - \delta_\ell(-\bar{\nu}_k)$, $\nu_1 = \tau$, $\nu_2 = \tau - s\omega$, $\nu_k = \sigma_k$, $k = 3, 4, \ldots, n$. Consequently,

$$\alpha_{12}^{(s)}(\sigma_1) = \frac{\sigma_1}{\sigma_2} \bar{\alpha}_{21}^{(s)}(\sigma_1)$$

qnd $H(s, \sigma_1) > 0$ for all s if $\alpha_{12}^{(s)}(\sigma_1) \neq 0$. Actually, all that we have required to obtain boundedness at $\sigma_1 - \sigma_2 = s\omega$ is $\varphi_{1k} = \varphi_{k1}$, $\varphi_{2k} = \varphi_{k2}$, $k = 1, 2, \ldots, n$. To obtain boundedness at all of the points $\sigma_j - \sigma_k = s\omega$, $j \neq k$, $j, k = 1, 2, \ldots, n$, the matrix Φ must be symmetric.

In case Φ is not symmetric, then some solution may not be bounded for an $\varepsilon \neq 0$. For example, consider the system (4.1), (4.2) with $\mu = n = 2$, $\Psi \equiv 0$, $\sigma_1 = \sigma_2(0)$, σ_1 independent of ε,

$$\Phi = \begin{bmatrix} 2p \cos t & 2 \cos t \\ 2q \cos t & 2r \cos t \end{bmatrix} .$$

From (5.12), $\alpha_{12}^{(0)}(\sigma_1) = \beta(p + r)$, $\alpha_{21}^{(0)}(\sigma_1) = \beta\, q(p + r)$, $\beta = \sigma_1^{-1}(4\sigma_1^2 - 1)^{-1}$ for $\varepsilon = 0$. Therefore, from Theorem 5.3 the AC solutions are bounded in a neighborhood of the point $(\sigma_1, 0)$ in the (σ_2, ε) plane for ε sufficiently small if $p + r \neq 0$, $q > 0$, and some solutions are unbounded in every neighborhood of this point if $p + r \neq 0$, $q < 0$.

As another example, consider system (4.1), (4.2) with $\mu = n = 2$, $\Psi \equiv 0$, $\sigma_1 - \sigma_2(0) = 1$, σ_1 independent of ε,

$$\Phi = \begin{bmatrix} 2p \cos t & 2 \cos 2t \\ 2q \cos 2t & 2r \cos 2t \end{bmatrix} .$$

From (5.12), $\alpha_{12}^{(1)}(\sigma_1) = \beta(\sigma_1 - 1)^{-1}$, $\alpha_{21}^{(1)}(\sigma_1) = \beta q \sigma_1^{-1}$, $\beta = 2p(1 + 2\sigma_1)^{-1}$ for $\varepsilon = 0$. Therefore, from Theorem (5.3) all of the AC solutions are bounded in a neighborhood of the point $(\sigma_1 - 1, 0)$ of the (σ_2, ε) plane for ε sufficiently small if $p \neq 0$, $\sigma_1 > 1$, $q > 0$, and some solutions are unbounded in every neighborhood of this point if $p \neq 0$, $\sigma_1 > 1$, $q < 0$.

The following theorem concerns the behavior of the solutions of (4.1), (4.2) with $\mu = n$ when one of the σ's approaches zero as ε approaches zero.

THEOREM (5.4). Consider the system of differential equations of order $2n$,

$$(5.15) \qquad y'' = Dy = \varepsilon\, \Phi(t) y + \varepsilon\, \Psi(t) y'$$

where $\varepsilon > 0$, $D = \mathrm{diag}(\varepsilon\sigma_1^2, \sigma_2^2, \ldots, \sigma_n^2)$, Φ Ψ are $n \times n$ real matrices whose elements $\varphi_{ij}(t)$, $\psi_{ij}(t)$ are periodic in t of period $T = 2\pi/\omega$, L-integrable in $[0, T]$ and have mean value zero; $\Phi = (\Phi_{jk})$, $\Psi = (\Psi_{jk})$, $j, k = 1, 2$, where Φ_{jk}, Ψ_{jk} are matrices of the same dimension and

$$(5.16) \qquad \Phi_{jk}(-t) = (-1)^{k+j}\Phi_{jk}(t), \quad \Psi_{jk}(-t) = (-1)^{k+j+1}\Psi_{jk}(t), \quad j, k = 1, 2 .$$

If $\sigma_j > 0$, $j = 1, 2, \ldots, n$,

$$(5.17) \qquad 2\sigma_k \not\equiv 0, \quad \sigma_j \pm \sigma_k \not\equiv 0, \quad j \neq k \pmod{\omega}, \quad j \neq k, \quad j, k = 2, 3, \ldots, n ,$$

[if the σ_j are analytic functions of ε at $\varepsilon = 0$, then σ_k in (5.17) is replaced by $\sigma_k(0)$], then there exists an $\varepsilon_0 > 0$ such that all of the absolutely continuous (AC) solutions of (5.15) are bounded for $0 < \varepsilon < \varepsilon_0$ (notice this is not true for $\varepsilon = 0$). Furthermore, the characteristic exponents of (5.15) are analytic functions of $\sqrt{\varepsilon}$ at $\varepsilon = 0$.

PROOF. By using an elementary argument, we show that there are $2n - 2$ linearly independent bounded AC solutions of (5.51). This same result could also be proved by the method of successive approximations. The characteristic exponents τ_k are only determined up to a multiple of ωi, but without loss of generality we may assume that the $\tau_k(\varepsilon)$, $k = 3, 4, \ldots, 2n$ are such that $\tau_{2j-1}(0) = i\sigma_j$, $\tau_{2j}(0) = -i\sigma_j$, $j = 2, 3, \ldots, n$. From the Floquet theory condition (5.17) assures us that these functions $\tau_k(\varepsilon)$, $k = 3, 4, \ldots, 2n$, are analytic in ε at $\varepsilon = 0$ and

$$\begin{aligned} &\tau_{2j-1}(\varepsilon) = \bar{\tau}_{2j}(\varepsilon), \ j = 2, 3, \ldots, n; \\ (5.19) \qquad &\tau_j \not\equiv 0, \ \tau_j \not\equiv \tau_k (\mathrm{mod}\ \omega i), \ j \neq k, \ j, k = 3, 4, \ldots, n; \\ &\qquad\qquad 0 \leqq \varepsilon \leqq \varepsilon_0, \ \varepsilon_0 > 0 \ , \end{aligned}$$

provided only that ε_0 is sufficiently small. If $y(t)$ is a solution of (5.15), there exist, from (5.16), a nonsingular matrix P such that $Py(-t)$ is also a solution of (5.15). This, together with (5.18) implies that each τ_k, $k = 3, 4, \ldots, 2n$, is purely imaginary and this obviously implies the existence of $2n - 2$ linearly independent solutions of (5.15).

It does not seem possible to obtain the other two characteristic exponents by such an elementary argument so we apply the above method of successive approximation.

The transformation of variables $y_j = (2i\sigma_j)^{-1}(z_{2j-1} + z_{2j})$, $y_j' = 2^{-1}(z_{2j-1} - z_{2j})$, $j = 2, 3, \ldots, n$, $y_1 = z_1$, $y_1' = \lambda z_2$, $\lambda = \sqrt{\varepsilon}$, leads to the equivalent system of first order equations

$$(5.19) \qquad z' = Az + \lambda C(t, \lambda)z$$

where $A = \mathrm{diag}(0, 0, i\sigma_2, -i\sigma_2, \ldots, i\sigma_n, -i\sigma_n)$, $C = (c_{jk})$, $j, k = 1, 2, \ldots, 2n$,

$$c_{11} = 0,\ c_{12} = 1,\ c_{1j} = 0,\ j = 2, 3, \ldots, 2n,\ c_{21} = \varphi_{11} - \sigma_1^2,$$
$$c_{22} = \psi_{11}$$

$$c_{2,2j-1} = (\frac{\varphi_{1j}}{i\sigma_j} + \psi_{1j}),\ c_{2,2j} = (\frac{\varphi_{1j}}{i\sigma_j} - \psi_{1j}),\ j = 2, 3, \ldots, n ,$$

(5.20) $$c_{2k,j} = - c_{2k-1,j},\ k = 2, 3, \ldots, n;\ j = 1, 2, \ldots, 2n ,$$

$$c_{2k-1,1} = \lambda\varphi_{k,1},\ c_{2k-1,2} = \lambda^2\psi_{k,1},\ k = 2, 3, \ldots, n ,$$

$$c_{2j-1,2k-1} = \lambda\ (\frac{\varphi_{jk}}{i\sigma_k} + \psi_{jk}),\ c_{2j-1,2k} = \lambda\ (\frac{\varphi_{jk}}{i\sigma_k} - \psi_{jk}),$$
$$j, k = 2, 3, \ldots, n .$$

To obtain the two characteristic exponents close to zero, we apply the first method of successive approximations to obtain a solution of the form $z = e^{i\tau t}p(t)$, $p(t + T) = p(t)$, where τ is a real number. The equation for $p(t)$ is

(5.21) $$p' = (A - i\tau I)p + \lambda\ C(t, \lambda)p .$$

If τ is close enough to zero so that

(5.22) $$\sigma_k \pm \tau \neq m\omega,\ k = 2, 3, \ldots, n;\ m = 0, 1, 2, \ldots ,$$

then the method of successive approximations may be applied to the auxiliary equation

(5.23) $$p' = Bp + \lambda\ C(t, \lambda)p$$

where $B = \text{diag}(0, 0, i(\sigma_2 - \tau), - i(\sigma_2 + \tau), \ldots, i(\sigma_n - \tau)), - i(\sigma_n + \tau)$ and the zeroth approximation is taken to be

(5.24) $$p^{(0)} = x^{(0)} = (1, ib, 0, \ldots, 0)$$

where b is a nonzero real number. The functions $x^{(r)}$ obtained in this manner are periodic of period $T = 2\pi/\omega$. If $x^{(r)}(t)$ has the Fourier series

(5.25) $$x^{(r)}(t) \sim \Sigma\ B_k^{(r)}\ e^{ik\omega t}$$

where $B_k^{(r)}$ is a $2n$ dimensional column vector, then we show by induction that

$$B_k^{(r)} = \text{col }(B_{k1}^{(r)}, iB_{k2}^{(r)}, iB_{k3}^{(r)}, B_{k4}^{(r)}),\ k = 0, \pm 1, \pm 2, \ldots ; \tag{5.26}$$
$$r = 0, 1, 2, \ldots .$$

where $B_{k1}^{(r)}, B_{k2}^{(r)}$ are real scalars, $B_{k3}^{(r)}$ is a $2\mu - 2$ dimensional real vector and $B_{k4}^{(r)}$ is a $2(n - \mu)$ dimensional real vector. The assertion is clearly true for $r = 0$. Assume that it is true for $r = 0, 1, 2, \ldots, v - 1$. Then from the definition of the c_{jk} in (5.20) and the assumptions on the φ_{jk}, ψ_{jk} of the theorem, it is clear that $C(t)x^{(r)}(t) \sim \Sigma D_k^{(r)} e^{ik\omega t}$ where

$$D_k^{(r)} = \text{col }(iD_{k1}^{(r)}, D_{k2}^{(r)}, D_{k3}^{(r)}, iD_{k4}^{(r)}) \tag{5.27}$$

and the real vectors $D_{kj}^{(r)}$ have the same dimension as the $B_{kj}^{(r)}$. From (3.6) and (5.24)

$$S_1^{(r)},\ S_2^{(r)} \text{ are purely imaginary, } r = 1, 2, \ldots, v . \tag{5.28}$$

It is now very easy to show from (3.7), (5.27), (5.28) that (5.26) is true for $r = v$ and the induction is completed. Consequently (5.28) is true for all r and the determining equations

$$\begin{aligned} -i\tau &= -\lambda S_1^{(1)} - \lambda^2 S_1^{(2)} \ldots \\ -i\tau &= -\lambda S_2^{(1)} - \lambda^2 S_2^{(2)} \ldots \end{aligned} \tag{5.29}$$

are purely imaginary. Furthermore, it is immediate from (3.6) and (5.24) that

$$S_1^{(1)} = ib,\ S_2^{(1)} = -\frac{\sigma_1^2}{ib}$$

and for $\lambda \neq 0$, the determining equations (5.29) become

$$\begin{aligned} \tau &= i\lambda b + O(\lambda^2) \\ b - \frac{\sigma_1{}^2}{b} &= O(\lambda^2) . \end{aligned} \tag{5.30}$$

For $\lambda = 0$, these equations have two solutions $\tau = 0$, $b = \pm \sigma_1$ and since the Jacobian of the functions τ, $b - \sigma_1^2 b^{-1}$, with respect to b, τ is $\neq 0$ at each of these solutions, it follows from the implicit function theorem that there are two solutions of (5.30) analytic in λ at $\lambda = 0$ and the solutions are given by

$$\tau = \pm i\lambda\sigma_1 + O(\lambda^2)$$

$$b = \pm \sigma_1 + O(\lambda) ,$$

where τ is purely imaginary and b is real. These two solutions lead to two linearly independent solutions of (5.15). These solutions together with the $2n - 2$ mentioned previously will form a fundamental system of bounded AC solutions of (5.15) and Theorem (5.4) is proved.

§6. ZONES OF INSTABILITY FOR SYSTEMS OF MATHIEU TYPE EQUATIONS

Consider the equation

$$(6.1) \qquad x_j'' + \sigma_j^2 x_j + 4\varepsilon(\cos 2t) \sum_{k=1}^{n} \sigma_k d_{jk} x_k = 0, \; j = 1, 2, \ldots, n ,$$

where ε is a real parameter and σ_j, d_{jk}, $j, k = 1, 2, \ldots, n$, are real constants. If $\tau_1, \ldots, \tau_{2n}$ are the characteristic exponents of (6.1) and $\rho_j = e^{\pi\tau_j}$, $j = 1, 2, \ldots, 2n$, are the characteristic multipliers, then the ρ_j are the solutions of a polynomial equation,

$$(6.2) \qquad F(\rho) = \prod_{j=1}^{n} (\rho^2 - 2A_j\rho + B_j) ,$$

where the A_j, B_j are real numbers and $\Pi_{j=1}^{n} B_j = 1$. If $\sigma_1, \ldots, \sigma_n$ are analytic functions of ε at $\varepsilon = 0$,

$$(6.3) \qquad \begin{aligned} &\sigma_1(0) = m, \quad m \text{ a positive integer} , \\ &2\sigma_j(0) \not\equiv 0, \; \sigma_j \pm \sigma_k \not\equiv 0 \pmod 2, \; j \neq k, \; j, k = 2, 3, \ldots, n , \end{aligned}$$

then it follows from Theorem (5.1) that $2n - 2$ of the characteristic exponents are purely imaginary, say $\tau_3, \ldots, \tau_{2n}$, and $\tau_{2j-1}(\varepsilon) = \bar{\tau}_{2j}(\varepsilon)$, $\tau_{2j-1}(0) = i\sigma_j(0)$, $j = 2, 3, \ldots, n$, for $|\varepsilon|$ sufficiently small. Therefore, $|\rho_j| = 1$, $j = 3, 4, \ldots, 2n$, and $B_j = 1$, $|A_j| < 1$, $j = 2, 3, \ldots, n$. Since $\Pi_{j=1}^{n} B_j = 1$, $B_1 = 1$ and

$$(6.4) \qquad F(\rho) = \prod_{j=1}^{n} (\rho^2 - 2A_j\rho + 1) ,$$

The remaining characteristic multipliers ρ_1, ρ_2 will have absolute value less than one if and only if $|A_1| < 1$ and have absolute value greater than one if and only if $|A| > 1$. Consequently, the transition curves in the (σ_1, ε)-plane from stability to instability will occur when $A_1 = +1$ or $A_1 = -1$. But this implies that $\rho_1 + \rho_2 = +2$ or $\rho_1 + \rho_2 = -2$, i.e., there is a periodic solution of (6.1) of period π or 2π. These comments are generalizations of the well known facts for the Mathieu equations [11].

The purpose of the present section is to determine the nature of these transition curves for system (6.1) satisfying condition (6.3).

To obtain these curves transform system (6.1) into an equivalent system [see (3.12)].

(6.5) $$z' = Az + i\varepsilon(e^{2it} + e^{-2it})Dz$$

where $A = \operatorname{diag}(i\sigma_1, -i\sigma_1, \ldots, i\sigma_n, -i\sigma_n)$, and

(6.6) $$D = \begin{bmatrix} d_{jk} & d_{jk} \\ -d_{jk} & -d_{jk} \end{bmatrix}, \; j, k = 1, 2, \ldots, n \; .$$

The alternative method of successive approximations is now applied to the auxiliary system

(6.7) $$z' = Bz + i\varepsilon(e^{2it} + e^{-2it})Dz$$

where $B = \operatorname{diag}(im, -im, i\sigma_2 - i\sigma_2, \ldots, i\sigma_n, -i\sigma_n)$, m a positive integer, and the zeroth approximation is given by

(6.8) $$z^{(0)} = x^{(0)} = (e^{imt}, be^{-imt}, 0, \ldots, 0)$$

and b is a real number. From Lemma (4.2), we know that the functions V_j obtained from the successive approximations are purely imaginary. Consequently, the determining equations for $\varepsilon \neq 0$ are equivalent to the equations

(6.9) $$\begin{aligned} & m - \varepsilon I\,[V_1(m, \sigma_1, b, \varepsilon)] = \sigma_1 \\ & I\,[V_1(m, \sigma_1, b, \varepsilon)] + I\,[V_2(m, \sigma_1, b, \varepsilon)] = 0 \; . \end{aligned}$$

Since m is fixed, equations (6.9) may be used to determine σ_1 and b as functions of ε for $|\varepsilon|$ sufficiently small and the corresponding function $\sigma_1 = \sigma_1(\varepsilon)$ will be a transition curve in the (σ_1, ε)-plane near the point $\varepsilon = 0$, $\sigma_1 = m$. Therefore, the only thing that remains to be done is to find the first nonzero terms of V_1 and V_2 containing b and solve equations (6.9).

First, it is clear that $x_j^{(\ell)}(t)$, $j = 1, 2, \ldots, 2n$, for all ℓ, contains only terms of the form $e^{i(m+2p)t}$, $e^{-i(m+2p)t}$, $p = 0, \pm 1, \ldots, \pm \ell$. Let

$$x_j^{(\ell)}(t) = \sum_{p=-\ell}^{+\ell} [a_{p,j}^{(\ell)} e^{i(m+2p)t} + c_{p,j}^{(\ell)} e^{-i(m+2p)t}] \quad , \tag{6.10}$$

$$j = 1, 2, \ldots, 2n \quad ,$$

for all ℓ, where $a_{p,j}^{(\ell)}$, $c_{p,j}^{(\ell)}$ are constants.

We now show by induction that

$$c_{p,2j}^{(\ell)} = ba_{p,2j-1}^{(\ell)}, \; c_{p,2j-1}^{(\ell)} = ba_{p,2j}^{(\ell)}, \quad j = 1, 2, \ldots, n; \tag{6.11}$$

$$p = 0, \pm 1, \ldots, \pm \ell; \; \ell = 0, 1, 2, \ldots, m - 1 \quad ,$$

where the constants $a_{p,2j-1}^{(\ell)}$, $a_{p,2j}^{(\ell)}$ are independent of b. For $k = 0$, the constants $a_{p,j}^{(o)}$, $c_{p,j}^{(o)}$ satisfy (6.11) for all p, j. If the constants $a_{p,j}^{(\ell)}$, $c_{p,j}^{(\ell)}$ satisfy (6.11) for all p, j and $\ell = 0, 1, 2, \ldots, v - 1$, then the terms $S_1^{(\ell)}$, $S_2^{(\ell)}$ in (3.6) are given by

$$S_1^{(\ell)} = i \sum_{k=1}^{n} d_{1k} \sum_{p=-1,1} (a_{p,2k-1}^{(\ell-1)} + a_{p,2k}^{(\ell-1)}) \quad ,$$

$$S_2^{(\ell)} = - ib^{-1} \sum_{k=1}^{n} d_{1k} \sum_{p=-1,1} (c_{p,2k-1}^{(\ell-1)} + c_{p,2k}^{(\ell-1)}) = - S_1^{(\ell)} \quad , \tag{6.12}$$

$$\ell = 1, 2, \ldots, v \quad ,$$

if $v < m$ and $S_1^{(\ell)}$, $S_2^{(\ell)}$ are independent of b. From the definition of $x_j^{(v)}$ in (3.7), it now follows immediately that the coefficients $a_{p,j}^{(v)}$, $c_{p,j}^{(v)}$ satisfy (6.11) for all $v < m$. Therefore, (6.12) is also satisfied for all $v < m$.

From (3.6) and (6.11),

$$S_1^{(m)} = i \sum_{k=1}^{n} d_{1k} \left[\sum_{p=-1,1} (a_{p,2k-1}^{(m-1)} + a_{p,2k}^{(m-1)}) + c_{-(m-1),2k-1}^{(m-1)} + c_{-(m-1),2k}^{(m-1)} \right]$$

(6.13)

$$S_1^{(m)} + S_2^{(m)} = i(b - b^{-1}) \sum_{k=1}^{n} d_{1k} \left[a_{-(m-1,2k-1}^{(m-1)} + a_{-(m-1),2k}^{(m-1)} \right]$$

where this last expression is independent of b. If we define γ_m by the relation $S_1^{(m)} + S_2^{(m)} = i(b - b^{-1})\gamma_m$, then it is a simple calculation to show that

$$(6.14) \qquad \gamma_m = \sum_{\substack{k_1,\ldots,k_{m-1}=1 \\ k_0=1}}^{n} \prod_{\ell=0}^{m-1} \frac{2d_{k_\ell k_{\ell+1}} \sigma_{k_{\ell+1}}}{(m-2\ell-2)^2 - \sigma_{k_{\ell+1}}^2} \cdot d_{k_{m-1},1}$$

where $\sigma_1 = m$ in this expression.

Consequently, for $\varepsilon \neq 0$, the determining equations (6.9) are

$$\sigma_1 = m - \varepsilon I\,[V_1(m, \sigma_1, b, \varepsilon)]$$

(6.15)

$$\gamma_m(b - b^{-1}) = O(\varepsilon) \quad .$$

If $\gamma_m \neq 0$ for $\varepsilon = 0$, then these two equations have two distinct solutions $\sigma_1 = m + O(\varepsilon)$, $b = \pm 1 + O(\varepsilon)$, where the two curves $\sigma_1 = \sigma_1(\varepsilon)$ coincide up to terms of order ε^{m-1}. We have proved the following theorem.

THEOREM (6.1). Consider the system of equations (6.1) satisfying (6.3). If γ_m defined by (6.14) is different from zero, then there are unbounded AC solutions of (6.1) in every neighborhood of the point $(m, 0)$ in the (σ_1, ε)-plane, where m is a positive integer. Furthermore, the transition curves in the (σ_1, ε)-plane from a region of stability to instability coincide up to terms of order ε^{m-1}.

COROLLARY (6.1). There are unbounded solutions of the Mathieu equation $x'' + \sigma^2 x + 4\varepsilon(\cos 2t)x = 0$ in every neighborhood of the point $(m, 0)$ in the (σ, ε)-plane for every positive integer m.

PROOF. In this case (6.14) implies $\gamma_m(0) = [-2^{m-1}m!(m-1)!]^{-1} \neq 0$ for every positive integer m.

COROLLARY (6.2). Consider the system of equations (6.1) satisfying condition (6.3) and suppose that $d_{jk}d_{rs} \geqq 0$ for every j, k, r, s, at least one d_{jk} is $\neq 0$, and $\sigma_k > M$, $k = 2, 3, \ldots, n$, M a positive integer. There are unbounded solutions of (6.1) in every neighborhood of the point $(m, 0)$ in the (σ_1, ε)-plane for every positive integer $m \leq M + 2$.

PROOF. In this case (6.14) implies $\gamma_m \neq 0$ if at least one d_{jk} is $\neq 0$.

In case $\gamma_m = 0$, then there still may be two distinct solutions to the determining equations (6.9). It should be clear from the proof of relations (6.11) and (6.12) that if the first $S_j^{(\ell)}$ which depends on b occurs in the m_1^{th} approximation, then relations (6.12) are valid for $v = m_1 - 1$, and (6.13), (6.14) will hold for $m = m_1$ and the determining equations (6.15) for $m = m_1$ will have two different solutions. Therefore, Theorem (6.1) will hold for $m = m_1$. In case $S_j^{(\ell)}$ is independent of b for every ℓ, then the determining equations (6.9) will always have only one solution $\sigma_1 = \sigma_1(\varepsilon)$ which is independent of b. Therefore, there are two linearly independent periodic solutions along this curve and the AC solutions are bounded in a neighborhood of the point $(m, 0)$ in the (σ_1, ε)-plane.

BIBLIOGRAPHY

[1] BAILEY, H. R. and CESARI, L., Boundedness of solutions of linear differential systems with periodic coefficients, Archive Rat. Mech. Ana. 3(1958), 246-271.

[2] BAILEY, H. R. and GAMBILL, R. A., On stability of periodic solutions of weakly nonlinear differential systems, Journ. Math. Mech. 6(1957), 655-668.

[3] CESARI, L., (a) Sulla stabilita delle soluzioni dei sistemi di equazioni differenziali lineari a coefficienti periodici, Atti. Accad. Italia, Mem. Cl. Fis. Mat. Nat. (6) 11(1940), 633-692; (b) Asymptotic behavior and stability problems in ordinary differential equations, Ergbn. d. Math. N.F. Heft 16, Springer 1959;

(c) Existence theorems for nonlinear Lipschitzian differential systems and fixed point theorems, this Study.

[4] CESARI, L. and HALE, J. K., (a) Second order linear differential systems with periodic L-integrable coefficients, Riv. Mat. Univ. Parma 5(1954), 55-61; (b) A new sufficient condition for periodic solutions of weakly nonlinear differential systems, Proc. Amer. Math. Soc. 8(1957), 757-764.

[5] GAMBILL, R. A., (a) Stability criteria for linear differential systems with periodic coefficients, Riv. Mat. Univ. Parma 5(1954), 169-181; (b) Criteria for parametric instability for linear differential systems with periodic coefficients, Riv. Mat. Univ. Parma. 6(1955), 37-43.

[6] GAMBILL, R. A. and HALE, J. K., Subharmonic and ultraharmonic solutions for weakly nonlinear differential systems, Journ. Rat. Mech. Ana. 5(1956), 353-394.

[7] GOLOMB, M., Expansion and boundedness theorems for solutions of linear differential systems with periodic or almost periodic coefficients, Archive Rat. Mech. Anal. 2(1958), 284-308.

[8] HAACKE, W., Über die Stabilität eines Systems von gewohnlichen linearen Differentialgleichungen zweiter Ordnung mit periodischen Koeffizienten, die von Parametern abhängen (I und II) Math. Z. 56(1952), 65-79, (1952), 34-45.

[9] HAUPT, O., Über lineare homogene Differentialgleichungen zweiter Ordnung mit periodischen Koeffizieten, Math. Ann. 79(1919), 278.

[10] HALE, J. K., (a) Evaluations concerning products of exponential and periodic functions, Riv. Mat. Univ. Parma, 5(1954), 63-81; (b) On boundedness of the solutions of linear differential systems with periodic coefficients, Riv. Mat. Univ. Parma, 5 (1954), 137-167; (c) Periodic solutions of nonlinear systems of differential equations, Riv. Mat. Univ. Parma, 5(1954), 281-311; (d) Sufficient conditions for the existence of periodic solutions of systems of weakly nonlinear first and second order differential equations, Journ. Math. Mech. 2(1958), 163-172; (e) Linear systems of first and second order differential equations with periodic coefficients, Ill. Journ. Math. 2, 586-591, (1958); (f) A short proof of a boundedness theorem for linear differential systems with periodic coefficients, Archive Rat. Mech. Ana. 2 (1959), 429-434; (g) On the stability of periodic solutions of weakly nonlinear periodic and autonomous differential systems. This Study.

[11] McLACHLAN, N. W., Theory and application of Mathieu functions, Clarendon Press, Oxford, 1947.

[12] METTLER, E., Allegemeine Theorie der Stabilität erzwungener Schwingungen elasticher Korper, Ing. Arch. 17(1949), 418-449.

[13] MOSER, J., New aspects in the theory of stability of Hamiltonian systems, Comm. Pure Appl. Math. 9(1958), 81-114.

[14] GEL'FAND, J. M. and LIDSKII, V. B., On the structure of the regions of stability of linear canonical systems of differential equations with periodic coefficients, Uspehi Mat. Nauk (N.S.), 10(1955), 3-40; Am. Math. Soc. Trans. (2) 8(1958), 143-182.

V. ON THE STABILITY OF PERIODIC SOLUTIONS OF WEAKLY NONLINEAR PERIODIC AND AUTONOMOUS DIFFERENTIAL SYSTEMS

J. K. Hale

§1. INTRODUCTION

Consider the system of differential equations

$$x'' + Ax = \varepsilon f(x, x', \varepsilon, t) \tag{1.1}$$

where ε is a small real parameter, $x = (x_1, \ldots, x_n)$, $A = \mathrm{diag}(\sigma_1^2, \ldots, \sigma_n^2)$, and the vector $f = (f_1, \ldots, f_n)$ is periodic in t of period $T = 2\pi/\omega$. For ε small, and certain conditions of smoothness on f, various methods have been given for the determination of periodic solutions, $x_0(\varepsilon, t)$ of (1.1) of the form

$$\begin{aligned} x_{jo}(0, t) &= a_j \cos(r_j \omega t + \varphi_j), \quad j = 1, 2, \ldots, \mu, \\ x_{jo}(0, t) &= 0, \quad j = \mu + 1, \ldots, n, \end{aligned} \tag{1.2}$$

where each a_j, φ_j is a real number and each $r_j = k_j/m_j$ is a rational number, $j = 1, 2, \ldots, \mu$ (for example, see [4a, Section (8.5)] and [4b]. If the partial derivatives of f_j with respect to x_k, x_k' exist, the standard procedure for determining whether or not such a solution is stable is to discuss the solutions of the linear variational equation

$$y'' + Ay = \varepsilon \frac{\partial f(x_0, x_0', \varepsilon, t)}{\partial x} y + \varepsilon \frac{\partial f(x_0, x_0', \varepsilon, t)}{\partial x'} y', \tag{1.3}$$

where $\partial f/\partial x$, $\partial f/\partial x'$ are matrices defined by $\partial f/\partial x = (\partial f_j/\partial x_k)$, $\partial f/\partial x' = (\partial f_j/\partial x_k')$, $j, k = 1, 2, \ldots, n$. The coefficients in (1.3) are periodic of period mT, $m = m_1 \ldots m_\nu$, $T = 2\pi/\omega$. If all of the characteristic exponents of system (1.3) have negative real parts, then the solution (1.2) of (1.1) is asymptotically stable for ε sufficiently small ([Liapunov, 9]).

From the Floquet theory [5, p. 78], the characteristic exponents of system (1.3) are obtained from the eigenvalues of a $2n \times 2n$ matrix, which depends upon a fundamental system of solutions of (1.3). Suppose that the coefficients in (1.3) are such that the characteristic exponents are continuous functions of ε at $\varepsilon = 0$. Some of the properties of the characteristic exponents of (1.3) can be obtained by using the fact that they are continuous functions of ε at $\varepsilon = 0$ and are only determined up to a multiple of $\omega i/m$. In fact, the characteristic exponents can be grouped into sets S_i, which are disjoint for all ε, $0 \leq \varepsilon \leq \varepsilon_0$, such that each S_i satisfies one of the following properties: (i) S_i contains more than one characteristic exponent all of which are equal to zero for $\varepsilon = 0$; (ii) S_i contains more than one characteristic exponent all of which are equal to a complex or real number $\neq 0$ for $\varepsilon = 0$; (iii) S_i contains only one characteristic exponent. Furthermore, if there exists an S_i which satisfies (ii) there is a set S_j (which may coincide with S_i) whose characteristic exponents are the complex conjugates of those characteristic exponents of S_i. Even with this more detailed information about the characteristic exponents, it does not seem possible to use the Floquet theory directly to reduce the order of the $2n \times 2n$ matrix from which the exponents are obtained.

For ε sufficiently small and using the method of successive approximations developed by L. Cesari, J. K. Hale and R. A. Gambill in a series of papers (see [4], Section 4.5 and 8.5), it was shown in [8] that the characteristic exponents of (1.3) which belong to any one of the sets S_j above, say S, are determined by the eigenvalues of a $p \times p$ matrix where p is the number of elements in S, provided that all of the eigenvalues of this matrix are distinct for $|\varepsilon|$ sufficiently small.[1] Furthermore, this matrix of order p can be written explicitly in terms of the coefficients in (1.3). In particular, for $p = 1$, this characteristic exponent can be obtained directly from the method. The purpose of the present paper is to illustrate the application of some of the results of [8] (see Lemmas 2.1 and 2.2) to the stability problem. We state some explicit stability theorems in §3, §4 and give a few examples.

In §3, under the assumption that each f_j in (1.1) is analytic

[1] Added in proofs: Since writing this paper, the author has become aware of some other recent literature on this same problem in which similar results have been obtained (see [13], [14]).

in x, x', ε and satisfies (2.2) these results are applied to systems (1.1) (and even more general systems) to determine a number of sufficient conditions for asymptotic stability (see Theorems 3.1, 3.2, 3.3). Using the same method, Theorem 3.1 had already been proved by H R. Bailey and R. A. Gambill [3] for the case where $\mu = n$ in (1.1), (1.2). As a corollary of Theorem 3.2 one obtains a result proved by L. Mandelstam and N. Papalexi [10] using another method for the case where $\mu = n = 1$ in (1.1), (1.2).

In §4, these same methods are applied to the linear variational equations for a periodic solution of an autonomous nonlinear differential system to obtain a number of sufficient conditions for asymptotic orbital stability (see Theorems 4.1, 4.2, 4.3). In his thesis at Purdue University, E. W. Thompson [12] has obtained Theorem 4.2 by first eliminating the zero characteristic root of the linear variational system and applying the method of successive approximations mentioned above; in particular, the results of H. R. Bailey and L. Cesari [3]. Theorem (4.2) is obtained in this paper by the above method in a more straightforward way. A special case of Theorem (4.2) is the following result which seems particularly significant. Consider the system of second order equations

$$(1.4)\quad x_j'' + \sigma_j^2 x_j = \varepsilon f_j(x_1, \ldots, x_n, x_1', \ldots, x_n'),\ j = 1, 2, \ldots, n \ ,$$

where $2\sigma_j \neq m\sigma_1$, $\sigma_j \pm \sigma_k \neq m\sigma_1$, $j \neq k$, $j, k = 2, 3, \ldots, n$; $m = 0, \pm 1, \pm 2, \ldots$ and each f_j is analytic in x_j, x_j' for $|x_j| \leq b$, $|x_j'| \leq b$, $b > 0$. and suppose that $x_{jo}(\varepsilon, t) = x_{jo}(\varepsilon, t + 2\pi/\omega)$, $j = 1, 2, \ldots, n$; $\omega = \sigma_1 + O(\varepsilon)$, is a periodic solution of (1.4) and is analytic in ε at $\varepsilon = 0$ with $x_{1o}(0, t) = a\cos(\sigma_1 t + \varphi)$, $x_{ko}(0, t) = 0$, $k = 2, 3, \ldots, n$, where a, φ are real numbers. If

$$(1.5)\quad \int_0^{2\pi} f_{jx_j'}(a\cos t, 0, \ldots, 0, -a\sigma_1 \sin t, 0, \ldots, 0)dt < 0 \ ,$$

$$j = 1, 2, \ldots, n \ ,$$

then this periodic solution of (1.4) is asymptotically orbitally stable in $[t_o + \infty]$, $0 \leq \varepsilon \leq \varepsilon_o$, $\varepsilon_o > 0$. This result was obtained by A. Andronov and A. Witt [1] for $n = 2$, by evaluating the characteristic polynomial of a matrix of order four. The result above for arbitrary n is proved by the above method and the properties of the characteristic exponents without evaluating any determinants.

In §3 and §4, the nonlinear functions in the differential equations are assumed to satisfy a certain condition of analyticity. It is shown in §5 that it is sufficient to require that these functions possess continuous second derivatives.

Throughout the present paper, the following notations are used: $i = \sqrt{-1}$; $\bar{a}$ is the complex conjugate of a; $R(a)$ is the real part of a; $I(a)$ is the imaginary part of a. Also ε_0 is used to denote a sufficiently small positive number.

§2. DEFINITIONS OF STABILITY

Consider a system of differential equations

$$x' = q(x, t) \tag{2.1}$$

where $q = (q_1, \ldots, q_n)$ is a continuous vector function of the real vector $x = (x_1, \ldots, x_n)$ and the time t and suppose that (2.1) satisfies a uniqueness condition in a region R of the $(n+1)$-dimensional (x, t) space. A solution $x^*(t) = x(t, x_0, t_0)$, $t_0 \leq t < \infty$, of (2.1) with $x^*(t_0) = x_0$ is said to belong to R if $[x^*(t), t]$ is an interior point of R for every $t \geq t_0$. A solution $x^*(t) = x(t; x_0, t_0)$ is said to be <u>asymptotically stable</u> (to the right) if there exists a $\delta > 0$ such that (i) every solution $x(t; x_1, t_0)$ exists for all $t \geq t_0$ and belongs to R if $\|x_1 - x_0\| < \delta$; (ii) $\|x(t; x_1, t_0) - x(t; x_0, t_0)\| \longrightarrow 0$ as $t \longrightarrow \infty$ if $\|x_1 - x_0\| < \delta$, where $\|x\| = \Sigma_{j=1}^n |x_j|$.

Suppose now that $q(x, t)$ in (2.1) is periodic in t and the solution $x^*(t)$ is also periodic and the second partial derivatives of the functions q_j, $j = 1, 2, \ldots, n$, with respect to x_k, $k = 1, 2, \ldots, n$, in the region R of (x, t) space. Then the linear variational equation for the solution x^* is given by

$$y' = Q(t)y, \quad Q(t) = (\partial q_j/\partial x_k)_{x=x^*}, \quad j, k = 1, 2, \ldots, n \ , \tag{2.2}$$

where $Q(t)$ is periodic in t. We need the following known theorem.

THEOREM 2.1. If $q(x, t)$ satisfies the above conditions and the characteristic exponents of (2.2) have negative real parts, then the periodic solution $x^*(t)$ of (2.1) is asymptotically stable to the right ([Liapunov, 9], see also [5, p. 314]).

If system (2.1) is autonomous, i.e., $x' = q(x)$, then any periodic solution $x^*(t)$ defines a closed curve C in the n-dimensional x-space, E_n. If $d(x, c)$ denotes the distance of a point x in E_n from the curve C in E_n, then the solution $x^*(t)$ is said to be <u>asymptotically orbitally stable</u> (to the right) provided, given $\varepsilon > 0$, there exists a $\delta > 0$ such that every solution $x(t)$ with $d(x(t_0), C) < \delta$ for some t_0 implies $d(x(t), C) < \varepsilon$ for $t \geqq t_0$ and $d(x(t), C) \longrightarrow 0$ as $t \longrightarrow +\infty$.

THEOREM 2.2. If $x^*(t)$ is a periodic solution of the autonomous system $x' = q(x)$ where each component of q_j has continuous second derivatives in a region R of E_n, and $n - 1$ of the characteristic exponents of the linear variarional equation for $x^*(t)$ have negative real parts, then the solution $x^*(t)$ is asymptotically stable to the right. [5, p. 323].

§3. PERIODIC DIFFERENTIAL SYSTEMS

Consider the periodic system of equations

$$(3.1)\qquad \begin{aligned} &x_j'' + \alpha_j x_j' + \sigma_j^2 x_j = \varepsilon\, f_j(x, x', \varepsilon, t),\ j = 1, 2, \ldots, \mu\ , \\ &x_j' + \beta_j x_j = \varepsilon\, f_j(x, x', \varepsilon, t),\ j = \mu + 1, \ldots, n;\ t_0 \leqq t < +\infty\ , \end{aligned}$$

where $\varepsilon > 0$ is a real parameter, $x = (x_1, \ldots, x_n)$, $x' = (x_1', \ldots, x_\mu)$, each f_j is real and analytic in a neighborhood U of the origin of the (x, x', ε) space; the power series expansion of each f_j in U has coefficients periodic in t of period $T = 2\pi/\omega$, L-integrable in $[0, T]$, and there exists a function $\eta(t)$, L-integrable in $[0, T]$, such that

$$(3.2)\qquad |f_j(x, x', \varepsilon, t)| < \eta(t),\ j = 1, 2, \ldots, n\ ,$$

for all (x, x', ε) in U. The parameters α_j, β_j, σ_j are real, analytic functions of ε, $0 \leqq \varepsilon \leqq \varepsilon_0$, $\varepsilon_0 > 0$; $\gamma_j(\varepsilon) = (4\sigma_j^2 - \alpha_j^2)^{\frac{1}{2}} > 0$, $j = 1, 2, \ldots, \mu$, $0 \leqq \varepsilon \leqq \varepsilon_0$. Furthermore, we define numbers ρ_j, $j = 1, 2, \ldots, n + \mu$, by the relations

$$(3.3)\qquad \begin{aligned} &\rho_{2j-1} = \tfrac{1}{2}(-\alpha_j + i\gamma_j) = \bar{\rho}_{2j},\ j = 1, 2, \ldots, \mu,\ \rho_{\mu+j} = -\beta_j\ , \\ &\qquad j = \mu + 1, \ldots, n\ . \end{aligned}$$

The system of periodic differential equations (3.1) satisfying all of the above conditions shall be referred to as system (3.1). Notice the vectors x, x' used above do not have the same dimension.

Suppose there exists a real periodic solution, $X(\varepsilon, t) = (X_1, \ldots, X_n)$, of system (3.1), analytic in ε for $0 \leq \varepsilon \leq \varepsilon_0$, $\varepsilon_0 > 0$ with

$$(3.4)\qquad \begin{aligned} X_{jo} &\equiv X_j(0, t) = a_j \cos(r_j \omega t + \varphi_j),\ j = 1, 2, \ldots, \nu;\ \nu \leq \mu\ , \\ X_{jo} &\equiv X_j(0, t) = 0,\ j = \nu + 1, \ldots, n\ , \end{aligned}$$

where a_j, φ_j are real numbers, $r_j = k_j/m_j$, k_j, m_j relatively prime positive integers, $j = 1, 2, \ldots, \nu$, and $X(\varepsilon, t + mT) = X(\varepsilon, t)$, $T = 2\pi/\omega$, $m = m_1 \cdots m_\nu$. Many methods have been given for obtaining such periodic solutions (for example, see [4a, Section 8.5] and [4b]). (3.1),

$$(3.5)\qquad x_j = X_j + y_j,\ j = 1, 2, \ldots, n\ ,$$

then the linear variational equation for y_j is

$$(3.6)\qquad \begin{aligned} y_j'' + \alpha_j y_j + \sigma_j^2 y_j &= \varepsilon \sum_{k=1}^{n} f^*_{jx_k} y_k + \varepsilon \sum_{k=1}^{\mu} f^*_{jx'_k} y'_k,\ j = 1, 2, \ldots, \mu\ , \\ y_j' + \beta_j y_j &= \varepsilon \sum_{k=1}^{n} f^*_{jx_k} y_k + \varepsilon \sum_{k=1}^{\mu} f^*_{jx'_k} y'_k,\ j = \mu + 1, \ldots, n\ , \end{aligned}$$

where

$$(3.7)\qquad f^*_{jx_k} = f_{jx_k}(X, X', \varepsilon, t),\ f^*_{jx'_k} = f_{jx'_k}(X, X', \varepsilon, t)\ ,$$

and $X = (X_1, \ldots, X_n)$, $X' = (X'_1, \ldots, X'_\mu)$. Each of the coefficients of y_k, y'_k in (3.7) is periodic in t of period $mT = 2\pi m/\omega$.

The transformation of variables

$$(3.8)\qquad \begin{aligned} y_j &= (2i\gamma_j)^{-1}(z_{2j-1} + z_{2j}),\ y'_j = (2i\gamma_j)^{-1}(\rho_{2j-1} z_{2j-1} + \rho_{2j} z_{2j})\ , \\ & j = 1, 2, \ldots, \mu,\ y_k = z_{\mu+k},\ k = \mu + 1, \ldots, n\ , \end{aligned}$$

leads to the equivalent system of first order equations

(3.9) $$z' = A(\varepsilon)z + \varepsilon\, C(t, \varepsilon)z$$

where $A = \mathrm{diag}(\rho_1, \ldots, \rho_{n+\mu})$, the ρ_j are defined by (3.3), and

$$C = (c_{jk}),\ j, k = 1, 2, \ldots, n + \mu ,$$

$$c_{2j-1,2k-1} = (2i\gamma_k)^{-1}(f^*_{jx_k} + \rho_{2k-1} f^*_{jx'_k});\ c_{2j-1,2k} = -\bar{c}_{2j-1,2k-1} ,$$

$$k = 1, 2, \ldots, \mu;$$

$$c_{2j-1,\mu+h} = f^*_{jx_k},\ h = \mu + 1, \ldots, n;\ c_{2j-1,\ell} = -c_{2j,\ell},\ \ell = 1, 2, \ldots, n + \mu,$$

(3.10) $$j = 1, 2, \ldots, \mu ;$$

$$c_{\mu+h,2k-1} = (2i\gamma_k)^{-1}(f^*_{hx_k} + \rho_{2k-1} f^*_{hx'_k});\ c_{\mu+h,2k} = -\bar{c}_{\mu+h,2k-1} ,$$

$$k = 1, 2, \ldots, \mu ;$$

$$c_{\mu+h,\mu+\ell} = f^*_{hx_\ell},\ \ell = \mu + 1, \ldots, n;\ h = \mu + 1, \ldots, n ,$$

and the functions $f^*_{jx_k}$, $f^*_{jx'_k}$ are defined by (3.7).

In the sequel, we need the following known results.

LEMMA 3.1. (J. K. Hale [8, Theorem 2.1]). Consider the system of differential equations (3.9) with $C(t + mT, \varepsilon) = C(t, \varepsilon)$, $T = 2\pi/\omega$, $m = m_1 \cdots m_\nu$. Suppose that $\omega' = \omega/m$,

$$\rho_j(0) - \rho_k(0) = m_{jk}\omega' i,\ m_{jk} \text{ an integer or zero,}$$

(3.11) $$j, k = 1, 2, \ldots, p;$$

$$\rho_j(0) \not\equiv \rho_k(0) \pmod{\omega' i},\ j = 1, 2, \ldots, p;\ k = p + 1, \ldots, n + \mu,$$

and define the $p \times p$ matrix $G = H - D$, $D = \mathrm{diag}(d_1, \ldots, d_p)$, $H = (h_{jk})$, $j, k = 1, 2, \ldots, p$, where

$$d_j = \lim_{\varepsilon \to 0} \varepsilon^{-1} [\rho_j(\varepsilon) - \rho_j(0)]$$

(3.12)

$$h_{jk} = (mT)^{-1} \int_0^{mT} e^{i(m_{k1} - m_{j1})\omega' t} c_{jk}(t, 0)dt,\ j, k = 1, 2, \ldots, p.$$

If λ_0 is a simple characteristic root of the matrix G and the corresponding eigenvector has no zero components, there is a characteristic exponent $\tau(\varepsilon)$ of (3.9) which is analytic in ε for $0 \leqq \varepsilon \leqq \varepsilon_0$, and

(3.13) $$\tau(\varepsilon) = \rho_1(0) + \varepsilon\, \lambda_0 + O(\varepsilon^2) \; .$$

COROLLARY 3.1. Consider the system of differential equations (3.9) with $C(t + mT, \varepsilon) = C(t, \varepsilon)$, $T = 2\pi/\omega$, $m = m_1 \cdots m_\nu$. If $\omega' = \omega/m$ and

(3.14) $$\rho_j(0) \not\equiv \rho_k(0) \pmod{\omega' i}, \; k \neq j, \; k = 1, 2, \ldots, n + \mu \; ,$$

where j is a fixed integer, $1 \leqq j \leqq n + \mu$, there is a characteristic exponent of (3.9), $\tau_j = \tau_j(\varepsilon)$, analytic in ε for $0 \leqq \varepsilon \leqq \varepsilon_0$, with

(3.15) $$\tau_j(\varepsilon) = \rho_j(\varepsilon) + \varepsilon T^{-1} \int_0^T c_{jj}(t, 0)dt + O(\varepsilon^2) \; .$$

LEMMA (3.2) (J. K. Hale, [8, Theorem 3.1]). Consider the system of differential equations (3.9) with $C(t + mT, \varepsilon) = C(t, \varepsilon)$, $T = 2\pi/\omega$, $m = m_1 \cdots m_\nu$. Suppose $\omega' = \omega/m$,

(3.16)
$$\rho_{2j-1} = \bar{\rho}_{2j}, \; \rho_{2j-1}(0) \not\equiv \rho_{2j}(0) \pmod{\omega' i},$$
$$\rho_{2j-1}(0) = \rho_{2k-1}(0) = m_{jk}\omega' i, \; m_{jk} \text{ and integer or zero},$$
$$j, k = 1, 2, \ldots, p \; ;$$
$$\rho_j(0) \not\equiv \rho_k(0) \pmod{\omega' i}, \; j = 1, 2, \ldots, 2p; \; k = 2p + 1, \ldots, n + \mu,$$

and define the $p \times p$ matrix $G = H - D$, $D = \operatorname{diag}(d_1, \ldots, d_p)$, $H = (h_{jk})$, $j, k = 1, 2, \ldots, p$, where

(3.17)
$$d_j = \lim_{\varepsilon \to 0} \varepsilon^{-1}[\rho_{2j-1}(\varepsilon) - \rho_{2j-1}(0)], \; j = 1, 2, \ldots, p \; ,$$
$$h_{jk} = (mT)^{-1} \int_0^{mT} c_{2j-1,2k-1}(t, 0)\, e^{i(m_{k1} - m_{j1})\omega' t} dt \; ,$$
$$j, k = 1, 2, \ldots, p \; .$$

If λ_o is a simple characteristic root of the matrix G and the corresponding eigenvector has no zero components, then there are two characteristic exponents τ, $\bar{\tau}$ of (3.9) analytic in ε for $0 \leqq \varepsilon \leqq \varepsilon_o$ and

(3.18) $$\tau(\varepsilon) = \rho_1(0) + \varepsilon \lambda_o + O(\varepsilon^2) .$$

From the Floquet theory [5, p. 78] the characteristic exponents of system (3.9) are obtained by finding the eigenvalues of a matrix of order $2n$. We shall illustrate by means of a few theorems how Lemmas 3.1 and 3.2 can be of assitance in reducing the order of this determinant. More specifically, if there are p of the characteristic exponents close to the origin in the complex plane, then they will be obtained from the eigenvalues of a determinant of order p. If the remaining characteristic exponents are isolated for all ε, $0 \leqq \varepsilon \leqq \varepsilon_o$, then they will be obtainable immediately from Corollary 3.1. If there are say $2p$ characteristic exponents which are two by two complex conjugate none of which are close to the origin in the complex plane, and p are equal for $\varepsilon = 0$, then they will be obtained from a determinant of order p.

THEOREM (3.1). Consider the system (3.1) and suppose that (3.4) is a periodic solution of (3.1). Suppose $\varepsilon > 0$, $\alpha_1 = \alpha_2 \cdots = \alpha_q = 0$, $\alpha_j > 0$, $j = q + 1, \ldots, \mu$; $\beta_j > 0$, $j = \mu + 1, \ldots, n$; for all ε, $0 \leqq \varepsilon \leqq \varepsilon_o$, and

$$\rho_{2j-1}(0) = i\, k_j\, \omega/m_j = \bar{\rho}_{2j}(0),\ j = 1, 2, \ldots, \nu;\ \nu \leqq \mu,$$

(3.19) $$\rho_j(0) \not\equiv \rho_k(0) \ (\mathrm{mod}\ \omega' i),\ j = 1, 2, \ldots, 2\nu;\ k = 2\nu + 1, \ldots, n + \mu;$$

$$\rho_j(0) \not\equiv \rho_k(0) \ (\mathrm{mod}\ \omega' i),\ j \neq k,\ j, k = 2\nu + 1, \ldots, n + \mu,$$

where $\omega' = \omega/m$, $m = m_1 \cdots m_\nu$, and the ρ_j are defined by (3.3). The periodic solution (3.4) of (3.1) is asymptotically stable to the right for $0 \leqq \varepsilon \leqq \varepsilon_o$, $\varepsilon_o > 0$, if the following conditions are satisfied:

(i) $$\int_0^{mT} f_{jx_j'}(X_o, X_o', 0, t)dt < 0,\ j = \nu + 1, \ldots, q ,$$

where $X_o = (X_{1o}, \ldots, X_{no})$, $X_o' = (X_{1o}', \ldots, X_{\mu o}')$ are defined by (3.4);

(ii) the eigenvalues $\lambda_1, \ldots, \lambda_{2\nu}$ of the $2\nu \times 2\nu$ matrix

$$G = H - D,\ D = \operatorname{diag}(d_1, \ldots, d_{2\nu}),\ H = (h_{jk}),$$
$$j,\ k = 1,\ 2,\ \ldots,\ 2\nu,$$

$$d_{2j-1} = i \lim_{\varepsilon \to 0} \varepsilon^{-1}[\sigma_j(\varepsilon) - \sigma_j(0)],\ d_{2j} = -d_{2j-1},$$
$$j = 1,\ 2,\ \ldots,\ \nu,$$

(3.20)
$$h_{2j-1,2k-1} = \frac{1}{2mT}\int_0^{mT} [(i\sigma_k)^{-1} f_{jx_k} + f_{jx'_k}]\, e^{i(m_{k1}-m_{j1})\omega' t}\, dt$$

$$h_{2j-1,2k} = \frac{1}{2mT}\int_0^{mT} [(i\sigma_k)^{-1} f_{jx_k} - f_{jx'_k}]\, e^{-i(m_{k1}+m_{j1})\omega' t}\, dt$$

$$h_{2j,2k-1} = \bar{h}_{2j-1,2k},\ h_{2j,2k} = \bar{h}_{2j-1,2k-1},\ j,\ k = 1,\ 2,\ \ldots,\ \nu,$$

where the functions under the integrals are evaluated at $(X_0, X'_0, 0, t)$, $\rho_{2j-1}(0) - \rho_{2k-1}(0) = m_{jk}\omega' i$, m_{jk} an integer or zero, $j,\ k = 1,\ 2,\ \ldots,\ \nu$, are distinct, have negative real parts and the corresponding eigenvectors have no zero components.

PROOF. From Corollary 3.1, it follows immediately that the characteristic exponents $\tau_j(\varepsilon)$, $j = 1, 2, \ldots, n + \mu$, $\tau_{2j-1}(0) = \bar{\tau}_{2j}(0) = \frac{1}{2}[-\alpha_j(0) = i\gamma_j(0)]$, $j = q + 1, \ldots, \mu$; $\tau_{\mu+j}(0) = -\beta_j(0)$ $j = \mu + 1, \ldots, n$, have negative real parts for $0 \leqq \varepsilon \leqq \varepsilon_0$. Furthermore, from the same corollary and (3.10),

$$R(\tau_{2j-1}) = R(\tau_{2j}) = \varepsilon(2mT)^{-1}\int_0^{mT} f_{jx'_j}(X_0, X'_0, 0, t)\,dt + o(\varepsilon^2)\ ,$$

for $j = \nu + 1, \ldots, q$. Therefore, from (i) of the theorem, the real parts of these characteristic exponents are negative for $0 < \varepsilon \leqq \varepsilon_0$. The other characteristic exponents $\tau_1, \ldots, \tau_{2\nu}$ are determined from Lemma 3.1 for $p = 2\nu$. The matrix G of the theorem is precisely the matrix G of Lemma 3.1. Therefore, from (ii) and (3.13), it follows that these characteristic exponents also have negative real parts for $0 < \varepsilon \leqq \varepsilon_0$. Theorem 2.1 now implies that Theorem 3.1 is true.

REMARK 3.1. Theorem 3.1 has been previously obtained by H. R. Bailey and R. A. Gambill [3] for the case where $\nu = \mu = q = n$; i.e., for systems of second order equations, $x''_j + \sigma_j^2 x_j = \varepsilon f_j(x_1, \ldots, x_n, x'_1, \ldots, x'_n, \varepsilon, t$

$j = 1, 2, \ldots, n$, where each $\sigma_j = \sigma_j(\varepsilon)$ satisfies the relation $\sigma_j(0) = k_j\omega/m_j$, $j = 1, 2, \ldots, n$. An interesting example is also given in [3] for $\nu = \mu = q = n = 2$ illustrating the application of this theorem.

REMARK 3.2. If any one of the roots of $|G - \beta I| = 0$ in (ii) have a positive real part or if any of the functions in (i) are positive, then the periodic solution (3.4) of (3.1) is unstable in $[t_0 + \infty]$ for every $\varepsilon \neq 0$.

THEOREM 3.2. Consider the system (3.1) and suppose that (3.4) with $\nu = 1$ is a periodic solution of (3.1). Suppose $\varepsilon > 0$ and the α_j, β_j, σ_j satisfy the conditions of Theorem 3.1 with $\nu = 1$; i.e.,

$$\rho_1(0) = i\, k_1\omega/m_1 = \bar{\rho}_2(0),$$

$$\rho_j(0) \not\equiv \rho_k(0) \pmod{\omega' i} \quad j = 1, 2;\ k = 3, 4, \ldots, n + \mu\ ,$$

(3.21)

$$\rho_j(0) \not\equiv \rho_k(0) \pmod{\omega' i} \quad j \neq k,\ j, k = 3, 4, \ldots, n + \mu\ ;$$

where $\omega' = \omega/m_1$. The periodic solution (3.4) of (3.1) is asymptotically stable to the right for $0 < \varepsilon \leqq \varepsilon_0$ if the following conditions are satisfied:

(i) $$\int_0^{mT} f_{jx_j'}(X_0, X_0', 0, t)dt < 0, \quad j = 1, 2, \ldots, q\ ;$$

where X_0, X_0' are defined by (3.4) for $\nu = 1$;

(ii) The numbers A, B defined by

$$(2m_1T)A = \int_0^{m_1T} f_{1x_1'}dt\ ,$$

(3.22)

$$(2m_1T)^2B = \left(\int_0^{m_1T} f_{1x_1'}dt\right)^2 + \left[2m_1Td_1 - \sigma_1^{-1}(0)\int_0^{m_1T} f_{1x_1}dt\right]^2$$
$$- \left[\sigma_1^{-1}(0)\int_0^{m_1T} f_{1x_1}\cos 2\sigma_1(0)t\,dt - \int_0^{m_1T} f_{1x_1'}\sin 2\sigma_1(0)t\,dt\right]^2$$
$$- \left[\sigma_1^{-1}(0)\int_0^{m_1T} f_{1x_1}\sin 2\sigma_1(0)t\,dt + \int_0^{m_1T} f_{1x_1'}\cos 2\sigma_1(0)dt\right]^2$$

$$d_1 = \lim_{\varepsilon \to 0} \varepsilon^{-1}\,[\sigma_1(\varepsilon) - \sigma_1(0)]\ ,$$

where all of the functions are evaluated at $(X_0, X_0', 0, t)$, satisfy the property $A \neq B$, $A^2 \neq B$, $B > 0$.

REMARK 3.3. This theorem has been proved by H. R. Bailey and R. A. Gambill [3] and L. Mandelstam and N. Papalexi [10] for the case where $q = \nu = \mu = n = 1$; i.e., a second order equation $x'' + \sigma^2 x = \varepsilon f(x, x', \varepsilon)$, where x, f are scalars and $\sigma = \sigma(\varepsilon)$, $\sigma(0) = k\omega/m$.

PROOF. The system under consideration in Theorem 3.2 is a special case of the system in Theorem 1 for $\nu = 1$. Condition (i) of Theorem 3.2, for $j = 2, 3, \ldots, q$ implies condition (i) of Theorem 1 is satisfied for $\nu = 1$. For $\nu = 1$, the characteristic equation of the matrix G in Theorem 3.1 is given by $\lambda^2 - 2A\lambda + B = 0$, where A, B are given in (3.22). The roots of this equation have negative real parts if and only if $A > 0$, $A^2 \neq B$, $B > 0$ which is assured by condition (i) and (ii) of Theorem 3.2. The condition $A \neq B$ assures that the corresponding eigenvectors of G have no zero components and Theorem 3.2 is proved.

EXAMPLE. Consider the system of equations

$$(3.23)\qquad \begin{aligned} x_1'' + x_1 &= \varepsilon(1 - x_1^2 - x_2^2)x_1' + \varepsilon\, p \cos t \equiv \varepsilon f_1(x_1, x_2, x_1', x_2', \varepsilon, t) \;, \\ x_2'' + \sigma_2^2 x_2 &= \varepsilon(1 - x_1^2 - x_2^2)x_2' \equiv \varepsilon f_2(x_1, x_2, x_1', x_2', \varepsilon, t) \;, \end{aligned}$$

where $\varepsilon > 0$, $p < 0$ and $\sigma_2 > 0$, $\sigma_2 \neq m$, m an integer. Applying the method of successive approximations [6], it is easy to see that (3.23) has a periodic solution $X_j(\varepsilon, t) = X_j(\varepsilon, t + 2\pi)$, $j = 1, 2$, analytic in ε for $0 \leqq \varepsilon \leqq \varepsilon_0$, $\varepsilon_0 > 0$,

$$(3.24)\qquad X_1(0, t) = a \sin t, \quad X_2(0, t) = 0 \;,$$

where $a > 2$ is the positive solution of the equation $1 - (\frac{a}{2})^2 + \frac{p}{a} = 0$. Furthermore,

$$\int_0^{2\pi} f_{jx_j'}(a \sin t, 0, a \cos t, 0, 0, t)dt = \pi(2 - a^2) < 0 \;,$$

$j = 1, 2$, and the numbers A, B of Theorem 3.2 are given by $4a = (2-a^2)$ $16B = (2 - a^2)^2 + 16a^2 - 9a^2/4$ and $A \neq B$, $A^2 \neq B$, $B > 0$ for every real $a \neq 0$. Therefore, from Theorem 3.2, the solution (3.24) of (3.23) is asymptotically stable to the right for $0 < \varepsilon \leqq \varepsilon_0$.

THEOREM 3.3. Consider the system (3.1) and suppose that (3.4) is a periodic solution of (3.1). Suppose $\varepsilon > 0$, $\alpha_1 = \alpha_2 = \dots = \alpha_q = 0$, $\alpha_j > 0$, $j = q + 1, \dots, \mu$, $\beta_j > 0$, $j = \mu + 1, \dots, n$, for all ε, $0 \leq \varepsilon \leq \varepsilon_0$, and

$$\rho_{2j-1}(0) = i\, k_j\omega/m_j = \bar{\rho}_{2j}(0),\ j = 1, 2, \dots, \nu;\ \nu \leq q\ ;$$

$$\rho_j(0) \not\equiv \rho_k(0) \pmod{\omega' i},\ j = 1, 2, \dots, 2\nu;$$
$$k = 2\nu + 1, \dots, n + \mu\ ;$$

(3.25) $$\rho_{2j-1}(0) \not\equiv \rho_{2j}(0) \pmod{\omega' i},\ \rho_{2j-1}(0) - \rho_{2k-1}(0) = m_{jk}\omega\, i\ ,$$

m_{jk} an integer or zero, $j, k = \nu + 1, \dots, r$; $\nu \leq r \leq q$;

$$\rho_j(0) \not\equiv \rho_k(0) \pmod{\omega' i},\ j = 2\nu + 1, \dots, 2r;$$
$$k = 2r + L, \dots, n + \mu\ ;$$

$$\rho_j(0) \not\equiv \rho_k(0) \pmod{\omega' i},\ j \neq k,\ j, k = 2r + 1, \dots, n + \mu\ ,$$

where $\omega' = \omega/m$, $m = m_1 \cdots m_\nu$ and the ρ_j are defined by (3.3). The periodic solution (3.4) of (3.1) is asymptotically stable to the right, $0 \leq \varepsilon \leq \varepsilon_0$, if condition (ii) of Theorem 3.1 is satisfied and, in addition,

(3.26) $$\int_0^{mT} f_{jx_j'}(X_0, X_0', 0, t)\,dt < 0,\ j = r + 1, \dots, q\ ,$$

where $X_0 = (X_{10}, \dots, X_{n0})$, $X_0' = (X_{10}', \dots, X_{\mu 0}')$ are defined by (3.4); and the eigenvalues $\lambda_1, \dots, \lambda_{r-\nu}$ of the $(r-\nu) \times (r-\nu)$ matrix $M = N - P$, $P = \mathrm{diag}(p_1, \dots, p_{r-\nu})$, $N = (n_{jk})$, $j, k = 1, 2, \dots, r - \nu$,

$$p_j = i \lim_{\varepsilon \to 0} \varepsilon^{-1}\,[\sigma_{\nu+j}(\varepsilon) - \sigma_{\nu+j}(0)],\ j = 1, 2, \dots, r - \nu\ ;$$

(3.27)
$$n_{j-\nu,k-\nu} = \frac{1}{mT} \int_0^{mT} c_{2j-1,2k-1}(t, 0)e^{i(m_{k1}-m_{j1})\omega' t}\,dt\ ,$$
$$j, k = \nu + 1, \dots, r\ ;$$

where the c_{jk} are defined by (3.10), are distinct, have negative real parts and the corresponding eigenvectors have no zero components.

PROOF. The proof that the characteristic exponents $\tau_1, \ldots, \tau_{2\nu}, \tau_{2r+1}, \ldots, \tau_{n+\mu}$ have negative real parts is exactly the same as the proof in Theorem 3.1. The remaining characteristic exponents $\tau_{2\nu+1}, \ldots, \tau_{2r}$ are obtained from Lemma 3.2. The matrix M of the theorem is precisely the matrix G of Lemma 3.2 if the equations (3.9) are reordered so that $A = \mathrm{diag}(\rho_{2\nu+1}, \ldots, \rho_{2r}, \rho_1, \ldots, \rho_{2\nu}, \rho_{2r+1}, \ldots, \rho_{n+\mu})$. Therefore, each of the eigenvalues $\lambda_{j-\nu}$ of the matrix M determines two characteristic exponents $\tau_{2j-1}, \tau_{2j} = \bar{\tau}_{2j-1}$, with $\tau_{2j-1} = i\rho_j(0) + \varepsilon\lambda_{j-\nu} + o(\varepsilon)$, $j = \nu + 1, \ldots, r$. The real parts of these characteristic exponents are also negative for $0 < \varepsilon \leqq \varepsilon_0$, and the theorem follows from Theorem 2.1.

REMARK 3.3. For $\nu = 1$ in Theorem 3.3, the condition (ii) referred to can be replaced by condition (ii) of Theorem 3.2.

This theorem should clarify the remark preceding Theorem 3.1. In fact, this theorem is mentioned only to illustrate the procedure for calculating the characteristic exponents of (3.9) which are two by two complex conjugate, $r - \nu$ which are equal for $\varepsilon = 0$ and none of which are close to the origin in the complex plane.

§4. AUTONOMOUS DIFFERENTIAL SYSTEMS

Consider the autonomous system of equations

$$
(4.1)\qquad
\begin{aligned}
&x_j'' + \alpha_j x_j' + \sigma_j^2 x_j = \varepsilon f_j(x, x', \varepsilon), \; j = 1, 2, \ldots, \mu, \\
&x_j' + \beta_j x_j = \varepsilon f_j(x, x', \varepsilon), \; j = \mu + 1, \ldots, n; \; t_0 \leq t < +\infty,
\end{aligned}
$$

where $\varepsilon > 0$ is a real parameter $x = (x_1, \ldots, x_n)$, $x' = (x_1', \ldots, x_\mu')$, each f_j is real and analytic in a neighborhood U of the origin of the (x, x', ε) space. The parameters $\alpha_j, \beta_j, \sigma_j$ are real analytic functions of ε, $0 \leqq \varepsilon \leqq \varepsilon_0$, $\varepsilon_0 > 0$; $\gamma_j(\varepsilon) = (4\sigma_j^2 - \alpha_j^2)^{\frac{1}{2}} > 0$, $j = 1, 2, \ldots, \mu$, $0 \leqq \varepsilon \leqq \varepsilon_0$. Furthermore, we define numbers ρ_j, $j = 1, 2, \ldots, n + \mu$, by the relations

$$
(4.2)\qquad \rho_{2j-1} = \tfrac{1}{2}(-\alpha_j + i\gamma_j) = \bar{\rho}_{2j}, \; j = 1, 2, \ldots, \mu, \; \rho_{\mu+j} = -\beta_j, \quad j = \mu + 1, \ldots, n.
$$

The system of autonomous differential equations (4.1) satisfying all of the above conditions shall be referred to as system (4.1). Notice that the vectors x, x' used above do not have the same dimension.

Suppose there exists a real periodic solution, $X(\varepsilon, t) = (X_1, \ldots, X_n)$, of system (4.1), analytic in ε for $0 \leqq \varepsilon \leqq \varepsilon_0$, $\varepsilon_0 > 0$ with

$$(4.3)\quad \begin{aligned} X_{jo} &\equiv X_j(0, t) = a_j \cos(r_j \omega t + \varphi_j),\ j = 1, 2, \ldots, \nu;\ \nu \leqq \mu , \\ X_{jo} &\equiv X_j(0, t) = 0,\ j = \nu + 1, \ldots, n , \end{aligned}$$

where a_j, φ_j are real numbers, $r_j = k_j/m_j$, k_j, m_j relatively prime positive integers $j = 1, 2, \ldots, \nu$, and $X(\varepsilon, t + mT) = X(\varepsilon, t)$, $T = 2\pi/\omega$, $m = m_1 \cdots m_\nu$. Many methods have been given for obtaining such periodic solutions (for example, see [4a, Section 8.5] and [4b]).

As in the previous section, the linear variational equation associated with the solution $X(\varepsilon, t)$ of (4.1) is given by

$$(4.4)\quad \dot{z'} = A(\varepsilon)z + \varepsilon\, C(t, \varepsilon)z$$

where $A = \mathrm{diag}(\rho_1, \ldots, \rho_{n+\mu})$, the ρ_j are defined by (4.2) and the matrix $C = (c_{jk})$, $j, k = 1, 2, \ldots, n + \mu$, is given by (3.10) with

$$(4.5)\quad f^*_{jx_k} = f_{jx_k}(X, X', \varepsilon),\ f^*_{jx'_k} = f_{jx'_k}(X, X', \varepsilon)$$

and $X = (X_1, \ldots, X_n)$, $X' = (X'_1, \ldots, X'_\mu)$. The matrix $C(t, \varepsilon)$, therefore, satisfies the relation $C(t + mT, \varepsilon) = C(t, \varepsilon)$.

We wish to apply Lemmas 3.1 and 3.2 to obtain sufficient conditions for the asymptotic orbital stability of the periodic solution $X(\varepsilon, t)$ of (4.1). Before proceeding to the formal discussion, observe that one expects each of the given numbers σ_j to satisfy a relation of the form $\sigma_j = r_j\sigma_j + O(\varepsilon)$, $j = 2, 3, \ldots, \nu$, to obtain a periodic solution of (4.1) of the form (4.3). The number ω (also the numbers r_j, φ_j, $j = 1, 2, \ldots, \nu$) is then determined as a function of ε and the numbers σ_j so that (4.3) satisfies (4.1). In particular $\omega = \sigma_1 + O(\varepsilon)$. As we know, one could just as well obtain σ_1 as a function of ω and ε simply by making a convenient change of scale in the variable t. In the present context, it is convenient to assume that the latter alternative is chosen, namely, ω is independent of ε and the numbers σ_j, $j = 1, 2, \ldots, \nu$, are convenient functions of ε so that (4.3) satisfies (4.1).

Another observation to be made is that the linear variational equation (4.4) always has a periodic solution of period mT [5, p. 322] and, therefore, one of the characteristic exponents (since they are only determined up to a multiple of $\omega i/m$) must be $k\omega i/m$, where k is an integer or zero.

THEOREM 4.1. Consider the system (4.1) and suppose that (4.3) is a periodic solution of (4.1). Suppose $\varepsilon > 0$, $\alpha_1 = \alpha_2 = \dots = \alpha_q = 0$, $\alpha_j > 0$, $j = q + 1, \dots, \mu$; $\beta_j > 0$, $j = \mu + 1, \dots, n$; for all ε, $0 \leqq \varepsilon \leqq \varepsilon_0$, $\varepsilon_0 > 0$, and

$$\rho_{2j-1}(0) = i\, k_j\omega/m_j = \bar{\rho}_{2j}(0),\ j = 1, 2, \dots, \nu,\ \nu \leqq \mu ,$$

(4.6) $$\rho_j(0) \not\equiv \rho_k(0) \pmod{\omega' i},\ j = 1, 2, \dots, 2\nu;\ k = 2\nu + 1, \dots, n + \mu ;$$

$$\rho_j(0) \not\equiv \rho_k(0) \pmod{\omega' i},\ j \neq k,\ j, k = 2\nu + 1, \dots, n + \mu ,$$

where $\omega' = \omega/m$, $m = m_1 \dots m_\nu$, and the ρ_j are defined by (4.2). The periodic solution (4.3) of (4.1) is asymptotically orbitally stable to the right, $0 < \varepsilon \leqq \varepsilon_0$, $\varepsilon_0 > 0$, if the following conditions are satisfied:

(i) $$\int_0^{mT} f_{jx_j'}(X_0, X_0', 0)dt < 0,\ j = \nu + 1, \dots, q ,$$

where $X_0 = (X_{10}, \dots, X_{n0})$, $X_0' = (X_{10}', \dots, X_{\mu 0}')$ are defined by (4.3);

(ii) $2\nu - 1$ of the eigenvalues of the $2\nu \times 2\nu$ matrix $G = H - D$, $D = \mathrm{diag}(d_1 \dots d_{2\nu})$, $H = (h_{jk})$, $j, k = 1, 2, \dots, 2\nu$,

$$d_{2j-1} = i \lim_{\varepsilon \to 0} \varepsilon^{-1}[\sigma_j(\varepsilon) - \sigma_j(0)],\ d_{2j-1} = -d_{2j},\ j = 1, 2, \dots, \nu ;$$

$$h_{2j-1,2k-1} = \frac{1}{2mT}\int_0^{mT} [(i\sigma_k)^{-1} f_{jx_k} + f_{jx_k'}] e^{i(m_{k1}-m_{j1})\omega' t} dt ,$$

(4.7)

$$h_{2j-1,2k} = \frac{1}{2mT}\int_0^{mT} [(i\sigma_k)^{-1} f_{jx_k} - f_{jx_k'}] e^{i(m_{k1}-m_{j1})\omega' t} dt ,$$

$$h_{2j,2k-1} = \bar{h}_{2j-1,2k},\ h_{2j,2k} = \bar{h}_{2j-1,2k-1};\ j, k = 1, 2, \dots, \nu ;$$

where the functions under the integrals are evaluated at $(X_o, X'_o, 0)$, $\rho_{2j-1}(0) - \rho_{2k-1}(0) = m_{jk}\omega' i$, m_{jk} an integer or zero, $j, k = 1, 2, \ldots, \nu$, are distinct, have negative real parts and the corresponding eigenvectors have no zero components.

PROOF. The proof of this theorem is exactly the same as the proof of Theorem 3.1 if we observe that one of the characteristic exponents of (4.4) being a multiple of $\omega i/m$ is equivalent to one of the roots β of the equation $|G - \beta I| = 0$, being zero. One then applies Theorem 2.2 rather than Theorem 2.1 to complete the proof.

THEOREM 4.2. Consider the system (4.1) and suppose that (4.3) with $\nu = 1$ is a periodic solution of (4.1). If $\varepsilon > 0$ and the α_j, β_j, σ_j satisfy the conditions of Theorem 4.1 with $\nu = 1$, i.e.,

$$\rho_1(0) = i\omega = \bar{\rho}_2(0),$$

$$\rho_j(0) \not\equiv \rho_k(0) \pmod{\omega i},\ j = 1, 2;\ k = 3, 4, \ldots, n + \mu\ ; \tag{4.8}$$

$$\rho_j(0) \not\equiv \rho_k(0) \pmod{\omega i},\ j \neq k,\ j, k = 3, 4, \ldots, n + \mu\ ;$$

the periodic solution (4.3) with $\nu = 1$ of (4.1) is asymptotically orbitally stable to the right, $0 < \varepsilon \leqq \varepsilon_o$, $\epsilon_o > 0$, if

$$\int_0^T f_{jx'_j}(X_o, X'_o, 0)dt < 0,\ j = 1, 2, \ldots, q\ ; \tag{4.9}$$

where $X_o = (X_{1o}, \ldots, X_{no})$, $X'_o = (X'_{1o}, \ldots, X'_{\mu o})$, $X_{1o} = a \cos(\omega t + \varphi)$, $X_{ko} = 0$, $k = 2, 3, \ldots, n$; a, φ real numbers.

REMARK 4.1. In his thesis at Purdue University, E. W. Thompson [12] has obtained this same result by another method. The method used by E. Thompson was to first eliminate the zero root of the variational equations and then apply some known results of H. R. Bailey and L. Cesari [2]. Also, A. Andronov and A. Witt [1], (see also [11, p. 153]) have obtained this result for the case $\nu = 1$, $q = \mu = n = 2$; i.e., for a system of two second order equations with no "large" damping terms. The method used by A. Andronov and A. Witt was to evaluate the characteristic polynomial of a matrix of order four. As we shall see in the proof of this theorem, no determinants are required.

PROOF. Exactly as in the proof of Theorem 4.1, we apply Corollary 3.1 to obtain the real parts of the characteristic exponents, $\tau_{2j-1} = \bar{\tau}_{2j}$, $j = 2, 3, \ldots, \mu$; $\tau_{\mu+j}$, $j = \mu + 1, \ldots, n$ as

$$
(4.10) \qquad
\begin{aligned}
R(\tau_{2j-1}) &= -\frac{\alpha_j}{2} + \frac{\varepsilon}{2T}\int_0^T f_{jx_j'}(X_0, X_0', 0)dt + O(\varepsilon^2), \; j = 1, 2, \ldots, \mu , \\
R(\tau_{\mu+j}) &= -\beta_j + \frac{\varepsilon}{T}\int_0^T f_{jx_j}(X_0, X_0', 0)dt + O(\varepsilon^2), \; j = \mu + 1, \ldots, n .
\end{aligned}
$$

We may assume that $\tau_1 = 0$, $\tau_2 = O(\varepsilon)$ and we know [5, p. 81] that

$$
T\sum_{j=1}^{n+\mu} \tau_j = \int_0^T \operatorname{tr}\,(A(\varepsilon) + \varepsilon\, C(t, \varepsilon))dt = -\sum_{j=1}^{\mu} \alpha_j - \sum_{j=\mu+1}^{n} \beta_j
$$

$$
+ \varepsilon \int_0^T \left(\sum_{j=1}^{\mu} f_{jx_j'} + \sum_{j=\mu+1}^{n} f_{jx_j} \right)dt .
$$

From (4.10) and the fact that $\tau_1 = 0$, we have

$$
R(\tau_2)T = \varepsilon \int_0^T f_{1x_1'}\,(X_0, X_0', 0)dt + O(\varepsilon^2) ,
$$

and Theorem 4.2 now follows immediately. One could prove this theorem in another way by evaluating directly the eigenvalues of the 2×2 matrix G in Theorem 4.1.

EXAMPLE. Consider the system of equations

$$
(4.11) \qquad
\begin{aligned}
x_1'' + \sigma_1^2 x_1 - \varepsilon(1 - x_1^2 - x_2^2)x_1' &= \varepsilon\, f(x_1, x_2, x_2', \varepsilon) \\
x_2'' + \sigma_2^2 x_2 - \varepsilon(1 - x_1^2 - x_2^2)x_2' &= \varepsilon\, g(x_1, x_1', x_2, \varepsilon), \; \varepsilon > 0 ,
\end{aligned}
$$

σ_1, σ_2 are analytic functions of ε at $\varepsilon = 0$ with $\sigma_2(0) = \sqrt{2}\sigma_1(0)$, $f(-x_1, x_2, x_2', \varepsilon) = -f(x_1, x_2, x_2', \varepsilon)$, $g(x_1, x_1', -x_2, \varepsilon) = -g(x_1, x_1', x_2, \varepsilon)$ are any analytic functions in a neighborhood U of the origin in

$(x_1, x_2, x_1', x_2', \varepsilon)$ space. It is known [7, p. 300] that there are two periodic solutions of this equation of the form

$$\text{(i)} \qquad x_1 = a_1 \sin(t + \varphi) + O(\varepsilon),\; x_2 = 0,\; a_1 = 2 + O(\varepsilon),\; \sigma_1 = 1 + O(\varepsilon)$$

$$\text{(ii)} \qquad x_1 = 0,\; x_2 = a_2 \sin(\sqrt{2}t + \varphi) + O(\varepsilon),\; a_2 = 2 + O(\varepsilon),\; \sigma_2 = \sqrt{2} + O(\varepsilon)$$

of periods 2π and $\sqrt{2}\pi$, respectively (we are choosing σ_1, σ_2 as functions of ϵ rather than choose the period). For (i) conditions (4.9) for $j = 1, 2,$ are equivalent to

$$\int_0^{2\pi} [1 - 4 \sin^2 t]dt = -2 \int_0^{2\pi} \sin^2 t\, dt < 0 \quad .$$

It follows from Theorem 4.2 that (i) is asymptotically orbitally stable to the right, $0 < \varepsilon \leqq \varepsilon_0$, $\varepsilon_0 > 0$. Similarly, one shows that (ii) is also asymptotically orbitally stable to the right for ε sufficiently small.

THEOREM 4.3. Consider the system (4.1) and suppose that (4.3) is a periodic solution of (4.1). Suppose $\varepsilon > 0$, $\alpha_1 = \alpha_2 = \dots = \alpha_q = 0$, $\alpha_j > 0$, $j = q + 1, \dots, \mu$; $\beta_j > 0$, $j = \mu + 1, \dots, n$, for all ε, $0 \leqq \varepsilon \leqq \varepsilon_0$, $\varepsilon_0 > 0$,

$$(4.12) \qquad \begin{aligned} &\rho_{2j-1}(0) = i\, k_j\omega/m_j = \bar{\rho}_{2j}(0),\; j = 1, 2, \dots, \nu;\; \nu \leqq q \;; \\ &\rho_j(0) \not\equiv \rho_k(0) \pmod{\omega' i},\; j = 1, 2, \dots, 2\nu; \\ &\qquad\qquad k = 2\nu + 1, \dots, n + \mu \;; \\ &\rho_{2j-1}(0) \not\equiv \rho_{2j}(0) \pmod{\omega' i}, \rho_{2j-1}(0) - \rho_{2k-1}(0) = m_{jk}\omega i \;, \\ &m_{jk} \text{ an integer or zero, } j, k = \nu + 1, \dots, r;\; \nu \leqq r \leqq q \;; \\ &\rho_j(0) \not\equiv \rho_k(0) \pmod{\omega' i},\; j = 2\nu + 1, \dots, 2r; \\ &\qquad\qquad k = 2r + 1, \dots, n + \mu \;; \\ &\rho_j(0) \not\equiv \rho_k(0) \pmod{\omega' i},\; j \neq k,\; j, k = 2r + 1, \dots, n + \mu \;, \end{aligned}$$

where $\omega' = \omega/m$, $m = m_1 \cdots m_\nu$ and the ρ_j are defined by (4.2). The periodic solution (4.3) of (4.1) is asymptotically orbitally stable to the right, $0 < \varepsilon \leqq \varepsilon_0$, $\varepsilon_0 > 0$ if condition (ii) of Theorem 4.1 is satisfied, and, in addition,

$$(4.13)\qquad \int_0^{mT} f_{jx_j'}(X_o, X_o', 0)dt < 0, \; j = r+1, \ldots, q ,$$

where $X_o = (X_{1o}, \ldots, X_{no})$, $X_o' = (X_{1o}', \ldots, X_{\mu o}')$ are defined by (4.3); and the eigenvalues $\lambda_1, \ldots, \lambda_{r-\nu}$ of the $(r-\nu) \times (r-\nu)$ matrix $M = N - P$, $P = \operatorname{diag}(p_1, \ldots, p_{r-\nu})$, $N = (n_{jk})$, $j, k = 1, 2, \ldots, r-\nu$;

$$p_j = i \lim_{\varepsilon \to 0} \varepsilon^{-1} \sigma_{\nu+j}(\varepsilon) - \sigma_{\nu+j}(0)], \; j = 1, 2, \ldots, r - \nu ,$$

$$(4.14)\qquad n_{j-\nu,k-\nu} = \frac{1}{mT}\int_0^{mT} c_{2j-1,2k-1}(t, 0)e^{i(m_{k1}-m_{j1})\omega' t}dt,$$

$$j, k = \nu+1, \ldots, r ,$$

where the functions c_{jk} are defined in (3.10) with the f^* given by (4.5) are distinct, have negative real parts and the corresponding eigenvectors have no zero components.

PROOF. The proof is exactly the same as the proof of Theorem 3.3, replacing everywhere the words Theorem 3.1, Theorem 2.1 by Theorem 4.1 and Theorem 2.2, respectively.

REMARK 4.2. For $\nu = 1$ in Theorem 4.3, the condition (ii) referred to can be replaced by

$$\int_0^{mT} f_{1x_1'}(X_o, X_o', 0)dt < 0 .$$

EXAMPLE. Consider the system of equations

$$x_1'' + \sigma_1^2 x_1 = \varepsilon(1 - x_1^2)x_1' + \varepsilon\, g_1(x_1, x_2, x_2', x_3, x_3') \equiv \varepsilon\, f_1(x, x')$$

$$(4.15)\qquad x_2'' + 2x_2 = \varepsilon(1 - x_1^2 - x_2^2 - x_3^2)x_2' + \varepsilon x_1^2 x_3 + \varepsilon g_2(x_1, x_1') \equiv \varepsilon f_2(x, x')$$

$$x_3'' + 2x_3 = \varepsilon(1 - x_1^2 - x_2^2 - x_3^2)x_3' + \varepsilon x_1^2 x_2 + \varepsilon g_3(x_1, x_1') \equiv \varepsilon f_3(x, x')$$

where $x = (x_1, x_2, x_3)$, $\varepsilon > 0$, $\sigma_1 = \sigma_1(\varepsilon)$, $\sigma_1(0) = 1$, $g_1(-x_1, x_2, x_2', x_3, x_3') = -g_1(x_1, x_2, x_2', x_3, x_3')$ and g_1, g_2, g_3 are analytic functions of their arguments for $|x_j|, |x_j'| \leq A$, $A > 0$. Using the method in [7], it is easy to see that there exists $\sigma_1 = \sigma_1(\varepsilon)$, $a = a(\varepsilon)$ analytic in ε for

$0 \leqq \varepsilon \leqq \varepsilon_0$, $\sigma_1(0) = 1$, $a(0) = 2$ such that there is a periodic solution $X(\varepsilon, t) = X(\varepsilon, t + 2\pi)$ analytic in ε, $0 \leqq \varepsilon \leqq \varepsilon_0$, $X = (X_1, X_2, X_3)$ where

$$(4.14) \qquad X_1(0, t) = 2 \sin(t + \varphi), \; X_2(0, t) = X_3(0, t) = 0 \quad ,$$

and φ is an arbitrary constant. This solution will be asymptotically orbitally stable if we can show that five of the six characteristic exponents of the linear variational equation associated with $X(\varepsilon, t)$ have negative real parts. Since this system is a special case of Theorem 4.3 for $\nu = 1$, $r = q = \mu = n = 3$, it follows from Remark 4.2 it is sufficient to show that

$$\int_0^{2\pi} f_{1x_1'} \, [(X(0, t), X'(0, t)]dt < 0$$

and the eigenvalues of the matrix M,

$$2\pi M = (m_{jk}), \; m_{jk} = \int_0^{2\pi} [- i2^{-\frac{1}{2}} f_{jx_k}(X(0, t), X'(0, t)) + f_{jx_k'}(X(0, t), X'(0, t))]dt \quad ,$$

$j, k = 2, 3$, have negative real parts. One finds immediately that

$$\int_0^{2\pi} f_{jx_j'}dt = - 2\pi, \; j = 1, 2, 3, \int_0^{2\pi} f_{jx_j}dt = 0 = \int_0^{2\pi} f_{jx_k'}dt, \int_0^{2\pi} f_{jx_k}dt = 4\pi \quad ,$$

$j \neq k$, $j, k = 2, 3$, where the functions under the integral sign are evaluated at $X(0, t)$, $X'(0, t)$. The characteristic equation for M is $\beta^2 + 2\beta + 3 = 0$ and both of the roots of this equation have negative real parts. Also the corresponding eigenvectors have no zero components and the solution (4.14) of (4.13) is asymptotically orbitally stable to the right, $0 < \varepsilon \leqq \varepsilon_0$, $\varepsilon_0 > 0$.

§5. GENERALIZATIONS

In the preceding sections, we have assumed that the f_j, α_j, β_j were analytic functions. This is not necessary but was assumed only to make the presentation simple. The basic Lemmas 3.1 and 3.2 hold under very general conditions on these functions [8], the most severe restriction arising from applying the implicit function theorem to show that the eigenvalues λ of the matrices G in these lemmas will yield a characteristic exponent of the form (3.13) or (3.18). One can show that the Lemmas are still valid if we assume only that the functions f_j have continuous second derivatives with respect to x_k, x_k', ε, $k = 1, 2, \ldots, n$, and the functions α_j, β_j, σ_j^2 have continuous second derivatives with respect to ε [8]. Therefore, the stability criteria in Sections 3 and 4 hold for this more general situation. For the existence of periodic solutions of systems (3.1), (4.1) under these and even weaker hypotheses, see L. Cesari [4b].

Another assumption that has been made in all of the theorems is $\beta_j(\varepsilon) > 0$, $j = \mu + 1, \ldots, n$, $0 \leqq \varepsilon \leqq \varepsilon_0$. This restriction was introduced only to simplify the notation. In fact, if one of the $\beta_j = 0$, say $\beta_{\mu+1} = 0$, then the peridoic solution (3.4) or (4.3) may have $X_{\mu+1}(0, t) \neq 0$. One then has one additional characteristic exponent which is close to the origin in the complex plane. There is a generalization of Lemma 3.1 [8] which will yield sufficient conditions for stability in these cases.

BIBLIOGRAPHY

[1] ANDRONOV, A. and WITT, A., On the mathematical theory of self-oscillatory systems with two degrees of freedom, J. Technical Physics (USSR), (Russian) 4(1933), 249-271.

[2] BAILEY, H. R. and CESARI, L., Boundedness of solutions for linear differential systems with periodic coefficients, Archive Rat. Mech. Ana. 3(1958), 246-271.

[3] BAILEY, H. R. and GAMBILL, R. A., On stability of periodic solutions of weakly nonlinear differential systems, Jour. Math. Mech. 6(1957), 655-668.

[4] CESARI, L., a) Asymptotic behavior and stability problems in ordinary differential equations, Ergbn. d. Math. N.F. Heft 16, Springer, 1959; b) Existence theorems for nonlinear Lipschitzian differential systems and fixed point theorems. This study.

[5] CODDINGTON, E. A. and LEVINSON, N., Theory of ordinary differential equations, McGraw-Hill, 1955.

[6] GAMBILL, R. A. and HALE, J. K., Subharmonic and ultraharmonic solutions for weakly nonlinear differential systems, Jour. Rat. Mech. Ana. 5(1956), 353-394.

[7] HALE, J. K., Periodic solutions of nonlinear systems of differential equations, Riv. Mat. Univ. Parma 5(1954), 281-311.

[8] HALE, J. K., On the behavior of the solutions of linear periodic differential systems near resonance points. This study.

[9] LIAPUNOV, A., Problème général de la stabilité du mouvement, Annals. Math. Studies No. 17, Princeton, 1947.

[10] MANDELSTAM, L. and PAPALEXI, N., Uber Resonanzerscheinungen bei Frequenzteilung, Zeit. Physik 73(1932), 223.

[11] MINORSKY, N., Introduction to nonlinear mechanics, J. W. Edwards, 1947.

[12] THOMPSON, E. W., On stability of periodic solutions of autonomous differential systems. To appear.

[13] BLEHMAN, I. I., On the stability of periodic solutions of quasi-linear systems with many degrees of freedom (Russian). Dokl. Akad. Nauk. SSSR 104(1955), 809-812; 112(1957), 183-186.

[14] NOHEL, J. A., Stability of perturbed motions. J. Reine Angew. Math. 1960.

VI. EXISTENCE THEOREMS FOR PERIODIC SOLUTIONS OF NONLINEAR LIPSCHITZIAN DIFFERENTIAL SYSTEMS AND FIXED POINT THEOREMS

Lamberto Cesari

INTRODUCTION

In the present paper we consider nonlinear systems of ordinary differential equations in the complex field containing a small parameter and satisfying merely conditions of continuity, or a Lipschitz condition. We prove existence theorems for periodic solutions, and families of periodic solutions. This is done by considering a convenient transformation $\psi = \mathfrak{T}\varphi$ from a space Ω of periodic vector functions into itself, and by proving that $\mathfrak{T}$ transforms a compact convex set $\Omega_0^* \subset \Omega$ into itself under mere hypotheses of continuity. Thus $\mathfrak{T}$ has at least a fixed element $y \in \Omega_0^*$, $y = \mathfrak{T}y$, and y is a periodic solution of the given differential system provided a (finite) equation is satisfied (determining equation). The analysis of $\mathfrak{T}$ and of the determining equation in the case of complex or real systems leads to final and simple existence theorems for periodic solutions (and families of periodic solutions). Under a Lipschitz condition $\mathfrak{T}$ is a contraction in a closed sphere $\Omega_0 \subset \Omega$ and thus $\mathfrak{T}$ has a unique fixed element $y \in \Omega$. Under a Lipschitz condition, the method of successive approximations $\varphi^{(k+1)} = \mathfrak{T}\varphi^{(k)}$, as usually associated with $\mathfrak{T}$, is uniformly convergent toward the periodic solution y.

This method of successive approximations, or variants of it, has been already studied and its convergence proved directly by the author, by J. K. Hale, and R. A. Gambill [4b, 8, 9c] in the analytic and in the linear case. In the present paper, therefore, a functional interpretation of it is given, and this assures the convergence of the method under a mere Lipschitz condition. Also, known existence theorems for periodic solutions

[1] This research was partially supported by AF Contract No. 49 (638)-382 at RIAS, Baltimore, Maryland.

and families of them are extended, particularly the ones recently proved by J. K. Hale [9 cf] in the analytic case using the same method. Also, other theorems are extended in various ways. The method just mentioned, or variants of it, has a bearing on two papers of J. K. Hale [9 hi, this volume of the Contributions], one dedicated to the study of the critical frequencies of linear systems with periodic coefficients, and the other to the proof of new and simple criteria for stability of periodic solutions of nonlinear systems. Finally, let it be mentioned that the same method of successive approximations was studied and applied by the author, J. K. Hale, R. A. Gambill, H. R. Bailey, W. R. Fuller in a series of papers listed in the references.[2]

All applications and results under consideration belong to the class of the problems of perturbation of linear differential systems. In successive papers I shall deal with straightforward nonlinear differential systems as well as with perturbation problems of nonlinear differential systems.

§1. THE CONCEPT OF MEAN VALUE AND OTHER REMARKS

(a). The concept of mean value. Consider the family C_ω of all functions $f(t)$, $-\infty < t < +\infty$, which are finite sums of functions of the form $e^{\sigma t}\varphi(t)$, where σ is any complex number, and $\varphi(t)$ is any complex-valued function, periodic of period $T = 2\pi/\omega$, L-integrable in $[0, T]$. Obviously C_ω is an additive class. If $\varphi(t)$ has the Fourier series

$$\varphi(t) \sim \sum_{n=-\infty}^{+\infty} c_m e^{im\omega t} \quad ,$$

then the series

$$f(t) = e^{\sigma t}\varphi(t) \approx \sum_{m=-\infty}^{+\infty} c_m e^{(im\omega + \sigma)t} \tag{1.1}$$

is said to be the series associated with $f(t) = e^{\sigma t}\varphi(t) \in C_\omega$. Note that the decomposition $e^{\sigma t}\varphi(t)$ is not unique since also $e^{(\sigma+ih\omega)t} \cdot e^{-ih\omega t}\varphi(t)$,

---

2 The present results contain some of those proved by W. R. Fuller [10] in his thesis at Purdue University. For instance, W. R. Fuller proved the convergence of the method of successive approximations in the autonomous case under a Lipschitz condition. Other results of this thesis, (e.g., the applications of the same method to systems containing differences) will appear independently.

h any integer, has the same properties. Nevertheless, the series (1.1) is uniquely determined. The mean value $\mathfrak{M}[f]$ of $f(t) = e^{\sigma t}\varphi(t)$ is defined by $\mathfrak{M}[f] = c_m$ if $im\omega + \sigma = 0$ for some m, by $\mathfrak{M}[f] = 0$ otherwise. For any function $f(t) \in C_\omega$, then $\mathfrak{M}[f]$ is defined as an additive linear functional. For periodic functions f, $\mathfrak{M}[f]$ is the usual mean value of Fourier series theory.

For vector functions $f = (f_1, \dots, f_n)$, all $f_j \in C_\omega$, we shall denote by $\mathfrak{M}[f]$ the vector $(m_1, \dots, m_n)$ with $m_j = \mathfrak{M}[f_j]$, $j = 1, \dots, n$.

This concept of mean value has been proposed by L. Cesari [4b] and further studied by J. K. Hale [9a]. It was proved first that $\mathfrak{M}[f]$ is uniquely defined in C_ω, and that, if $f = f_1 + \dots + f_N$ is any finite decomposition of f in C_ω, then $\mathfrak{M}[f] = \mathfrak{M}[f_1] + \dots + \mathfrak{M}[f_N]$. Also the following simple statement holds:

(1.i). If $f(t) \in C_\omega$ then any primitive $F(t)$ of $f(t)$ belongs to C_ω if and only if $\mathfrak{M}[f] = 0$. In addition, if $\mathfrak{M}[f] = 0$, there exists one and only one primitive $F(t) \in C_\omega$ with $\mathfrak{M}[F] = 0$. We shall denote this primitive by

$$\int f(t)dt$$

[see 4b and 9a].

If $f(t) = e^{(\alpha+i\beta)t}\varphi(t) \in C_\omega$, α, β real, $\mathfrak{M}[f] = 0$,

$$\varphi(t) \sim \sum c_m e^{im\omega t} \quad ,$$

the unique primitive $F(t)$ of $f(t)$ of mean value zero is given by

$$F(t) = e^{(\alpha+i\beta)t}\Phi(t), \quad \Phi(t) = \sum c_m(\alpha + i\beta + im\omega)^{-1}e^{im\omega t} \quad , \tag{1.2}$$

where in the last series the term with $\alpha + i\beta + im\omega = 0$ (if any) is omitted, the series is convergent for all t, and $\Phi(t)$ is absolutely continuous in each finite interval [see 9a]. The following simple examples may be considered, all with $\omega = 1$, $\mathfrak{M}[f] = 0$:

$$\int \cos t\, dt = \sin t, \quad \int \sin t\, dt = -\cos t$$

$$\int e^{(\alpha+i\beta)t}dt = (\alpha + i\beta)^{-1}e^{(\alpha+i\beta)t}, \quad \alpha + i\beta \neq 0 \quad .$$

REMARK. Though not needed in the sequel the following considerations may be of some interest. Given a function $f(t) \in C_\omega$ with $\mathfrak{M}[f] = 0$, it is of interest to know whether its unique primitive $F(t) \in C_\omega$ with $\mathfrak{M}[F] = 0$ is a definite integral of the form

$$\int_{\xi}^{t} f(u)du$$

with $0 \leq \xi < T$. For functions $f(t) = e^{(\alpha+i\beta)t}\varphi(t) \in C_\omega$, $\mathfrak{M}[f] = \mathfrak{M}[\varphi] = 0$, this is true, for instance, if the function $\Phi(t)$ above happens to be real, and then we may take for ξ any zero of $\Phi(t)$. In particular this is true for $\alpha + i\beta = 0$, $f = \varphi$ real. For $A + i\beta = 0$, $f = \varphi = \varphi_1 + i\varphi_2$, φ_1, φ_2 real, $\mathfrak{M}[f] = \mathfrak{M}[\varphi_1] = \mathfrak{M}[\varphi_2] = 0$, there are two points $0 \leq \xi_1$, $\xi_2 < T$, such that

$$F(t) = \int_{\xi_1}^{t} \varphi_1 \, du + \int_{\xi_2}^{t} \varphi_2 \, du \quad ,$$

and the points ξ_1, ξ_2 may be distinct, as for $f(t) = e^{it} = \cos t + i \sin t$, $F(t) = - ie^{it} = \sin t - i \cos t$, $\xi_1 \equiv 0$, $\xi_2 \equiv \pi/2$, $\xi_1 \not\equiv \xi_2$, (mod π). If $f(t) = e^{(\alpha+i\beta)t}\varphi(t) \in C_\omega$, $\mathfrak{M}[f] = \mathfrak{M}[\varphi] = 0$, then there is always a decomposition $f = f_1 + f_2$ and two points $0 \leq \xi_1$, $\xi_2 < T$ such that $f_j \in C_\omega$, $\mathfrak{M}[f_j] = 0$,

$$\int f_j(t) \, dt = \int_{\xi_j}^{t} f_j \, du \quad ,$$

$j = 1, 2$ [J. K. Hale, 9a, proved this by the use of faltung integrals]. If $\mathfrak{M}[\varphi] \neq 0$ then not even this is true as the example shows: $f(t) = e^{\sigma t}$ $\mathfrak{M}(f) = 0$, $\mathfrak{R}(\sigma) \neq 0$,

$$F(t) = \sigma^{-1}e^{\sigma t} = \sigma^{-1} \int_{\pm\infty}^{t} e^{\sigma u} \, du \quad ,$$

$\pm$ according as $\mathfrak{R}(\sigma) < 0$, or $\mathfrak{R}(\sigma) > 0$. Note that for $f(t) = e^{\sigma t}\varphi(t) \in C_\omega$, $\sigma \not\equiv 0$ (mod ωi), we have

$$F(t) = (e^{\sigma T} - 1)^{-1} \int_{t}^{t+T} f(u) \, du \quad .$$

For $\sigma \equiv 0$ (mod ωi) we have

$$F(t) = T^{-1} \int_t^{t+T} uf(u)\, du$$

[J. Moser].

(1.ii). If $f(t) = e^{\sigma t}\varphi(t) \in C_\omega$, $\mathfrak{M}[f] = 0$, and $F(t) \in C_\omega$ is its unique primitive with $\mathfrak{M}[F] = 0$, if $V \geq T$ is any constant, then there is a constant N, depending only on σ, T, V, such that

$$(1.3) \qquad |F(t)| \leq N \int_0^T |\varphi(u)|\, du, \qquad 0 \leq t \leq V \;.$$

If $\sigma \equiv 0 \pmod{\omega i}$ we may take $N = 2$, if $|\sigma + im\omega| \geq \delta$ for all $m = 0, \pm 1, \pm 2, \dots$, and some $\delta > 0$, we may take N depending only on δ, T, V. If $F(t)$ is periodic then (1.3) holds for all t [J. K. Hale, 9a; see also H. R. Bailey and L. Cesari, 2].

Given $\sigma = \alpha + i\beta$, α, β real, with $\sigma + im\omega \neq 0$, $m = 0, \pm 1, \pm 2, \dots$, let $\delta = \min\,[|\sigma + im\omega|, m = 0, \pm 1, \dots,]$. For every $\sigma' = \alpha' + i\beta'$, α', β' real, with $|\sigma' - \sigma| \leq \delta/2$ we have $|\sigma' + im\omega| \geq \delta/2$ for the same m.

(1.iii). Let $\sigma = \alpha + i\beta$, α, β real, be any complex number with $\sigma + im\omega \neq 0$, $m = 0, \pm 1, \dots$, and let $0 < \delta \leq \min\,[|\sigma + im\omega|, m = 0, \pm 1, \dots]$, $0 < \delta \leq \omega$, let σ' be any other number with $|\sigma' - \sigma| \leq \delta/2$, let $\varphi(t)$ be any complex valued function periodic of period $T = 2\pi/\omega$, L-integrable in $[0, T]$, let $f(t) = e^{\sigma t}\varphi(t)$, $f_1(t) = e^{\sigma' t}\varphi(t)$, hence $\mathfrak{M}[f] = \mathfrak{M}[f_1] = 0$, and let $V \geq T$ be any given constant. Then the unique primitives F, F_1 of f, f_1 of class C_ω and mean values zero verify the relation

$$(1.5) \qquad |F(t) - F_1(t)| \leq |\sigma - \sigma'|\, N' \int_0^T |\varphi(u)|\, du, \qquad 0 \leq t \leq V \;,$$

where N' is a constant depending only on δ, T, V (σ bounded).

PROOF. A simple proof can be given by making use of the last lines of the remark above. By taking

$$\alpha = (e^{\sigma T} - 1)^{-1}(e^{\sigma' T} - 1)^{-1} \;,$$

we have

$$F(t) - F_1(t) = \int_t^{t+T} [(e^{\sigma T} - 1)^{-1} e^{\sigma u} - (e^{\sigma' T} - 1)^{-1} e^{\sigma' u}] \varphi(u)\, du =$$

$$= \alpha \int_t^{t+T} [e^{\sigma' T}(1 - e^{(\sigma'-\sigma)(u-T)}) + (e^{(\sigma'-\sigma)u} - 1)] e^{\sigma u} \varphi(u)\, du.$$

For $0 \leq t \leq V$, $T \leq V$, we have also

$$|F(t) - F_1(t)| \leq |\sigma' - \sigma| N' \int_0^T |\varphi(u)|\, du \quad .$$

This proves (1.iii).

(1.iv). If A is a constant $n \times n$ matrix with characteristic roots $\rho_1, \ldots, \rho_n$, if $f(t) = (f_1, \ldots, f_n)$ is a vector function whose components f_j are periodic of period $T = 2\pi/\omega$ and L-integrable in $[0, T]$, if $|\rho_j + im\omega| \geq \delta > 0$ for some $\delta > 0$, all $j = 1, \ldots, n$, and $m = 0, \pm 1, \pm 2, \ldots$, then the differential system $y' = Ay + f(t)$, $y = (y_1, \ldots, y_n)$, has exactly one solution $y(t) = (y_1, \ldots, y_n)$ whose components are periodic of period T. Also, there is a constant N depending only on A and T and not on $f(t)$ such that

$$(1.6) \qquad |y_j(t)| \leq N \sum_{h=1}^{n} \int_0^T |f_h(t)|\, dt, \; j = 1, \ldots, n, \; -\infty < t < +\infty \quad .$$

PROOF. It is not restrictive to suppose that $A = [a_{jh}]$ is given in triangular form, i.e., $a_{jh} = 0$, $j > h$, $a_{jj} = \rho_j$, $j = 1, \ldots, n$. Let $\rho_j = \alpha_j + i\beta_j$, α_j, β_j real, $j = 1, \ldots, n$, $\gamma = \max |\alpha_j|$, $M = e^{\gamma T}$, Q the sum in (1.6), $a = \max |a_{jh}|$. The last equation of the system,

$$y_n' = \rho_n y_n + f_n(t) \quad ,$$

has a unique periodic solution of period $T = 2\pi/\omega$, which, by using the notation of this Section, can be written as

$$y_n(t) = e^{\rho_n t} \int e^{-\rho_n u} f_n(u)\, du \quad ,$$

where the integrand is of class C_ω and has mean value zero, the integral is the unique primitive of class C_ω and mean value zero, has the form

$e^{-\rho_n t}\Phi(t)$ considered in (1.2), and hence $y_n = \Phi(t)$ is periodic of period T. Let N be the constant of (1.ii) depending only on δ, T, and V = T. Then by (1.ii) for all $0 \leq t \leq T$, we have

$$|y_n(t)| \leq MN \int_0^T |f_n(t)|\, dt \leq MNQ \quad .$$

The last but one equation of the system is

$$y'_{n-1} = \rho_{n-1} y_{n-1} + [a_{n-1,n} y(t) + f_{n-1}(t)] \quad ,$$

where the expression in brackets is periodic of period T. Hence this equation has a unique periodic solution given by

$$y_{n-1}(t) = e^{\rho_{n-1} t} \int e^{-\rho_{n-1} u} [a_{n-1,n} y_n(u) + f_{n-1}(u)]\, du \quad ,$$

and we have, for $0 \leq t \leq T$,

$$|y_{n-1}(t)| \leq MN \int_0^T [\,|a_{n-1,n}|\; |y_n(u)| + |f_{n-1}(u)|\,]\, du$$

$$\leq MNQ(aMNT + 1) \quad .$$

By repeating this procedure n times we find, for $0 \leq t \leq T$,

$$|y_j(t)| \leq MNQ(aMNT + 1)^{n-j}, \quad j = 1, \ldots, n \; ,$$

and these relations hold for all t since $y_j(t)$ is periodic.

(b). A form of Brouwer's fixed point theorem. For the use in §5 we mention here the following statement which has been shown to be equivalent to the Brouwer fixed point theorem for a cell in the Euclidean space E_n [C. Miranda, Un'osservazione su un teorema di Brouwer, Boll. Unione Mat. Ital. 3, 1941, 5-7].

(1.v). If $K \subset E_n$ is a cube whose 2n opposite (closed) faces are K'_j, K''_j, $j = 1, \ldots, n$, if $f(x) = (f_1, \ldots, f_n)$, $x \in K$, $x = (x_1, \ldots, x_n)$, is a real vector function, continuous in K, if $f_j(x)$ has opposite constant signs on the two faces K'_j, K''_j, $(j = 1, \ldots, n)$, then there is at least one point $x_0 \in K^0$, such that $f(x_0) = 0$, i.e., $f_j(x_0) = 0$, $j = 1, \ldots, n$.

An immediate corollary of (1.v) is

(1.vi). If $K \subset E_n$ is a cube whose $2n$ opposite (closed) faces are K_j', K_j'', $j = 1, \ldots, n$, if M is a compact topological space, and $f(x, m) = (f_1, \ldots, f_n)$, $x \in K$, $m \in M$, is a real vector function, continuous in $K \times M$, if for some $m_0 \in M$, $f_j(x, m_0)$ has opposite signs on the two faces K_j', K_j'', $(j = 1, \ldots, n)$, then there is a neighborhood N of m_0 in M and, for each $m \in N$, at least one $x_0 = x_0(m)$, $m \in N$, $x^0 \in K^0$, such that $f[x_0(m), m] = 0$ for every $m \in N$.

Indeed, by continuity, there is a neighborhood N of m_0 such that, for every $m \in N$, $f_j(x, m)$ has the same constant sign on $K_j'[K_j'']$ as $f_j(x, m_0)$, and thus $f(x, m)$ has opposite constant signs on K_j', K_j'', $(j = 1, \ldots, n)$.

A slightly different form of (1.vi) is the following one:

(1.vii). If $K \subset E_n$ is a cube whose $2n$ opposite (closed) faces are K_j', K_j'', $j = 1, \ldots, n$, if M_1, M_2 are compact topological spaces and $f(x, m_1, m_2) = (f_1, \ldots, f_n)$, $x \in K$, $m_1 \in M_1$, $m_2 \in M_2$, is a real vector function, continuous in $K \times M_1 \times M_2$, if for some $m_{20} \in M_2$, $f_j(x, m_1, m_{20})$ has opposite constant signs on the two faces K_j', K_j'', $(j = 1, \ldots, n)$, independently from $m_1 \in M_1$, then there is a neighborhood N_2 of m_{20} in M_2 and for each $m_1 \in M_1$, and $m_2 \in N_2$, at least one $x_0 = x_0(m_1, m_2)$, $m_1 \in M_1$, $m_2 \in N_2$, $x^0 \in K^0$, such that $f[x_1(m_1, m_2), m_1, m_2] = 0$ for every $m_1 \in M_1$, $m_2 \in N_2$.

(c). Schauder's fixed point theorem. We shall need in §3 the following form of Schauder's fixed point theorem:

(1.viii). Any continuous mapping $f : K \longrightarrow K$, from a convex, closed, compact subset K of a linear space M has at least one fixed point $y \in K$, $fy = y$.

[See J. Schauder, Der Fixpunkt in Funktionalraümen, Studia Math. 2, 1930, 171-180. Also, S. Lefschetz, Topics in Topology, Annals of Math. Studies, No. 10, 1942.] We do not repeat all definitions, but only the following ones: K convex means that $x, y \in K$ implies $tx + (1 - t)y \in K$ for all $0 \leq t \leq 1$; K closed means that $M - K$ is open, and hence $x_n \longrightarrow x$, $x_n \in K$, $x \in M$, implies $x \in K$; K compact means that every sequence $[x_n]$, $x_n \in K$, possesses a convergent subsequence.

We shall need in §3 also the following statement:

(1.ix). Any continuous mapping $f : K \longrightarrow K$, from a complete subspace K of a metric space M, which is a contraction in K, has exactly one fixed point in K (Banach's fixed point theorem).

If $d(x, y)$ is the metric in M, then f a contraction means that $d(fx, fy) \leq md(x, y)$ for all $x, y \in K$, and a constant $m < 1$. That K is complete means that every Cauchy sequence in K has its limit in K. Thus K is closed [H. Hahn, Reelle Funktionen I, Chelsea 1948, p. 118]. The uniqueness follows by the obvious remark that $fx = x$, $fy = y$, $x, y \in K$, $x \neq y$, implies $0 < d(x, y) = d(fx, fy) \leq md(x, y)$ with $m < 1$, a contradiction. The existence follows by proving that any sequence $[x_n]$ with $x_{n+1} = f(x_n)$, $n = 0, 1, \ldots$, $x_0 \in K$ arbitrary, is a Cauchy sequence [Cf. E. A. Coddington and N. Levinson, Ordinary differential equations, McGraw-Hill, 1955, p. 41].

§2. A TRANSFORMATION $\mathfrak{T}$ IN FUNCTIONAL SPACES

We shall consider first, for the sake of simplicity, a differential system of the form

$$dy/dt = Ay + \varepsilon q(y, t, \varepsilon), \quad -\infty < t < +\infty \quad , \tag{2.1}$$

where $y = (y_1, \ldots, y_n)$, $A = A(\varepsilon)$ is a constant real or complex $n \times n$ matrix, $\varepsilon \geq 0$ is a real parameter which will be supposed to be sufficiently small, $q = (q_1, \ldots, q_n)$, and each $q_j(y, t, \varepsilon)$ is a real or complex function of t, y, ε. We shall make on A and q the following assumptions:

(α). There are numbers $\omega > 0$, $\delta > 0$, $\varepsilon_0 > 0$, and integers $0 \leq \nu \leq n$, $a_j \gtrless 0$, $b_j > 0$, $j = 1, \ldots, \nu$, such that the n characteristic roots $\rho_j(\varepsilon)$, $j = 1, \ldots, n$, of A are continuous functions of ε in $0 \leq \varepsilon \leq \varepsilon_0$, and verify the relations

$$\begin{aligned} &\rho_j(0) = i\, a_j\omega/b_j, && j = 1, \ldots, \nu \quad , \\ &|\rho_j(0) - im\omega/b_0| > \delta > 0, && j = \nu + 1, \ldots, n, \quad b_0 = b_1 \cdots b_\nu \quad , \\ & && m = 0, \pm 1, \pm 2, \ldots \quad . \end{aligned} \tag{2.2}$$

In addition we assume $A = \mathrm{diag}(A_1, A_2)$, where A_1, A_2 are $\nu \times \nu$ and $(n-\nu) \times (n-\nu)$ matrices, $A = \mathrm{diag}[\rho_1(\varepsilon), \ldots, \rho_\nu(\varepsilon)]$, and (*) $A_2 = \mathrm{diag}[\rho_{\nu+1}(\varepsilon), \ldots, \rho_n(\varepsilon)]$.

Actually, the last requirement (*) is unnecessarily restrictive, and all of the present results are valid under a much weaker assumption replacing (*); namely (**) B_2 is any matrix whose coefficients are continuous functions of ε in $0 \leq \varepsilon \leq \varepsilon_0$. It is only for the sake of simplicity that we use condition (*) [See remarks at the end of §2]. Also, b_0 in (2.2) need only be any common multiple of $b_1, \ldots, b_\nu$.

Finally, $A(\varepsilon)$ could actually be any $n \times n$ matrix whose elements are continuous functions of ε in $0 \leq \varepsilon \leq \varepsilon_0$, having a matrix A_0

as above for its canonical form, i.e., provided there exists a matrix $P(\varepsilon)$ whose coefficients are continuous functions of ε in $0 \leq \varepsilon \leq \varepsilon_0$, with $\det P(\varepsilon) \neq 0$ and $PAP^{-1} = A_0$.

(K) There exists a number $R > 0$ and a function $\psi(t) > 0$, $-\infty < t < \infty$, L-integrable in every finite interval such that $|y_\ell| \leq R$, $\ell = 1, \ldots, n$, $-\infty < t < \infty$, implies $|q_j(y, t, \varepsilon)| \leq \psi(t)$, $j = 1, \ldots, n$. Given $\zeta > 0$ there exists $\xi > 0$ such that $|y_\ell^1|, |y_\ell^2| \leq R$, $0 \leq \varepsilon^1, \varepsilon^2 \leq \varepsilon_0$, $|y_\ell^1 - y_\ell^2| \leq \xi$, $\ell = 1, \ldots, n$, $|\varepsilon^1 - \varepsilon^2| \leq \xi$, implies

$$|q_j(y^1, t, \varepsilon^1) - q_j(y^2, t, \varepsilon^2)| \leq \zeta \psi(t), \; j = 1, \ldots, n, \; -\infty < t < \infty .$$

Finally, (p) the functions $q_j(y, t, \varepsilon)$ are periodic in t of period $2\pi/\omega$, or are independent of t and then $\psi(t)$ is replaced by a constant $K > 0$.

Sometimes we shall replace (K) by the stronger assumption:

(L). Condition (K) holds, and, in addition, y_ℓ^1, $y_\ell^2 \leq R$, $\ell = 1, \ldots, n$, $0 \leq \varepsilon \leq \varepsilon_0$, implies

$$|q_j(y^1, t, \varepsilon) - q_j(y^2, t, \varepsilon)| \leq \psi(t) \sum_{\ell=1}^{n} |y_\ell^1 - y_\ell^2|, \; j = 1, \ldots, n, - \infty < t < \infty.$$

Condition (p) will be replaced by a condition of quasi periodicity in some theorems. Also, it may well occur that we can take $R = \infty$, or that we may take R any arbitrary constant and $\psi(t)$ depends on R, all with obvious simplifications in the statements of the present paper.

REMARK 1. Let us give here in a few words the actual meaning of condition (α). In most cases (2.1) will be the canonical form of a real system containing at least one equation of the form $x_1'' + \sigma^2 x_1 = \varepsilon f_1$, and then $\rho_1 = i\sigma$, $\rho_2 = - i\sigma$. Then the first condition (2.2) is satisfied by simply taking $\nu = 2$, $\omega = \sigma$, $a_1 = 1$, $a_2 = -1$, $b_1 = b_2 = 1$, while the second condition (2.2) requires that the remaining characteristic roots are not "close" to any one of the numbers $im\sigma$, $m = 0, \pm 1, \pm 2, \ldots$. If the real system contains, besides the equation above, a first order equation of the form $x_3' = \varepsilon f_3$, then we have $\rho_3 = 0$, and the first condition (2.2) is satisfied by simply taking $\nu = 3$, $\omega = \sigma$, $a_1 = 1$, $a_2 = - 1$, $a_3 = 0$, $b_1 = b_2 = b_3 = 1$, while the second condition (2.2) requires again that the remaining characteristic roots are not "close" to any one of the same numbers above. In §5 we shall actually consider these and more general situations. Integers $a_j, b_j \neq \pm 1$ certainly occur in connection with subharmonic and ultraharmonic solutions [cf. 8]. As a further informal comment on the general process discussed rigorously below we wish to mention that periodic solutions of system (2.1) can be expected to be of the form $y_j = c_j e^{i\tau_j t} + O(\varepsilon)$, $j = 1, \ldots, \nu$, $y_j = O(\varepsilon)$, $j = \nu + 1, \ldots, n$, where $i\tau_j = \rho_j(0) + O(\varepsilon)$ in the autonomous case, and where the numbers $c_j = c_j(\varepsilon)$ do not necessarily approach zero as $\varepsilon \longrightarrow 0$. This is actually the case in innumerable known examples [see, e.g., 8, 9c for references]. Thus it is natural to look for periodic solutions of (2.1) in the "neighborhood" of solutions of (2.1) for $\varepsilon = 0$ of the form $y_j = c_j e^{\rho_j(0)t}$, $j = 1, \ldots, \nu$, $y_j = 0$, $j = \nu + 1, \ldots, n$.

Note that, by force of the continuity requirement in condition (α), we may assume $\varepsilon_0 > 0$ and $\delta > 0$ sufficiently small so as

$$(2.3)\qquad \begin{aligned} &|\rho_j(\varepsilon) - ia_j\omega/b_j| < \delta, \ j = 1, \ldots, \nu, \ 0 \le \varepsilon \le \varepsilon_0 , \\ &|\rho_j(\varepsilon) - im\omega/b_0| > \delta, \ j = \nu + 1, \ldots, n, \ 0 \le \varepsilon \le \varepsilon_0, \\ &\qquad\qquad\qquad m = 0, \pm 1, \ldots . \end{aligned}$$

Note that ω appears in both conditions (α) and (K). In case the functions q_j do not depend on t (autonomous case), then we may assume that relations (2.3) hold for all ω of a sufficiently small neighborhood U of the number, say ω_0, for which (2.2) holds.

Note that the points $z = \rho_j(0)$, $j = 1, \ldots, \nu$, are apart from all points $z = \rho_j(0)$, $j = \nu + 1, \ldots, n$, and $z = im\omega/b_0$, $m/b_0 \ne a_j/b_j$, $m = 0, \pm 1, \ldots,$ in the complex plane. Then we can take δ in (2.3) sufficiently small in such a way that (2.3) still holds, and the δ-neighborhoods of the points $z = \rho_j(0)$, $j = 1, \ldots, \nu$, are δ apart from all remaining points.

Let us consider the auxiliary matrices $B = \text{diag}\,(B_1, B_2)$, $B_1 = A_1(0) = \text{diag}\,(i\tau_1, \ldots, i\tau_\nu)$, $i\tau_j = \rho_j(0) = ia_j\omega/b_j$, $j = 1, \ldots, \nu$, and $B_2 = A_2(\varepsilon) = (\rho_j(\varepsilon), j = \nu + 1, \ldots, n)$. Thus $e^{Bt} = \text{diag}(e^{B_1 t}, e^{B_2 t})$, $e^{B_1 t} = \text{diag}\,(e^{i\tau_1 t}, \ldots, e^{i\tau_\nu t})$, $e^{-B_1 t} = \text{diag}\,(e^{-i\tau_1 t}, \ldots, e^{-i\tau_\nu t})$, $e^{B_2 t} = \text{diag}\,(e^{\rho_j(\varepsilon)t}, j = \nu + 1, \ldots, n)$. Let us denote by $z(t)$ any particular solution of system (2.1) for $\varepsilon = 0$ of the form

$$(2.4)\qquad z(t) = (c_1 e^{i\tau_1 t}, \ldots, c_\nu e^{i\tau_\nu t}, 0, \ldots, 0) = e^{Bt}C ,$$

where $c_j \ne 0$, $j = 1, \ldots, \nu$, are constants, $C = \text{col}\,(c_1, \ldots, c_\nu, 0, \ldots, 0)$. Let $\Omega = \Omega(c_1, \ldots, c_\nu, b_0, \omega)$ be the class of all continuous periodic vector functions $\varphi(t) = (\varphi_1, \ldots, \varphi_n)$ of period $T = 2\pi b_0/\omega$, such that

$$(2.5) \qquad \mathfrak{M}\left[e^{-i\tau_j t}\varphi_j(t)\right] = c_j, \qquad j = 1, \ldots, \nu ,$$

that is, the first ν component φ_j of φ have the form

$$(2.6) \qquad \varphi_j(t) = e^{i\tau_j t}\left[c_j + \varphi_j^*(t)\right], \qquad j = 1, \ldots, \nu ,$$

with $\mathfrak{M}[\varphi_j^*] = 0$. Let us denote by Ω_R the subclass of Ω of all $\varphi = (\varphi_1, \ldots, \varphi_n)$ with $|\varphi_j(t)| \leq R$, $j = 1, \ldots, n$. Let $\mathfrak{T}$ be the transformation of Ω_R into Ω defined by $\psi(t) = (\mathfrak{T}\varphi)(t)$ with

$$(2.7) \qquad \psi(t) = z(t) + \varepsilon e^{Bt}\int e^{-Bu}\{q[\varphi(u), u, \varepsilon] - D[\varphi]\varphi(u)\}\, du ,$$

$$(2.8) \qquad D[\varphi]C = \mathfrak{M}\{e^{-Bu}q[\varphi(u), u, \varepsilon]\} ,$$

where $D = D[\varphi] = \operatorname{diag}(d_1, \ldots, d_\nu, 0, \ldots, 0)$. The following remarks are essential:

(a) As we shall prove, each component of the integrand is of class $C_{\omega'}$ with $\omega' = \omega/b_0$ and mean value zero;

(b) The integral in (2.7) denote the unique primitive of class $C_{\omega'}$ and mean value zero;

(c) For $\varphi \in \Omega_R$ we have $\psi \in \Omega$, i.e., $\mathfrak{T}$ transforms Ω_R into Ω [see proof below].

In component form relations (2.7), (2.8) become

$$(2.9) \qquad \psi_j(t) = c_j e^{i\tau_j t} + \varepsilon e^{i\tau_j t}\int e^{-i\tau_j u}\{q_j[\varphi(u), u, \varepsilon] - d_j\varphi_j(u)\}\, du , \qquad j = 1, \ldots, \nu ,$$

$$(2.10) \qquad \psi_j(t) = \varepsilon e^{\rho_j t}\int e^{-\rho_j u}q_j[\varphi(u), u, \varepsilon]\, du , \qquad j = \nu + 1, \ldots, n ,$$

$$(2.11) \qquad c_j d_j = \mathfrak{M}\{e^{-i\tau_j u}q_j[\varphi(u), u, \varepsilon]\} , \qquad j = 1, \ldots, \nu ,$$

$$(2.12) \qquad 0 = \mathfrak{M}\{e^{-\rho_j u}q_j[\varphi(u), u, \varepsilon]\} , \qquad j = \nu + 1, \ldots, n,$$

and we shall analyze these relations below.

Obviously the products $e^{-i\tau_j u} q_j[\varphi(u), u, \varepsilon]$, $j = 1, \ldots, \nu$, are periodic of period $T = 2\pi b_0/\omega$, and are L-integrable in $[0, T]$. Thus $\mathfrak{M}$ in (2.11) is the usual mean value, and since $c_j \neq 0$, d_j is defined, $j = 1, \ldots, \nu$. By (2.6) we have

$$(2.13) \qquad \begin{aligned} & e^{-i\tau_j u}\{q_j[\varphi(u), u, \varepsilon] - d_j\varphi_j(u)\} = \\ & = \{e^{-i\tau_j u} q_j[\varphi(u), u, \varepsilon] - c_j d_j\} - d_j\varphi_j^*(u) \quad , \end{aligned}$$

and, by (2.11), these functions are periodic of period $T = 2\pi b_0/\omega$, of class $C_{\omega'}$ and mean value zero. Hence, by §1, there is one and only one primitive, say $\psi_j^*(t)$, of (2.13), also periodic and of mean value zero, and finally, by (2.9), we have

$$(2.14) \qquad \psi_j(t) = [c_j + \varepsilon\psi_j^*(t)]e^{i\tau_j t}, \qquad j = 1, \ldots, \nu \quad ,$$

that is, the functions $\psi_j(t)$ satisfy (2.5). Also, the same functions $\psi_j(t)$ are absolutely continuous. The functions $e^{-\rho_j u} q_j[\varphi(u), u, \varepsilon]$, $j = \nu + 1, \ldots, n$, are obviously of class $C_{\omega'}$ with $|-\rho_j + im\omega/b_0| > \delta > 0$ for all $j = \nu + 1, \ldots, n$, $m = 0, \pm 1, \pm 2, \ldots,$ and hence have mean value zero according to §1. Thus (2.12) is justified. Also, there is one and only one primitive of class $C_{\omega'}$ and of mean value zero, and of the form $e^{-\rho_j t}\psi_{jo}(t)$, $\psi_{jo}(t)$ periodic of period T (ψ_{jo} of mean value not necessarily zero), and then $\psi_j(t) = \varepsilon\psi_{jo}(t)$, $j = \nu + 1, \ldots, n$. Also, the same functions $\psi_{jo}(t)$ are periodic and absolutely continuous. Thus (a), (b), (c) are completely proved, and $\mathfrak{T}\,\Omega_R \subset \Omega$.

We shall take in Ω the uniform topology (§3) and we shall prove that, for ε sufficiently small, $\mathfrak{T}$ transforms a closed sphere Ω_0 around $z(t)$, $\Omega_0 \subset \Omega_R$, into itself under the hypotheses of continuity (K). Also, (§3), there is a compact convex set $\Omega_0^* \subset \Omega_0 \subset \Omega_R$ which is transformed into itself by $\mathfrak{T}$. This implies (§3) that $\mathfrak{T}$ has in Ω_0^* at least one fixed element $y(t)$, $y = \mathfrak{T}\, y$, by Schauder's fixed point theorem. Under the Lipschitz condition (L), $\mathfrak{T} \mid \Omega_0$ is a contraction (§3) and thus the fixed element $y = \mathfrak{T}\, y$ is unique in Ω_0. In any case, $y(t)$ satisfies the integral equation

$$(2.15) \qquad \begin{aligned} & y(t) = z(t) + \varepsilon e^{Bt}\int e^{-Bu}\{q[y(u), u, \varepsilon] - D[y]y(u)\}\,du \quad , \\ & DC = D[y]C = \mathfrak{M}\{e^{-Bu}q[y(u), u, \varepsilon]\} \quad . \end{aligned}$$

Since both $y(t)$ and D depend upon the integers $a_j \gtrless 0$, $b_j > 0$, and the complex constants $c_j \neq 0$, $j = 1, \ldots, \nu$, besides ω and ε, we shall write $y(t, a, b, c, \omega, \varepsilon)$, $D(a, b, c, \omega, \varepsilon)$, where $a = (a_1, \ldots, a_\nu)$, $b = (b_1, \ldots, b_\nu)$, $c = (c_1, \ldots, c_\nu)$. By (2.15), $y(t)$ is absolutely continuous, has first derivative $y'(t)$ a.e., and, by differentiation, we obtain

$$y'(t) = (B - \varepsilon D)y(t) + \varepsilon q[y(t), t, \varepsilon] \quad ; \tag{2.16}$$

in other words, $y(t)$ satisfies a differential equation analogous to (2.1). The vector function $y(t)$ will be a solution of (2.1) provided the equation $B - \varepsilon D = A$ is satisfied, or, in component form,

$$ia_j\omega/b_j - \varepsilon d_j(a, b, c, \omega, \varepsilon) = \rho_j(\varepsilon), \qquad j = 1, \ldots, \nu \ . \tag{2.17}$$

These equations are called the determining equations of system (2.1).

Under the Lipschitz condition (L), $\mathfrak{T} \mid \Omega_0$ is a contraction, and the actual determination of the unique fixed element $y(t) \in \Omega_0$ and of the numbers $d_1, \ldots, d_\nu$ can be obtained by a method of successive approximations, namely

$$y^{(o)}(t) = z(t) = (c_1 e^{i\tau_j t}, \ldots, e^{i\tau_\nu t}, 0, \ldots, 0) \ ,$$

$$y^{(m)}(t) = z(t) + \varepsilon e^{Bt} \int e^{-Bu}\{q[y^{(m-1)}(u), u, \varepsilon] - D^{(m-1)}y^{(m-1)}\} \, du \ ,$$

$$D^{(m-1)}C = D[y^{m-1}]C = \mathfrak{M}\{e^{-Bu}q[y^{(m-1)}(u), u, \varepsilon]\}, \quad m = 1, 2, \ldots,$$

$$C = \mathrm{col}(c_1, \ldots, c_\nu, 0, \ldots, 0), \ D^{(m-1)} = \mathrm{diag}(d_1^{(m-1)}, \ldots, d_\nu^{(m-1)}, 0, \ldots, 0),$$

where each integrand is of class C_ω and mean value zero, and each integral is the unique primitive of class C_ω and mean value zero. The uniform convergence of the process, i.e.,

$$y(t) = \lim y^{(m)}(t) \ , \qquad D[y] = \lim D[y^{(m)}] \ ,$$

as $m \longrightarrow \infty$, is a consequence of the fact that $\mathfrak{T}$ is a contraction in Ω_0. This is the method of successive approximations mentioned in the Introduction and used in most papers listed in the references.

REMARK 2. A few informal words on the definition (2.7) of the transformation $\mathfrak{T}$ may be of interest. First, the term $D[\varphi]\varphi(u)$, which is subtracted under $\{ \ \}$ in (2.7) has the effect to make the integrand of mean value zero in the class C_ω, so as the integral is also in C_ω. Otherwise, there could be constant terms $c \neq 0$ in the integrand, and terms ct in the integral (the so called secular terms of classical analysis). Secondly, we actually subtract more than secular terms, but, by this device [cf. Cesari 4b] it is assured that the fixed element $y(t)$ of $\mathfrak{T}$, or the limit $y(t)$ in the method of successive approximations, satisfies the differential equation (2.16) which is similar to (2.1) and in fact has exactly the same terms εq as (2.1).

§3. PROOF THAT $\mathfrak{T}$ HAS A FIXED POINT. EXTENSIONS

(a). Proof that $\mathfrak{T}$ has a fixed point under condition (K). We shall introduce a uniform topology in Ω by defining as norm $\mathfrak{N}[\varphi]$ of an element $\varphi(t) = (\varphi_1, \ldots, \varphi_n)$ of Ω (φ_j periodic of period $T = 2\pi b_0/\omega$ and continuous, $j = 1, \ldots, n$), the number $\mathfrak{N}[\varphi] = \max |\varphi_j(t)|$, where the maximum is taken with respect to all $-\infty < t < +\infty$, and $j = 1, \ldots, n$. Thus Ω_R is the closed set of all $\varphi \in \Omega$ with $\mathfrak{N}[\varphi] \leq R$.

(3.i). Under hypotheses (α) and (K), given $a = (a_1, \ldots, a_\nu)$, $b = (b_1, \ldots, b_\nu)$, and any two numbers r_1, r_2, $0 < r_1 < r_2 < R$, there is an ε_1, $0 < \varepsilon_1 \leq \varepsilon_0$, such that, for every $c = (c_1, \ldots, c_\nu)$ and ε with $r_1 \leq |c_j| \leq r_2$, $j = 1, \ldots, \nu$, $0 \leq \varepsilon \leq \varepsilon_1$, the transformation $\mathfrak{T}$ is continuous, maps the closed sphere Ω_0 into itself, $\Omega_0 = \{\mathfrak{N}[\varphi(t) - z(t)] \leq r\}$, $r = R - r_2$, and has at least a fixed point in Ω_0.

PROOF. Obviously $z(t) = (c_1 e^{i\tau_1 t}, \ldots, c_\nu e^{i\tau_\nu t}, 0, \ldots, 0) \in \Omega$, and $\mathfrak{N}[z(t)] = \max |c_j| \leq r_2 < R$. Hence, $z(t) \in \Omega_R$. Also, for every $\varphi(t) \in \Omega_0$ we have $\mathfrak{N}[\varphi] \leq \mathfrak{N}[z] + r \leq r_2 + r = R$, and $\Omega_0 \subset \Omega_R \subset \Omega$. Let K be any number such that

$$\int_0^T \psi(t)\, dt \leq KT \quad . \tag{3.1}$$

For every $\varphi \in \Omega_0$ and $j = 1, \ldots, \nu$, we have, by (2.11),

$$c_j d_j = \mathfrak{M}\left\{e^{-i\tau_j u} q_j[\varphi(u), u, \varepsilon]\right\} = T^{-1} \int_0^T e^{-i\tau_j u} q_j[\varphi(u), u, \varepsilon]\, du \quad ,$$

(3.2)

$$|d_j| \leq |c_j|^{-1} T^{-1} \int_0^T |q_j[\varphi(u), u, \varepsilon]|\, du \leq |c_j|^{-1} T^{-1} \int_0^T \psi(u)\, du \leq r_1^{-1} K \; .$$

For $j = 1, \ldots, \nu$, we have, by (2.4), (2.9),

$$\psi_j(t) - z_j(t) = \varepsilon e^{i\tau_j t} \int e^{-i\tau_j u} \left\{q_j[\varphi(u), u, \varepsilon] - d_j\varphi_j(u)\right\} du ,$$

and, by (1.ii) with $V = T$, (K), and (3.2), also

$$(3.3) \quad \begin{aligned} |\psi_j(t) - z_j(t)| &\leq 2\varepsilon \int_0^T [|q_j[\varphi(u), u, \varepsilon]| + |d_j| \, |\varphi_j(u)|] \, du \\ &\leq 2\varepsilon[KT + Tr_1^{-1}KR] = 2\varepsilon(1 + r_1^{-1}R)KT , \end{aligned}$$

for all $-\infty < t < +\infty$, since ψ_j, z_j are periodic of period T, $j = 1, \ldots, \nu$. For $j = \nu + 1, \ldots, n$, we have, by (2.11), (2.10),

$$\psi_j(t) - z_j(t) = \psi_j(t) = e^{\rho_j t} \int e^{-\rho_j u} q_j[\varphi(u), u, \varepsilon] \, du ,$$

with $|-\rho_j + im\omega/b_0| \geq \delta > 0$, $j = \nu + 1, \ldots, n$, $m = 0, \pm 1, \ldots$. Let $\gamma = \max |\Re \rho_j|$, $j = \nu + 1, \ldots, n$, and $H = e^{\gamma T}$. By (1.ii) there is a constant $N = N(\delta, \omega, b)$ such that

$$(3.4) \quad |\psi_j(t) - z_j(t)| \leq \varepsilon HN \int_0^T |q_j[\varphi(u), u, \varepsilon]| \, du \leq \varepsilon HNKT, \quad j = \nu + 1, \ldots, n$$

By (3.3) and (3.4), for $M' = \max[2(1 + r_1^{-1}R)KT, HNKT]$, $\varepsilon' = \min[\varepsilon_0, r/M]$, $0 \leq \varepsilon \leq \varepsilon'$, we have

$$\Re[\varphi(t) - z(t)] \leq \varepsilon M' \leq \varepsilon' M' \leq r .$$

Thus $\mathfrak{T}\varphi \in \Omega_0$ for every $\varphi \in \Omega_0$, i.e., $\mathfrak{T}\Omega_0 \subset \Omega_0$.

Let $L > 0$ denote any number with $\rho_j(\varepsilon) \leq L$, $j = 1, \ldots, n$. Let us consider the subset $\Omega_0^* \subset \Omega_0$ of all $\varphi(t) \in \Omega_0$ whose components $\varphi_j(t)$ are absolutely continuous, with first derivatives (a.e.) satisfying the relations

$$|\varphi_j'(t)| \leq 2LR + \varepsilon_0\psi(t), \; j = 1, \ldots, n, \; -\infty < t < \infty .$$

Obviously Ω_0^* is convex. Note that the functions $\varphi_j(t)$, $j = 1, \ldots, n$, with $\varphi(t) \in \Omega_0^*$ are equibounded, equiabsolutely continuous, and equicontinuous. Thus, by Ascoli's thecrem, Ω_0^* is a compact family.

Finally, if $y^n = (y_{jn}, j = 1, \ldots, n) \in \Omega_0^*$, $n = 1, \ldots,$ and

$y_{jn}(t) \rightrightarrows y_j(t)$ as $n \longrightarrow \infty$, the equiabsolute continuity of the functions y_{jn} implies that $y_j(t)$ is absolute continuous, $j = 1, \dots, n$. If v_j, v_{jn} denote the total variations of y_j, y_{jn} in any interval $[t', t'']$, by the lower semicontinuity of the total variations we have

$$\int_{t'}^{t''} |y_j'(t)|\, dt = v_j \le \varliminf_{n\to\infty} v_{jn} = \varliminf_{n\to\infty} \int_{t'}^{t''} |y_{jn}'(t)|\, dt \ .$$

Hence

$$\int_{t'}^{t''} |y_j'(t)|\, dt \le \int_{t'}^{t''} [2LR + \varepsilon_0\psi(t)]\, dt \ ,$$

and, by a limit, also $|y_j'(t)| \le 2LR + \varepsilon_0\psi(t)$ for almost all t. Thus Ω_0^* is a closed set.

Let us prove that $\mathfrak{T}\,\Omega_0^* \subset \Omega_0^*$. By (2.9), (2.10), by differentiation, we have

$$\psi_j'(t) = \rho_j(0)\psi_j(t) - \varepsilon d_j\varphi_j(t) + \varepsilon q_j[\varphi(t), t, \varepsilon]\ , \qquad j = 1, \dots, \nu\ ,$$

$$\psi_j'(t) = \rho_j(\varepsilon)\psi_j(t) + \varepsilon q_j[\varphi(t), t, \varepsilon], \qquad j = \nu + 1, \dots, n,$$

and hence

$$|\psi_j'(t)| \le LR + \varepsilon r_1^{-1}KR + \varepsilon\psi(t), \qquad j = 1, \dots, n\ ,$$

Thus, for $\varepsilon'' = \min\,[\varepsilon', r_1LK^{-1}]$, $0 \le \varepsilon \le \varepsilon''$, we have

$$|\psi_j'(t)| \le 2LR + \varepsilon_0\psi(t), \qquad j = 1, \dots, n\ .$$

This proves that $\psi = \mathfrak{T}\varphi \in \Omega_0^*$ for every $\varphi \in \Omega_0^*$, i.e., $\mathfrak{T}\,\Omega_0^* \subset \Omega_0^*$.

We must now prove that, for every ε, $0 \le \varepsilon \le \varepsilon''$, $\mathfrak{T}$ is a continuous transformation in Ω_0. Given $\zeta > 0$, let $\xi > 0$ be the number defined in (K) relative to ζ. For every two elements $\varphi^s = (\varphi_{s1}, \dots, \varphi_{s2}) \in \Omega_0$, $s = 1, 2$, let $\psi^s = \mathfrak{T}\varphi^s$, $\psi^s = (\psi_{s1}, \dots, \psi_{sn})$, and $D_s = D[\varphi^s] = (d_{s1}, \dots, d_{s\nu}, 0, \dots, 0)$. By (3.2) we have $|d_{sj}| \le r_1^{-1}K$, $j = 1, \dots, \nu$, $s = 1, 2$. We have also, by (K) and (3.1), for $\mathfrak{N}[\varphi^1 - \varphi^2] < \xi$, and by taking $\xi \le \zeta$,

$$c_j d_{1j} - c_j d_{2j} = \mathfrak{M}\left\{e^{-i\tau_j u}\left[q_j[\varphi^1(u),u,\varepsilon] - q_j[\varphi^2(u),u,\varepsilon]\right]\right\},$$

$$(3.5)\quad |d_{1j}-d_{2j}| \le |c_j|^{-1}T^{-1} \int_0^T \left|q_j[\varphi^1(u),u,\varepsilon] - q_j[\varphi^2(u),u,\varepsilon]\right| du \quad .$$

$$\le |c_j|^{-1}T^{-1}\zeta \int_0^T \psi(u)\, du \le r_1^{-1}K\zeta \quad .$$

For $j = 1, \ldots, \nu$, we have now, by (2.4), (2.9),

$$\psi_{1j}(t) - \psi_{2j}(t) = \varepsilon\, e^{i\tau_j t} \int e^{-i\tau_j u}\Big\{q_j[\varphi^1(u),u,\varepsilon] -$$

$$- q_j[\varphi^2(u),u,\varepsilon] - d_{1j}\varphi_{1j} + d_{2j}\varphi_{2j}\Big\}\, du ,$$

and, by (1.ii), (K), (3.1), (3.2), (3.5), also

$$|\psi_{1j}(t) - \psi_{2j}(t)| \le 2\varepsilon \int_0^T \Big\{|q_j[\varphi^1(u),u,\varepsilon] - q_j[\varphi^2(u),u,\varepsilon]| +$$

$$(3.6)\qquad + |d_{1j}|\; |\varphi_{1j}(u) - \varphi_{2j}(u)| + |d_{1j} - d_{2j}|\; |\varphi_{2j}(u)|\Big\}\, du$$

$$\le 2\varepsilon[KT + r_1^{-1}KT + r_1^{-1}KTR]\zeta \le 2\varepsilon KT(1 + r_1^{-1} + r_1^{-1}R)\zeta \, .$$

For $j = \nu + 1, \ldots, n$, we have, by (2.10),

$$\psi_{1j}(t)-\psi_{2j}(t) = \varepsilon e^{\rho_j t} \int e^{-\rho_j u} \Big\{q_j[\varphi^1(u),u,\varepsilon] - q_j[\varphi^2(u),u,\varepsilon]\Big\}\, du \quad ,$$

and by (1.ii) for the same constants $H = e^{\gamma T}$, $N = N(\delta, \omega, b)$, we have

$$|\psi_{1j}(t) - \psi_{2j}(t)| \le \varepsilon HN \int_0^T |q_j[\varphi^1(u), u, \varepsilon] - q_j[\varphi^2(u), u, \varepsilon]|\, du$$

$$(3.7)\qquad \le \varepsilon HNKT\, \zeta, \qquad\qquad j = \nu + 1, \ldots, n \quad .$$

By (3.6), (3.7), for $M'' = \max\,[2KT(1 + r_1^{-1} + r_1^{-1}R),\ HNKT]$ we have

$$\mathfrak{N}[\psi^1 - \psi^2] \le \varepsilon M''\zeta \le (\varepsilon_0 M'')\zeta ,$$

and this assures that $\mathfrak{T}$ is a continuous transformation in Ω_0.

Since $\mathfrak{T}$ is a continuous transformation and $\mathfrak{T}$ maps the convex closed compact set Ω_0^* into itself, where $\Omega_0^* \subset \Omega_0 \subset \Omega_R \subset \Omega$, and Ω is linear, by Schauder's fixed point theorem we conclude that $\mathfrak{T}$ has a fixed element in Ω_0^*.

(b). <u>Proof that $\mathfrak{T}$ is a contraction in Ω_0 under condition (L).</u>

(3.ii). Under hypotheses (α) and (L), we can take ε_1 sufficiently small in (3.i) so that $\mathfrak{T} \mid \Omega_0$ is a contraction, namely $\mathfrak{N}[\mathfrak{T}\varphi^1 - \mathfrak{T}\varphi^2] \le 1/2 \ \mathfrak{N}[\varphi^1 - \varphi^2]$ for every $\varphi^1, \varphi^2 \in \Omega_0$. Thus, the fixed element $y(t) \in \Omega_0^* \subset \Omega_0$ is unique in Ω_0.

PROOF. We have already proved that $\mathfrak{T}\Omega_0 \subset \Omega_0$. We need not consider Ω_0^*. Relations (3.1), (3.2), (3.3), (3.4) will be needed. Relation (3.5) can be replaced by

$$(3.5\text{L}) \qquad \begin{aligned} |d_{1j} - d_{2j}| &\le |c_j|^{-1}T^{-1} \sum_{\ell=1}^{n} \int_0^T \psi(u)|\varphi_{1\ell}(u) - \varphi_{2\ell}(u)|\,du \\ &\le r_1^{-1}nK \ \mathfrak{N}[\varphi^1 - \varphi^2], \qquad j = 1, \ldots, \nu \ . \end{aligned}$$

Relation (3.6) can be replaced by

$$(3.6\text{L}) \qquad \begin{aligned} |\psi_{1j}(t) - \psi_{2j}(t)| &\le 2\varepsilon[nKT + r_1^{-1}KT + r_1^{-1}nKTR] \ \mathfrak{N}[\varphi^1 - \varphi^2] \\ &\le 2\varepsilon nKT[1 + r_1^{-1}(1 + R)] \ \mathfrak{N}[\varphi^1 - \varphi^2], \\ & \qquad j = 1, \ldots, \nu \ . \end{aligned}$$

Relation (3.7) is replaced by

$$(3.7\text{L}) \qquad |\psi_{1j}(t) - \psi_{2j}(t)| \le \varepsilon HNnKT \ \mathfrak{N}[\varphi^1 - \varphi^2], \qquad j = \nu + 1, \ldots, n \ .$$

By (3.6L) and (3.7L), for $M''' = \max\,[2nKT\,[1 + r_1^{-1}(1 + R)],\ nHNKT]$, $\varepsilon_1 = \min\,[\varepsilon'', 1/2M''']$, $0 \le \varepsilon \le \varepsilon_1$, we have

$$(3.8\text{L}) \qquad \begin{aligned} \mathfrak{N}[\psi^1 - \psi^2] &\le \varepsilon M''' \ \mathfrak{N}[\varphi^1 - \varphi^2] \le \varepsilon_1 M''' \ \mathfrak{N}[\varphi^1 - \varphi^2] \le \\ &\le (1/2)\ \mathfrak{N}[\varphi^1 - \varphi^2] \ . \end{aligned}$$

Thus $\mathfrak{T} \mid \Omega_0$ is a contraction and (3.ii) is proved.

REMARK 1. If we denote by $\|D\|$, or norm of $D = (d_1, \ldots, d_\nu, 0, \ldots, 0)$ the number $\|D\| = \max |d_j|$, $j = 1, \ldots, \nu$, and by M_1 the constant $M = nr_1^{-1}K$, then, under the same conditions of (3.ii), by (3.5L) and (3.8L) and for $0 \leq \varepsilon \leq \varepsilon_1$ we have

$$(3.9)\quad \begin{aligned} &\mathfrak{M}[\mathfrak{T}\varphi^1 - \mathfrak{T}\varphi^2] \leq (1/2)\ \mathfrak{M}[\varphi^1 - \varphi^2] \quad , \\ &\|D[\varphi^1] - D[\varphi^2]\| \leq M_1\ \mathfrak{M}[\varphi^1 - \varphi^2] \quad . \end{aligned}$$

Note that under the present hypotheses, i.e., (L) instead of (K), the existence of a fixed element in Ω_0 could be deduced by the fact that $\mathfrak{T} \mid \Omega_0$ is a contraction and Ω_0 is a complete set.

REMARK 2. The fixed element $y(t) = (y_1, \ldots, y_n)$, $y = \mathfrak{T}y$, whose existence has been proved above (3.i) satisfies the integral equation

$$(3.10)\qquad y(t) = z(t) + \varepsilon e^{Bt} \int e^{-Bu} \{q[y(u), u, \varepsilon] - D[y]y(u)\}\, du \quad .$$

As noted in §2, the components $y_j(t)$ are absolutely continuous and thus, by differentiation of (3.10), we conclude that the differential equation (2.16) is satisfied for almost all t. By (2.6), (2.9), (2.10), relation (3.10) in component form is

$$(3.11)\quad \begin{aligned} y_j(t) &= c_j e^{i\tau_j t} + \varepsilon e^{i\tau_j t} \int e^{-i\tau_j u} \{q_j[y(u),u,\varepsilon] - d_j y_j(u)\}\, du, \\ &\qquad\qquad j = 1, \ldots, \nu \quad , \\ y_j(t) &= \varepsilon e^{\rho_j(\varepsilon)t} \int e^{-\rho_j(\varepsilon)u} q_j[y(u),u,\varepsilon]\, du, \quad j = \nu + 1, \ldots, n \quad . \end{aligned}$$

By (3.3), (3.4), (2.4) we know that, for all t, we have

$$(3.12)\quad \begin{aligned} &|y_j(t) - c_j e^{i\tau_j t}| \leq 2\varepsilon KT(1 + r_1^{-1}R), \qquad j = 1, \ldots, \nu \quad . \\ &|y_j(t)| \leq \varepsilon HNKT, \qquad j = \nu + 1, \ldots, n \quad . \end{aligned}$$

Having already denoted by L any number such that $|\rho_j(\varepsilon)| \leq L$, $j = 1, \ldots, n$, $0 \leq \varepsilon \leq \varepsilon_0$, by (2.16) and (3.2) we have

$$(3.13)\qquad |y'_j(t)| \leq LR + \varepsilon r_1^{-1}KR + \varepsilon\psi(t) \quad .$$

(c). Case of large forcing terms. We shall consider the system analogous to (2.1):

$$dy/dt = Ay + F(t) + \varepsilon q(y, t, \varepsilon) ,$$

where $F(t) = (F_1, \ldots, F_n)$ is a periodic vector function of period $2\pi/\omega$, whose components are L-integrable in $[0, 2\pi/\omega]$. For what all previous considerations are concerned we may actually replace (3.14) by the system

$$(3.14) \qquad dy/dt = By + F(t) + \varepsilon q(y, t, \varepsilon) .$$

We must suppose that

$$\mathfrak{M}[e^{-i\tau_j t} F_j(t)] = T^{-1} \int_0^T e^{-i\tau_j t} F_j(t)\, dt = 0, \qquad j = 1, \ldots, \nu .$$

Under this condition, system

$$dy/dt = By + F(t)$$

has a particular solution given by

$$Y(t) = e^{Bt} \int e^{-Bu} F(u)\, du$$

with the usual conventions. Let $y = Y(t) + z$, $z = (z_1, \ldots, z_n)$. Then (3.14) is transformed into the system

$$dz/dt = Bz + \varepsilon q[Y(t) + z, t, \varepsilon]$$

which is of the type (2.1).

(d). Extensions.

(d_1). The considerations above can be repeated for systems much more general than (2.1). For instance, we may consider systems of the form

$$(3.15) \qquad dy/dt = Ay + q(y, t, \varepsilon), \qquad -\infty < t < +\infty ,$$

$y = (y_1, \ldots, y_n)$, $q = (q_1, \ldots, q_n)$, under conditions (α) and (K) provided $q_j(y, t, 0) = 0$, $j = 1, \ldots, n$. Given $\zeta > 0$ there is a $\xi > 0$ such that

$$(3.16) \qquad |q_j(y, t, \varepsilon)| < \zeta\psi(t), \qquad j = 1, \ldots, n ,$$

for all $|y_j| \leq R$, $-\infty < t < +\infty$, $0 \leq \varepsilon \leq \zeta$.

(d_2). The considerations above can be repeated for systems of the form

$$(3.17) \qquad dy/dt = Ay + q(y, t, \varepsilon) + g(y, t, \varepsilon), \quad -\infty < t < +\infty \quad ,$$

under conditions (α) and (K) (both q and g) provided q is as above in (d_1) and the following holds: given $\zeta > 0$ there is a $\xi > 0$ such that, for all $0 \leq \varepsilon \leq \varepsilon_0$, $-\infty < t < +\infty$, $|y_j| \leq R$, we have

$$(3.18) \qquad |g_j(y, t, \varepsilon)| \leq \zeta[|y_1| + \dots + |y_n|]\psi(t) \quad .$$

In this case we must take Ω_R with R sufficiently small. The last condition is certainly satisfied if $|g_j(y, t, \varepsilon)| \leq M[|y_1|^2 + \dots + |y_n|^2]$ for some constant M. Finally the case

$$(3.19) \qquad dy/dt = Ay + F(t) + q(y, t, \varepsilon) + g(y, t, \varepsilon) \quad ,$$

where q, g are as above and $F(t)$ is a given periodic function of period $2\pi/\omega$, can be reduced to the previous ones by the remark (c) above.

(d_3). Note that systems of the form

$$dy/dt = A(t)y + \varepsilon q(y, t, \varepsilon) \quad ,$$

or the similar ones, where $A(t)$ is a matrix whose elements are continuous periodic functions of t of period $2\pi/\omega$, can be reduced to systems (2.1) by a change of coordinates $z = Py$ where P is a matrix whose elements are continuous periodic functions of t, $\det P \neq 0$ for all t, such that PAP^{-1} is a constant matrix.

(d_4). Another important extension concerns the case where $B_2 = B_2(\varepsilon)$ is an arbitrary $(n-\nu) \times (n-\nu)$ matrix (§2, condition (α)) whose elements are continuous functions of ε for $0 \leq \varepsilon \leq \varepsilon_0$, and whose characteristic roots $\rho_j(\varepsilon)$, $j = \nu + 1, \dots, n$, verify the relations $\rho_j(0) \neq im\omega/b_0$, $b_0 = b_1 \dots b_\nu$, $m = 0, \pm 1, \pm 2, \dots$. Then the coefficients of the characteristic polynomials of $B_2(\varepsilon)$ are also continuous functions of ε, and finally the characteristic roots $\rho_j(\varepsilon)$ are continuous functions of ε (the latter statement can be proved in terms of theory of functions of one complex variable and the use of the logarithmic indicator. See, e.g., [4a]). Also, it is known that there exists a $(n-\nu) \times (n-\nu)$ matrix $P = P(\varepsilon)$ whose elements are continuous functions of ε in $0 \leq \varepsilon \leq \varepsilon_0$ for some $\varepsilon_0 > 0$, with $\det P(\varepsilon) \neq 0$, and PB_2P^{-1}

is a triangular matrix. See, e.g., for analogous considerations [4a]. See also, S. P. Diliberto, On systems of ordinary differential equations. These Contributions, Vol. 1, 1950, 1-38. Thus we can suppose $B_2 = B_2(\varepsilon)$ a triangular matrix, say $B_2 = [b_{jh}]$, $j, h = \nu + 1, \dots, n$, such that $b_{jh} = 0$ for $j > h$, $b_{jj} = \rho_j(\varepsilon)$. Note that the least $n - \nu$ equations (2.1), written in inverse order, are

$$y_n' = \rho_n(\varepsilon)y_n + \varepsilon q_n \quad ,$$

$$y_{n-1}' = \rho_{n-1}(\varepsilon)y_{n-1} + b_{n-1,n}y_n + \varepsilon q_{n-1} \quad ,$$

$$y_{n-2}' = \rho_{n-2}(\varepsilon)y_{n-2} + b_{n-2,n-1}y_{n-1} + b_{n-2,n}y_n + \varepsilon q_{n-2} \quad ,$$

. . . .

$$y_{\nu+1}' = \rho_{\nu+1}(\varepsilon)y_{\nu+1} + \sum_{\ell=\nu+2}^{n} b_{\nu+1,\ell}y_\ell + \varepsilon q_{\nu+1} \quad .$$

The transformation $\psi = \mathfrak{T}\varphi$ can now be defined by the same relations (2.9), (2.11) for $j = 1, \dots, \nu$, and, for $j = \nu + 1, \dots, n$, by

$$\psi_n(t) = \varepsilon e^{\rho_n t} \int e^{-\rho_n u} q_n[\varphi(u), u, \varepsilon] \, du \quad ,$$

$$\psi_{n-1}(t) = e^{\rho_{n-1} t} \int e^{-\rho_{n-1} u} \left\{\varepsilon q_{n-1}[\varphi(u), u, \varepsilon] + b_{n-1,n}\psi_n(u)\right\} du \quad ,$$

$$\psi_{n-2}(t) = e^{\rho_{n-2} t} \int e^{-\rho_{n-2} u}\left\{\varepsilon q_{n-2}[\varphi(u),u,\varepsilon] + b_{n-2,n-1}\psi_{n-1}(u) + b_{n-2,n}\psi_n(u)\right\} du,$$

. . . .

$$\psi_{\nu+1}(t) = e^{\rho_{\nu+1} t} \int e^{-\rho_{\nu+1} u}\left\{\varepsilon q_{\nu+1}[\varphi(u),u,\varepsilon] + \sum_{\ell=\nu+2}^{n} b_{\nu+1,\ell}\psi_\ell(u)\right\} du \quad .$$

All considerations of the present paper extend to this case without difficulties. Note that for $\nu = 0$ no condition is imposed on $A(\varepsilon)$ but $\rho_j(0) \neq im\omega$, $j = 1, \dots, n$, $m = 0, \pm 1, \pm 2, \dots$. For the sake of simplicity we shall refer in the following mainly to systems of the form (2.1). All considerations of §3 extend with obvious changes to the more general situations mentioned above.

(e). <u>The elementary case</u> $\nu = 0$.

(3.iii). System

$$y' = A(\varepsilon)y + \varepsilon q(y, t, \varepsilon) \quad , \tag{3.20}$$

$y = (y_1, \dots, y_n)$, $q = (q_1, \dots, q_n)$, satisfying (α) and (K) with $\nu = 0$, has always a periodic solution for all $0 \leq \varepsilon \leq \varepsilon_1$ and some $\varepsilon_1 > 0$ sufficiently small.

Indeed, for $\nu = 0$, there are no a_j, b_j, c_j, and (α) requires only that $\rho_j(0) \neq im\omega$, $j = 1, \dots, n$, $m = 0, \pm 1, \dots,$ where $2\pi/\omega$ is a period of q. Then, for $0 \leq \varepsilon \leq \varepsilon_0$ and some $\varepsilon_0 > 0$, $\delta > 0$, we have also $|\rho_j(\varepsilon) - im\omega| > \delta > 0$ for the same j's and m's. The transformation $\mathfrak{T}$, or $\psi = \mathfrak{T}\varphi$, of §2, is now the obvious one

$$\psi(t) = e^{iAt} \int e^{-iAu} q[\varphi(u), u, \varepsilon]\, du \quad ,$$

where the integrands are of class C_ω and mean value zero and the integrals are the unique primitives of class C_ω and mean value zero. Actually $\psi(t)$ is the unique periodic solution of the linear system $y' = Ay + \varepsilon q[\varphi(t), t, \varepsilon]$. Note that we have here $z(t) = (0, \dots, 0)$, and $A = B$. Also Ω_0 is a sphere about $\varphi = 0$, $\mathfrak{T}\Omega_0 \subset \Omega_0$, and $\mathfrak{T}\Omega_0 \subset \Omega_0$ for $0 \leq \varepsilon \leq \varepsilon_1$ and some $\varepsilon_1 > 0$. Finally $\mathfrak{T}$ has a fixed element $y(t) \in \Omega_0$. The determining equations (2.16) are nonexistent (thus certainly satisfied) and $y(t)$ is a periodic solution of (3.20). Under condition (L) the corresponding method of successive approximations is the usual one $y^{(o)} = 0$, $y^{(m)} = \mathfrak{T} y^{(m-1)}$, $m = 1, 2, \dots$ [See A. Lyapunov, Problème général de la stabilité du mouvement. Ann. of Math. Studies, 17, 1949].

(3.iv). System

$$y' = A(\varepsilon)y + F(t) + \varepsilon q(y, t, \varepsilon) \quad , \tag{3.21}$$

where y, q, F are n-vectors, where A and q satisfy (α) and (K) with $\nu = 0$, and $F(t)$ is periodic of period $2\pi/\omega$, L-integrable in $[0, 2\pi/\omega]$, $|F_j(t)| \leq r < R$, $j = 1, \dots, n$, has always a periodic solution for all $0 \leq \varepsilon \leq \varepsilon_1$ and some ε_1 sufficiently small.

Indeed the condition $\mathfrak{M}[e^{-\rho_j t} F_j(t)] = 0$, $j = 1, \dots, n$, is necessarily satisfied since $|\rho_j(\varepsilon) - im\omega| > \delta > 0$, for all j's and m's as before. Hence, system (3.21) can be reduced to another analogous to (3.20). A little reflection shows that the method of successive approximations can actually be defined in terms of system (3.21) by taking for $y^{(o)}(t) = z(t)$ the unique periodic solution of the linear system $y' = A(\varepsilon)y + F(t)$, and for $y^{(m)}(t)$ the unique periodic solution of the linear system $y' = A(\varepsilon)y + F(t) + \varepsilon q[y^{(m-1)}(t), t, \varepsilon]$, $m = 1, 2, \ldots,$ in other words

$$y^{(o)}(t) = z(t) = e^{iAt} \int e^{-iAu} F(u)\, du \quad ,$$

$$y^{(m)}(t) = z(t) + \varepsilon e^{iAt} \int e^{-iAu} q[y^{(m-1)}(u), u, \varepsilon]\, du \quad .$$

Extension of (3.iv) in the lines of the remarks d_1, d_2 are the following ones.

(3.v). Consider the system

$$dy/dt = A(\varepsilon)y + q(y, t, \varepsilon) + g(y, t, \varepsilon) \quad , \tag{3.22}$$

where y, q, g are n-vectors, where A, q, g satisfy conditions (α) and (K) with $\nu = 0$, and, in addition, q satisfies condition (3.16) and g condition (3.18). Then there is a periodic solution $y(t)$ of (3.22) for all $0 \leq \varepsilon \leq \varepsilon_1$ and some $\varepsilon_1 > 0$ sufficiently small.

Finally, we may replace $A(\varepsilon)$ in (3.22) by a $n \times n$ matrix $A(t, \varepsilon)$ whose elements are continuous functions of t and ε, $-\infty < t < +\infty$, $0 \leq \varepsilon \leq \varepsilon_0$, periodic in t of period $2\pi/\omega$ (the same ω as for q and g), and whose characteristic exponents, say $\lambda_j(\varepsilon)$, $j = 1, \ldots, n$, (all defined mod ωi) verify the relations $\lambda_j(0) \not\equiv 0 \pmod{\omega i}$, $j = 1, \ldots, n$.

The last extension contains, as particular cases, results of Antosiewicz and Diliberto, since the former assumed $R(\lambda_j) < 0$, and the latter $R(\lambda_j) \neq 0$, $j = 1, \ldots, n$. Furthermore we assume that the functions q and g are periodic and L-integrable in the period (in the sense of condition K) instead of periodic and continuous (S. P. Diliberto and M. D. Marcus, On systems of ordinary differentiable equations. These Contributions, Vol. 3, 1956, 237-241).

An extension in a different direction can be obtained as follows. Suppose that system (3.21) satisfies conditions (α) and (L) with $\nu = 0$ and $|R(\rho_j)| \geq \delta > 0$, $j = 1, \ldots, n$, suppose that $F(t) \in QP$, where QP is the class of all vector functions whose components $F_j(t)$ are L^2-quasi periodic, i.e.,

$$F_j(t) \sim \sum c_{jm} e^{i\lambda_m t} \quad , \qquad \sum |c_{jm}|^2 < +\infty, \quad j = 1, \ldots, n.$$

We shall suppose also that $q_j(y, t, \varepsilon) \in QP$ and $q_j[y(t), t, \varepsilon] \in QP$ This condition replaces the periodicity condition (p) of (K) as mentioned in §2.

Note that we have now $z(t) \equiv 0$ and

$$y_j^o(t) \sim \sum (\rho_j + i\lambda_m)^{-1} c_{jm} e^{i\lambda_m t}, \quad \sum |(\rho_m + i\lambda_m)^{-1} c_{jm}|^2 \leq \delta^{-2} \sum |c_{jm}|^2 < +\infty ,$$

i.e., $y^o(t) \in QP$. The same argument proves that system (3.21) has a solution $y(t)$ of class QP. [See J. J. Stoker, Nonlinear vibrations, App. 2., Interscience, 1950].

(f). <u>The analytic case</u>. Though not needed in the sequel the following remark may be of some interest. The method of successive approximations defined by relations (2.18) is the one initially proposed by the author [4b] in 1940 for linear systems, and successively used by J. K. Hale, R. A. Gambill [9b, 7abc] for linear systems, and by W. R. Fuller [10] for Lipschitzian nonlinear systems. The corresponding series

$$(3.23) \qquad y_j(t) = y_j^{(o)}(t) + \sum_{m=o}^{\infty} [y_j^{(m+1)}(t) - y_j^{(m)}(t)], \qquad j = 1, \ldots, n ,$$

are not power series in ε even in the case where the functions $q_j(y, t, \varepsilon)$ are analytic in $y_1, \ldots, y_n, \varepsilon$. For this case a modification of the process (2.18) has been used in [9c, 8, 6b]. In harmony with these papers, suppose

$$q_j(y, t, \varepsilon) = \sum_{k=o}^{\infty} \varepsilon^k q_j^{(k)}(y, t), \; q_j^{(k)}(y, t) \in C_\omega, \qquad 0 \leq \varepsilon \leq \varepsilon_o ,$$

$q^{(k)}(y, t)$ analytic in $y_1, \ldots, y_n$, and let $s_j^{(r,k)}(t)$ denote the coefficient of ε^{r-1}, $r = 1, 2, \ldots,$ when y in $q_j^{(k)}(y, t)$ is replaced by

$$(3.24) \qquad y = z^{(o)}(t) + \varepsilon z^{(1)}(t) + \varepsilon^2 z^{(2)}(t) + \ldots ,$$

where each $z^{(k)}(t)$ is independent of ε. Moreover, let the corresponding coefficient of ε^{r-1} in $q_j(y, t, \varepsilon)$ be denoted by $s_j^{(r)}(t)$. Let

$$c_j S_j^{(r,k)} = \mathfrak{M}\left\{e^{-i\tau_j t} s_j^{(r,k)}(t)\right\}, \qquad j = 1, \ldots, \nu ,$$

$$c_j S_j^{(r)} = \mathfrak{M}\left\{e^{-i\tau_j t} s_j^{(r)}(t)\right\} , \qquad j = 1, \ldots, \nu , \tag{3.25}$$

$$S^{(r)} = \mathrm{diag}\ (S_1^{(r)}, \ldots, S_\nu^{(r)}, 0, \ldots, 0) ,$$

where

$$s_j^{(r)} = s_j^{(r,o)} + s_j^{(r-1,1)} + \ldots + s_j^{(1,r-1)} ,$$

$$S_j^{(r)} = S_j^{(r,o)} + S_j^{(r-1,1)} + \ldots + S_j^{(1,r-1)}, \quad j = 1, \ldots, \nu.$$

The modified method of successive approximations is then defined by

$$y^{(o)}(t) = z(t) = z^{(o)}(t) = (c_1 e^{i\tau_1 t}, \ldots, c_\nu e^{i\tau_\nu t}, 0, \ldots, 0) ,$$

$$y^{(r)}(t) = z^{(o)}(t) + e^{Bt} \int e^{-Bu}\Big[\sum_{k=1}^{r} \varepsilon^k s^{(k)}(u) -$$

$$- \Big(\sum_{k=1}^{r} \varepsilon^k S^{(k)}(u) \Big) z^{(k-1)}(u)\Big] du , \tag{3.26}$$

$$\bmod \varepsilon^{r+1}, \qquad r = 1, 2, \ldots ,$$

with $s^{(k)} = (s_1^{(k)}, \ldots, s_\nu^{(k)}, 0, \ldots, 0)$, $y^{(r)} = z^{(o)} + \varepsilon z^{(1)} + \ldots + \varepsilon^r z^{(r)}$, and exactly the same conventions above. In [8, 9c] it was proved directly that this method converges and that $y(t) = \lim y^{(r)}(t)$ as $r \longrightarrow \infty$, satisfies the differential system $dy/dt = (B - \varepsilon h)y + \varepsilon q(y, t, \varepsilon)$, where $h = h(a, b, c, \omega, \varepsilon) = S^{(1)} + \varepsilon S^{(2)} + \varepsilon^2 S^{(3)} + \ldots$. Thus h may replace d in the analytic case.

All the proofs in the present paper are given in terms of the transformation $\mathfrak{T}$ and a fixed point theorem. In the Lipschitzian case, some could be given in terms of the process of successive approximations (2.18) and, in the analytic case, even in terms of the process (3.24, 25, 26). The differences are mostly notational.

(g). <u>A remark</u>. In this section we have supposed all $c_j \neq 0$ and between positive bounds, $0 < r_1 \leq |c_j| \leq r_2 < R$, $j = 1, \ldots, \nu$. There are cases where this requirement can be weakened or omitted.

A first situation where we may consider some c_j arbitrarily small, say $0 \leq |c_j| \leq r_2 < R$, is when, in a convenient class Ω, the corresponding mean value (2.11) is identically zero. Then we may assume $d_j \equiv 0$, c_j arbitrary. This situation will occur in §6.

A second analogous situation will occur when we can prove that the corresponding mean value (2.11) is $O(c_j)$ as $|c_j| \longrightarrow 0$. Then $d_j = O(1)$, $|d_j| \leq M$, and the previous evaluations are simplified. This situation will occur in §4 with applications in §6.

Another analogous situation would occur if for $c_j = O(\varepsilon^k)$ as $\varepsilon \longrightarrow 0$ for some j and $k > 0$, we could prove that also the corresponding mean values (2.11) are $O(\varepsilon^k)$. We do not discuss this case in the present paper.

§4. PROPERTIES OF THE FIXED ELEMENT

(a). The general case. Given system (2.1) under hypotheses (α) and (K), and the two sets of integers $a = (a_1, \ldots, a_\nu)$, $b = (b_1, \ldots, b_\nu)$, $a_j \gtrless 0$, $b_j > 0$, $j = 1, \ldots, \nu$, we have determined the n-vector $y(t) = y(t, a, b, c, \omega, \varepsilon) = (y_1, \ldots, y_n)$ and the relative ν-vector of numbers d_j, $D = D[y] = D(a, b, c, \omega, \varepsilon) = (d_1, \ldots, d_\nu, 0, \ldots, 0)$ with $-\infty < t < +\infty$, $0 \leq \varepsilon \leq \varepsilon_1$, $c = (c_1, \ldots, c_\nu)$, $0 < r_1 \leq |c_j| \leq r_2 < R$, $j = 1, \ldots, \nu$. As mentioned in §2, in the periodic case ω is fixed, since the relations

$$\rho_j(0) = ia_j\omega/b_j, \qquad j = 1, \ldots, \nu ,$$

are satisfied, and the functions q_j are periodic functions of t of period $T = 2\pi/\omega$. In the autonomous case we shall denote by $\omega = \omega_0$ the value of ω for which the ν relations above are verified and then ω is thought of as a parameter varying in an interval $U = [\omega_{01}, \omega_{02}]$, $0 < \omega_{01} < \omega_0 < \omega_{02} < +\infty$. In this section we shall discuss the properties of y as a function of t, as well as the properties of y and D as functions of c, ε and (in the autonomous case) of ω. For the sake of brevity we must treat the two cases together: system (2.1) periodic, and system (2.1) autonomous. To do this we shall suppose, in all that follows, that in the periodic case the interval U is reduced to the single point ω_0, thus $\omega_{01} = \omega_0 = \omega_{02} = \omega$. We shall take for the ν-vectors c the norm $\|c\| = \max |c_j|$, $j = 1, \ldots, \nu$. The vector function $y(t)$ is periodic in t of period $2\pi/\omega$ and verifies the integral equation (2.15) and the differential equation (2.16), thus

$$y'(t) = (B - \varepsilon D)y(t) + \varepsilon q(y, t, \varepsilon) ,$$

where $A = \text{diag}\ [\rho_j(\varepsilon),\ j = 1, \ldots, n]$, and $B = \text{diag}\ [\rho_j(0),\ j = 1, \ldots, \nu;\ \rho_j(\varepsilon),\ j = \nu + 1, \ldots, n]$. Let M be large enough in order that $|\rho_j(\varepsilon)| \leq M$, $j = 1, \ldots, n$, $0 \leq \varepsilon \leq \varepsilon_1$. Since $|y_j| \leq R$, $|q_j| \leq \psi(t)$, and $|d_j| \leq r_1^{-1}K$ (§3) we have, as in §3,

$$|y_j'(t)| \leq MR + \varepsilon_1 r_1^{-1}KR + \varepsilon_1\psi(t) \quad . \tag{4.1}$$

Thus, we have, for all t^1, t^2,

$$|y_j(t^2) - y_j(t^1)| \leq (M + \varepsilon_1 r_1^{-1}K)R\ |t^2 - t^1| + \varepsilon_1 \int_{t^1}^{t^2} \psi(t)\, dt \quad ,$$

i.e., the functions $y_j(t)$, $j = 1, \ldots, n$, are equiabsolutely continuous in t (with respect to c and ε). In the autonomous case, we have $\psi(t) = K$, and

$$|y_j'(t)| \leq MR + \varepsilon_1 r_1^{-1}KR + \varepsilon_1 K \quad , \tag{4.2}$$

i.e., the functions $y_j(t)$, $j = 1, \ldots, n$, are uniformly Lipschitzian in t (with respect to c, ε, and ω).

We could now prove a number of theorems stating that, under condition (L), $y(t, a, b, c, \omega, \varepsilon)$, $D(a, b, c, \omega, \varepsilon)$ are continuous functions of c, ε, and (in the autonomous case) of ω, and that further smoothness conditions on the q_j imply analogous properties for y and D. We will prove only the following theorem of a mixed type (which cannot be deduced from general theorems. Cfr. J. Schauder and J. Leray, loc. cit. in §5).

(4.i). Suppose hypotheses (α) and (L) are satisfied, that $a = (a_1, \ldots, a_\nu)$, $b = (b_1, \ldots b_\nu)$, $0 < r_1 < r_2 < R$, $0 \leq \varepsilon_1 \leq \varepsilon_0$, and Ω_0 are given as in (3.i), and let $V \geq 2\pi b_0/\omega_{01}$, $b_0 = b_1 \cdots b_\nu$ be any number. Then, there are numbers ε_2, M_2, $0 < \varepsilon_2 \leq \varepsilon_1$, $M_2 > 0$, and, given $\zeta > 0$ arbitrary, there is another number $\xi > 0$ such that, for all $c^s = (c_{js},\ j = 1, \ldots, \nu)$, $r_1 \leq |c_{js}| \leq r_2$, $j = 1, \ldots, \nu$, $s = 1, 2$, for all ω^1, ω^2, $\omega_{01} \leq \omega^1, \omega^2 \leq \omega_{02}$, for all ε^1, ε^2, $0 \leq \varepsilon^1, \varepsilon^2 \leq \varepsilon_2$, $|\varepsilon^2 - \varepsilon^1| \leq \xi$, we have

$$\left.\begin{array}{l} |y_j(t, a, b, c^1, \omega^1, \varepsilon^1) - y_j(t, a, b, c^2, \omega^2, \varepsilon^2)| \\ |d_h(a, b, c^1, \omega^1, \varepsilon^1) \quad - d_h(a, b, c^2, \omega^2, \varepsilon^2)| \end{array}\right\} \leq$$

$$\leq M_2\ \|c^1 - c^2\| + |\omega^1 - \omega^2| + \zeta \quad ,$$

$j = 1, \ldots, n$, $h = 1, \ldots, \nu$, $0 \leq t \leq V$. Note that for $\omega^1 = \omega^2$, the functions y_j above are all periodic of the same period $2\pi b_0/\omega' \leq V$, and hence, the relations above hold for all t, $-\infty < t < +\infty$. This occurs, in particular, in the periodic case, since then $\omega^1 = \omega^2 = \omega_0$.

PROOF. Take $\zeta' = \zeta/N$, where N is a constant which will be determined later, and let $\xi > 0$ be the number defined in (K) in relation to ζ'. We may assume $\xi \leq \zeta'$. Also, we may assume ξ so small that $|\rho_j(\varepsilon^1) - \rho_j(\varepsilon^2)| \leq \zeta'$, $j = 1, \ldots, n$, for all $0 \leq \varepsilon^1, \varepsilon^2 \leq \varepsilon_1$, $|\varepsilon^1 - \varepsilon^2| \leq \xi$.

We may suppose $\omega_{01} \leq \omega^1 \leq \omega^2 \leq \omega_{02}$. Take $V_1 = 2\pi b_0/\omega_{02}$, $V_2 = 2\pi b_0/\omega_{01}$, $V_1 \leq V_2 \leq V$. Since $\omega_{01} \leq \omega \leq \omega_{02}$ and $T = 2\pi b_0/\omega$, we have $V_1 \leq T \leq V_2 \leq V$. Note that $|q_j| \leq \psi(t)$, $j = 1, \ldots, n$, and that we took K in such a way that

$$\int_0^T \psi(t)\, dt \leq KT \quad .$$

In the autonomous case $\psi(t) \equiv K$ and this requirement is certainly satisfied. We shall denote by y^s, D^s, y_{js}, d_{js}, τ_{js}, ρ_{js}, T_s, etc., $s = 1, 2$, the values of y, D, y_j, d_j, τ_j, ρ_j, T for $c = c^1$, $\omega = \omega^1$, $\varepsilon = \varepsilon^1$, and $c = c^2$, $\omega = \omega^2$, $\varepsilon = \varepsilon^2$. We shall omit to indicate all the parameters which are not essential. Note that $V_1 \leq T_2 \leq T_1 \leq V_2 \leq V$.

For $j = 1, \ldots, \nu$, we have by (2.11)

$$d_{j2} - d_{j1} = c_{j2}^{-1} T_2^{-1} \int_0^{T_2} e^{-i\tau_{j2} t} q_{j2}[y^2(t)]dt - c_{j1}^{-1} T_1^{-1} \int_0^{T_1} e^{-i\tau_{j1} t} q_{j1}[y^1(t)]\, dt \; ,$$

where $q_{js}[y^s(t)]$ denotes $q_j[y(t, c^s, \omega^s, \varepsilon^s), t, \varepsilon^s]$, $s = 1, 2$. By manipulation, by introducing the new variable of integration $u = \omega_1 t/\omega_2$ $'\omega_2$ in the second integral, and changing u into t we have

$$d_{j2} - d_{j1} = (c_{j2}^{-1} - c_{j1}^{-1}) T_2^{-1} \int_0^{T_2} e^{i\tau_{j2} t} q_{j2}[y^2(t)]\, dt \; +$$

$$(4.3) \qquad + \, c_{j1}^{-1} T_2^{-1} \int_0^{T_2} e^{-i\tau_{j2} t} \left\{ q_{j2}[y^2(t)] - q_{j1}[y^1(\omega^2 t/\omega^1)] \right\} dt \; .$$

Note that in the periodic case $T_1 = T_2$, $\tau_{j1} = \tau_{j2}$, $\omega_1 = \omega_2$, and thus the expression in $\{\ \}$ actually is

$$q_j[y(t, c^2, \omega, \varepsilon^2), t, \varepsilon^2] - q_j[y(t, c^1, \omega, \varepsilon^1), t, \varepsilon^1] \quad , \tag{4.4}$$

while in the autonomous case it is

$$q_j[y(t, c^2, \omega^2, \varepsilon^2), \varepsilon^2] - q_j[y(\omega^2 t/\omega^1, c^1, \omega^1, \varepsilon^1), \varepsilon^1] \; . \tag{4.5}$$

We shall denote by Δ the number $\Delta = \max |y_{j2}(t) - y_{j1}(t)|$ for all $0 \leq t \leq V$, $j = 1, \ldots, n$. For every $0 \leq t \leq V$ we have $0 \leq \omega^2 t/\omega^1 \leq \omega^2 V/\omega^1$ and hence $|t - \omega^2 t/\omega^1| \leq \omega_{01}^{-1} V \, |\omega^2 - \omega^1|$. Because of (4.2) [only in the autonomous case we may take $\omega^1 \neq \omega^2$], we have

$$\begin{aligned} &|y_j(t, c^2, \omega^2, \varepsilon^2) - y_j(\omega^2 t/\omega^1, c^1, \omega^1, \varepsilon^1)| \leq \\ &\leq |y_j(t, c^2, \omega^2, \varepsilon^2) - y_j(t, c^1, \omega^1, \varepsilon^1)| + \\ &+ |y_j(t, c^1, \omega^1, \varepsilon^1) - y_j(\omega^2 t/\omega^1, c^1, \omega^1, \varepsilon^1)| \leq \\ &\leq \Delta + \omega_{01}^{-1} V |\omega^2 - \omega^1| \, [MR + \varepsilon_1 K(1 + r_1^{-1} R)] = \\ &= \Delta + P|\omega^2 - \omega^1|, \quad 0 \leq t \leq V \quad , \end{aligned} \tag{4.6}$$

where P is a constant.

By (L) we can now evaluate the expression in { } of (4.3), that is, either (4.4) or (4.5):

$$\begin{aligned} |q_{j2}[y^2(t)] - q_{j1}[y^1(t)]| &\leq \left[\sum_{\ell=1}^{n} |y_{\ell 2}(t) - y_{\ell 1}(t)| + \zeta' \right] \psi(t) \\ &\leq [n\Delta + nP \, |\omega^2 - \omega^1| + \zeta'] \, \psi(t) \end{aligned} \tag{4.7}$$

By (4.3) we have then

$$|d_{j2} - d_{j1}| \leq r_1^{-1} K (r_1^{-1} \|c^2 - c^1\| + n\Delta + nP \, |\omega^2 - \omega^1| + \zeta') \quad . \tag{4.8}$$

For $j = 1, \ldots, \nu$, by (2.9) we have

$$\begin{aligned} y_{j2}(t) - y_{j1}(t) &= c_{j2} e^{i\tau_{j2} t} - c_{j1} e^{i\tau_{j1} t} + \\ &+ \varepsilon^2 e^{i\tau_{j2} t} \int e^{-i\tau_{j2} u} \{q_{j2}[y^2(u)] - d_{j2} y_{j2}(u)\} \, du - \\ &- \varepsilon^1 e^{i\tau_{j1} t} \int e^{-i\tau_{j1} u} \{q_{j1}[y^1(u)] - d_{j1} \bar{y}_{j1}(u)\} \, du \quad . \end{aligned}$$

By manipulation, by introducing the variable $v = \omega^1 u/\omega^2$ in the second integral and replacing v by u, we have

$$y_{j2}(t) - y_{j1}(t) = (c_{j2} - c_{j1})e^{i\tau_{j1}t} + (1 - e^{i(\tau_{j1}-\tau_{j2})t})y_{j2}(t) +$$

$$+ (\varepsilon^2 - \varepsilon^1)e^{i\tau_{j1}t}\int e^{-i\tau_{j2}u}\left\{q_{j2}[y^2(u)] - d_{j2}y_{j2}(u)\right\} du +$$

$$+ \varepsilon^1 e^{i\tau_{j1}t}\int e^{-i\tau_{j2}u}\left\{q_{j2}[y^2(u)] - \frac{\omega^2}{\omega^1} q_{j1}\left[y^1(\frac{\omega^2}{\omega^1} u)\right] - d_{j2}y_{j2}(u) + \right.$$

$$+ \frac{\omega^2}{\omega^1} d_{j1}y_{j1}(\frac{\omega^2}{\omega^1} u)\Big\} du + \varepsilon^1 \frac{\omega^2}{\omega^1} e^{i\tau_{j1}t}\int_{u=\omega^1 t/\omega^2}^{u=t} e^{-i\tau_{j2}u}\left\{q_{j1}\left[y^1 (\frac{\omega^2}{\omega^1} u)\right] - \right.$$

$$(4.9) \qquad - d_{j1}y_{j1}(\frac{\omega^2}{\omega^1} u)\Big\} du = s_1 + \dots + s_5 .$$

For $0 \leq t \leq V$ we have $|s_1| \leq \|c^2 - c^1\|$, and

$$|s_2| = \left|y_{j2}(t)\int_{s=\tau_{j1}-\tau_{j2}}^{s=0} (it)e^{ist}ds\right| \leq RV|\tau_{j1} - \tau_{j2}| \leq RVQ|\omega^2 - \omega^1| ,$$

since $\tau_{js} = ia_j\omega^s/b_j$, $s = 1, 2$, and we take $Q = \max a_j/b_j$, $j = 1, \dots, \nu$. By $|\varepsilon^2 - \varepsilon^1| \leq \zeta \leq \zeta'$, and by (1.ii) taking $N = 2$, and by (3.2), we have

$$|s_3| \leq \xi \cdot 2\int_0^{T_2} |q_{j2}[y^2(u)] - d_{j2}y_{j2}(u)| \leq 2KT_2(1 + r_1^{-1}R)\zeta' .$$

In s_4 the integrand is periodic of period T_2, of class C_{ω^2} and mean value zero. By (1.ii), taking $N = 2$, we have

$$|s_4| \leq 2\varepsilon^1\int_0^{T_2}\left\{q_{j2}[y^2(u)] - (\omega^2/\omega^1)q_{j1}y^1(\omega^2 u/\omega^1) + \right.$$

$$+ |d_{j2} - (\omega^2/\omega^1)d_{j1}|\,|y_{j2}(u) + (\omega^2/\omega^1)|d_{j1}|\,|y_{j2}(u) - y_{j1}(\omega^2 u/\omega^1)|\Big\} du ,$$

and, by (4.6), (4.7), (4.8), also

$$|q_{j2}[y^2(u)] - (\omega^2/\omega^1)q_{j1}y^1(\omega^2 u/\omega^1)|$$

$$\leq |q_{j2}[y^2(u)] - q_{j1}[y^1(\omega^2 u/\omega^1)]| + |1 - \omega^2/\omega^1|\ |q_{j1}[y^1(\omega^2 u/\omega^1)]|$$

$$\leq [n\Delta + nP|\omega^2 - \omega^1| + \zeta']\psi(t) + \omega_{01}^{-1}\ |\omega^2 - \omega^1|\psi(t),\ j = 1, \ldots, n\ ;$$

$$|d_{j2} - (\omega^2/\omega^1)d_{j1}| \leq r_1^{-1}K(r_1^{-1}\|c^2 - c^1\| + n\Delta + nP|\omega^2 - \omega^1| + \zeta') + \omega_{01}^{-1}r_1^{-1}K|\omega^2 - \omega^1|,$$

$$|(\omega^2/\omega^1)|d_{j1}|\ |y_{j2}(u) - y_{j1}(\omega^2 u/\omega^1)| \leq \omega_{01}^{-1}\omega_{02}r_1^{-1}K[\Delta + P|\omega^2 - \omega^1|],$$

$$j = 1, \ldots, \nu .$$

Thus

$$|s_4| \leq \varepsilon^1 P'\Delta + P''|\omega^2 - \omega^1| + P'''\ \|c^2 - c^1\| + P^{iv}\zeta' ,$$

for convenient constants P', P'', P''', P^{iv}. Finally, s_5 may be not zero only in the autonomous case, and

$$|s_5| \leq \varepsilon^1\omega_{01}^{-1}\omega_{02}|t - \omega^1 t/\omega^2|\ (K + r_1^{-1}KR) .$$

By (4.9) and the present evaluations, we have

$$|y_{j2}(t) - y_{j1}(t)| \leq |s_1| + |s_2| + |s_3| + |s_4| + |s_5|$$

$$\leq \varepsilon^1 Q'\Delta + Q''|\omega^2 - \omega^1| + Q'''\|c^2 - c^1\| + Q^{iv}\zeta' ,\tag{4.10}$$

$$j = 1, \ldots, \nu ,$$

for convenient constants Q.

For $j = \nu + 1, \ldots, n$, by (2.10) we have

$$y_{j2}(t)-y_{j1}(t) = \varepsilon^2 e^{\rho_{j2}t}\int e^{-\rho_{j2}u}q_{j2}[y^2(u)]du - \varepsilon^1 e^{\rho_{j1}t}\int e^{-\rho_{j1}u}q_{j1}[y^1(u)]du =$$

$$= (\varepsilon^2-\varepsilon^1)e^{\rho_{j2}t}\int e^{-\rho_{j2}u}q_{j2}[y^2(u)]du + \varepsilon^1(e^{\rho_{j2}t}-e^{\rho_{j1}t})\int e^{-\rho_{j2}u}q_{j2}[y^2(u)]\,du +$$

$$+ \varepsilon^1 e^{\rho_{j1}t}\left\{\int e^{-\rho_{j2}u}q_{j2}[y^2(u)]\,du - \int e^{-\rho_{j1}u}q_{j2}[y^2(u)]\,du\right\} +$$

$$+ \varepsilon^1 e^{\rho_{j1}t}\int e^{-\rho_{j1}u}\left\{q_{j2}[y^2(u)] - q_{j1}[y^1(u)]\right\}du = \sigma_1 + \ldots + \sigma_4 .$$

For $0 \leq t \leq V$, by (1.iii), (4.7), and for convenient constants N, $H = e^{\gamma V}$, $\gamma = \max|\Re \rho_j(\varepsilon)|$, $0 \leq \varepsilon \leq \varepsilon_1$, $j = 1, \ldots, n$, we have

$$|\sigma_1| \leq \xi HNKT,$$

$$|\sigma_4| \leq \varepsilon^1 HNKT\ [n\Delta + nP\ |\omega^2 - \omega^1| + \zeta']\ ,$$

$$|\sigma_2| \leq \varepsilon^1 NKT\ |e^{\rho_{j2}t} - e^{\rho_{j1}t}| \leq \varepsilon^1 HNKTV\ |\rho_{j2} - \rho_{j1}|$$

$$\leq \varepsilon^1 HNKTV\zeta'\ .$$

Finally, by (1.iv), we have

$$|\sigma_3| \leq \varepsilon^1 HC|\rho_{2j} - \rho_{1j}| \int_0^T |q_{j2}[y^2(u)]|\ du \leq \varepsilon^1 HCKT\zeta'\ .$$

Thus, for $0 \leq t \leq V$, we have

$$(4.11) \quad |y_{j2}(t) - y_{j1}(t)| \leq \varepsilon^1 N'\Delta + N''\ |\omega^2 - \omega^1| + N'''\zeta',\ j = \nu + 1, \ldots, n,$$

for convenient constants N', N'', N'''.

By (4.10) and (4.11) we have now

$$|\ y_{j2}(t) - y_{j1}(t)| \leq \varepsilon^1 M'\Delta + M''\ |\omega^2 - \omega^1| + M'''\ \|c^2 - c^1\| + M^{iv}\zeta'\ ,$$

for all $j = 1, \ldots, n$, $0 \leq t \leq V$, and hence, if we take $\varepsilon_2 = \min[\varepsilon_1, 1/2M']$, for $0 \leq \varepsilon^1, \varepsilon^2 \leq \varepsilon_2$, we have

$$\Delta \leq (1/2)\Delta + M''\ |\omega^2 - \omega^1| + M'''\ \|c^2 - c^1\| + M^{iv}\zeta'$$

and finally

$$\Delta \leq 2M''\ |\omega^2 - \omega^1| + 2M'''\ \|c^2 - c^1\| + 2M^{iv}\zeta'\ .$$

If we suppose $N = 2M^{iv}$ and $M = \max\ [2M'', 2M''']$, we have $2M^{iv}\zeta' = N\zeta' = \zeta$ and

$$(4.12) \quad |y_{j2}(t) - y_{j1}(t)| \leq \Delta \leq M[|\omega^2 - \omega^1| + \|c^2 - c^1\|] + \zeta$$

for all $0 \leq t \leq V$, $j = 1, \ldots, n$. This proves (4.i).

(c). Further properties of y in the autonomous case and $\nu = 2 \leq n$. The theorems (3.i), (4.i) admit of a stronger form under additional hypotheses. We consider here only the autonomous case with $\nu = 2$, $-a_2 = a_1 = a = 1$, $b_1 = b_2 = b = 1$, $-c_2 = c_1 = c > 0$, $-\tau_2 = \tau_1 = \tau = a\omega/b$. We need the following additional hypothesis:

$$(K') \qquad q_j(0, 0, x_3, \ldots, x_n, \varepsilon) = 0, \qquad j = 1, 2 .$$

The following statement is stronger than (3.i):

(4.ii). In the situation above and under hypotheses (α), (L), (K'), given a, b, and any r_2, $0 < r_2 < R$, there is an ε_1, $0 < \varepsilon_1 \leq \varepsilon_0$, such that for every c, $0 < c \leq r_2$ and $0 \leq \varepsilon \leq \varepsilon_1$, $\mathfrak{T} \mid \Omega_0$ is a contraction and into, where Ω_0 is a convenient closed subset of Ω.

PROOF. The proof is the same as for (3.i) with the following modifications. First, by virtue of (L) and (K'), we have

$$|q_j(x, \varepsilon) - 0| \leq K[|x_1| + |x_2|], \qquad j = 1, 2 .$$

For $r_2/3 \leq c \leq r_2$ Theorem (3.i) holds. Suppose $0 < c \leq r_2/3$, and define Ω_0 as the closed subset of Ω with $|\varphi_j(t) - z_j(t)| \leq 2c$, $j = 1, 2$, $|\varphi_j(t)| \leq r_2$, $j = 3, \ldots, n$, where $z(t) = (ce^{i\tau t}, -ce^{-i\tau t}, 0, \ldots, 0)$. Thus $|\varphi_j(t)| \leq |z_j(t)| + 2c \leq 3c \leq r_2 < R$, $j = 1, 2$. Instead of (3.2) we have now

$$|d_j| \leq c^{-1}T^{-1} \int_0^T |q_j[\varphi(u), \varepsilon]| \, du$$

$$\leq c^{-1}T^{-1}K \int_0^T [|\varphi_1(u)| + |\varphi_2(u)|] \, du \leq 6K, \qquad j = 1, 2 .$$

Relation (3.3) is now replaced by

$$|\psi_j(t) - z_j(t)| \leq 2\varepsilon[6KTc + 6KT3c] = 48\varepsilon KTc, \qquad j = 1, 2 ,$$

while relation (3.4) remains the same

$$|\psi_j(t)| \leq \varepsilon HNKT, \qquad j = 3, \ldots, n .$$

If $M = \max\,[24KT, HNKT]$, $\varepsilon_1 = \min\,[\varepsilon_0, 1/M, r_2/M]$, $0 \leq \varepsilon \leq \varepsilon_1$, we have

$$|\psi_j(t) - z_j(t)| \leq 2\varepsilon Mc \leq 2c, \qquad j = 1, 2 \;,$$

$$|\psi_j(t)| \leq \varepsilon M \leq r_2, \qquad j = 3, \ldots, n \;,$$

i.e., $\mathfrak{T}\,\Omega_0 \subset \Omega_0$. For any two $\varphi^s \in \Omega_0$, $s = 1, 2$, we have, instead of (3.5L)

$$c(d_{1j} - d_{2j}) = \mathfrak{M}\left[e^{i\tau_j u}\left\{q_j[\varphi^1(u), \varepsilon] - q_j[\varphi^2(u), \varepsilon]\right\}\right] \;,$$

$$c|d_{1j} - d_{2j}| \leq T^{-1}\int_0^T K\sum_{\ell=1}^{n}|\varphi^1_\ell(u) - \varphi^2_\ell(u)|\,du \leq nK \cdot \mathfrak{N}[\varphi^1 - \varphi^2] \;,$$

$$|(d_{1j} - d_{2j})\varphi_{2j}(u)| \leq 3c\,|d_{1j} - d_{2j}| \leq 3nK \cdot \mathfrak{N}[\varphi^1 - \varphi^2], \quad j = 1, 2 \,.$$

Finally (3.6L) is replaced by

$$|\psi_{1j}(t) - \psi_{2j}(t)| \leq 2\varepsilon[nKT + 3nKT + 3KnT]\,\mathfrak{N}[\varphi^1 - \varphi^2]$$

$$\leq 14\varepsilon nKT\,\mathfrak{N}[\varphi^1 - \varphi^2], \qquad j = 1, 2 \;,$$

while (3.7L) remains the same. For $M_2 = \max[14nKT, nHNKT]$, $\varepsilon_1^1 = \min[\varepsilon_1, 1/2M_2]$, $0 \leq \varepsilon \leq \varepsilon_1^1$, we have

$$\mathfrak{N}[\psi^1 - \psi^2] \leq \varepsilon M_2\,\mathfrak{N}[\varphi^1 - \varphi^2] \leq 2^{-1}\,\mathfrak{N}[\varphi^1 - \varphi^2] \;,$$

and thus $\mathfrak{T} \mid \Omega_0$ is a contraction. Thereby, (4.ii) is proved.

In analogy with (3.12) we have now

$$|y_1(t)|,\ |y_2(t)| \leq 3c, \ |y_j(t)| \leq r_2, \qquad j = 3, \ldots, n \;,$$

$$|y_1(t) - ce^{i\tau t}|,\ |y_2(t) + ce^{-i\tau t}| \leq 2c \;,$$

$$|y_j(t)| \leq \varepsilon HNKT, \qquad j = 3, \ldots, n \,.$$

If we suppose: (K") the requirement (K') holds for all q_j, $j = 1, \ldots, n$, then we have also

$$|y_j(t)| \leq 3HNKnc, \qquad j = 3, \ldots, n \,.$$

The statements of this section become, under hypotheses (α), (L), (K'):

(4.iii). Under hypotheses (α), (L) and (K'), given a, b, r_2, and Ω_0 as in (4.ii), there is an ε_2, $0 < \varepsilon_2 < \varepsilon_1$, and $M > 0$ such that, for all c^s, $0 < c^s \leq r_2$, $s = 1, 2$, $0 \leq \varepsilon \leq \varepsilon_2$, the conclusions of (4.i) hold.

§5. FIRST EXISTENCE THEOREMS FOR REAL SYSTEMS

Consider the real system, of order $N = \mu + n$,

$$\text{(5.1)}\qquad \begin{aligned} &x_j'' + 2\alpha_j x_j' + \sigma_j^2 x_j = \varepsilon f_j(x, x', t, \varepsilon) \qquad && j = 1, \ldots, \mu, \\ &x_j' + \beta_j x_j = \varepsilon f_j(x, x', t, \varepsilon), && j = \mu + 1, \ldots, n, \end{aligned}$$

where $(') = d/dt$, $-\infty < t < +\infty$, $0 \leq \mu \leq n$, $x = (x_1, \ldots, x_n)$, $x' = (x_1', \ldots, x_\mu')$, $\sigma_j > 0$, α_j, β_j real constants, or continuous functions of ε, $0 \leq \varepsilon \leq \varepsilon_0$, and where $f_j(u, v, t, \varepsilon)$, $j = 1, \ldots, n$, are real functions of the n-vector u, of the μ-vector v, of $-\infty < t < +\infty$, and of $\varepsilon \geq 0$.

For the sake of simplicity we shall assume $\alpha_j^2 < \sigma_j^2$, $j = 1, \ldots, \mu$, and put $\gamma_j = (\sigma_j^2 - \alpha_j^2)^{1/2} > 0$, $\rho_{j1}(\varepsilon) = \rho_{j2}(\varepsilon) = -\alpha_j \pm i\gamma_j$, $j = 1, \ldots, \mu$.

We shall order the first μ equations (5.1) in such a way that for some fixed $\omega > 0$ we have $\alpha_j(0) = 0$, $\sigma_j(0) = a_j\omega/b_j$, a_j, $b_j > 0$ integers, $j = 1, \ldots, \nu$, and $\rho_{j1}(0)$, $\rho_{j2}(0) \neq im\omega/b_0$, $j = \nu + 1, \ldots, \mu$ $m = 0, \pm 1, \pm 2, \ldots$. The latter can be expressed by saying that either $\alpha_j(0) \neq 0$, or $\alpha_j(0) = 0$ and $\sigma_j \neq m\omega/b_0$, $b_0 = b_1 \cdots b_\nu$, $m = 0, 1, 2, \ldots$, $(j = \nu + 1, \ldots, \mu)$. We shall order the last $n - \mu$ equations (5.1) in such a way that $\beta_j(0) \neq 0$ for $j = \mu + 1, \ldots, r$, and $\beta_j(0) = 0$ for $j = r + 1, \ldots, n$, with $\mu \leq r \leq n$.

Thus we have $0 \leq \nu \leq \mu \leq r \leq n$.

We shall finally assume that the functions f_j, $j = 1, \ldots, n$, satisfy condition (L) (with respect to the $n + \mu$ vector $(x_1, \ldots, x_n, x_1', \ldots, x_\mu')$).

Let us introduce in (5.1) the new variables $y_1, \ldots, y_N$, $N = n + \mu$, by the usual relations

$$y_{2j-1} = -\rho_{j2}x_j + x_j', \quad y_{2j} = \rho_{j1}x_j - x_j', \qquad j = 1, \ldots, \mu ,$$

(5.2)

$$y_{\mu+j} = x_j , \qquad j = \mu + 1, \ldots, n,$$

and hence

$$(5.3) \quad x_j = (2i\gamma_j)^{-1}(y_{2j-1} + y_{2j}), \quad x_j' = (2i\gamma_j)^{-1}(\rho_{j1}y_{2j-1} + \rho_{j2}y_{2j}), \qquad j = 1, \ldots, \mu .$$

Then system (5.1) is transformed into the system of $N = \mu + n$ first order differential equations

$$(5.4) \qquad y' = Ay + \varepsilon q(y, t, \varepsilon) ,$$

where $y = (y_1, \ldots, y_N)$, $q = (q_1, \ldots, q_N)$, A is the diagonal matrix of the N elements

$$A = \text{diag}\ (\rho_{j1}, \rho_{j2}, j = 1, \ldots, \mu, -\beta_j, j = \mu + 1, \ldots, n) ,$$

and

$$q_{2j-1} = f_j, \ q_{2j} = -f_j, \ j = 1, \ldots, \mu, \ q_{\mu+j} = f_j, \ j = \mu + 1, \ldots, n ,$$

where the $N + 2$ arguments of the q's are, in order, $y_1, \ldots, y_N; t, \varepsilon$, and the $N + 2$ arguments of the f's are, in order

$$(2i\gamma_j)^{-1}(y_{2j-1} + y_{2j}), \ j = 1, \ldots, \mu; \ y_{\mu+j}, \ j = \mu + 1, \ldots, n ;$$

$$(2i\gamma_j)^{-1}(\rho_{j1}y_{2j-1} + \rho_{j2}y_{2j}), \ j = 1, \ldots, \mu; \ t, \varepsilon .$$

System (5.4) is of the type considered in §2, where the first 2ν equations and the last $n - r$ equations replace the first ν equations of §2; hence $2\nu + (n - r)$ replaces ν, $N = n + \mu$ replaces n, and $n + \mu - [2\nu + (n - r)] = \mu + r - 2\nu$ replaces $n - \nu$. The numbers $\rho_j(0) = i\tau_j$, $j = 1, \ldots, \nu$, of §2 are now replaced by the following $2\nu + (n - r)$ numbers

$$i\tau_1, -i\tau_1, \ldots, i\tau_\nu, -i\tau_\nu, 0, \ldots, 0 ,$$

(the zero repeated $n - r$ times), where $\tau_j = a_j\omega/b_j$, $j = 1, \ldots, \nu$. In other words, for the numbers a_j, b_j of §2, say a_j', b_j', we have now the integers $a'_{2j-1} = a_j$, $a'_{2j} = -a_j$, $b'_{2j-1} = b'_{2j} = b_j$, $j = 1, \ldots, \nu$, $a'_j = 0$, $b'_j = 1$, $j = r+1, \ldots, n$. Let us take for the ν numbers c of §2 the following $2\nu + (n-r)$ numbers

$$c'_{2j-1} = c_j,\ c'_{2j} = -\bar{c}_j,\ c_j \neq 0, \text{ complex},\ j = 1, \ldots, \nu \ ,$$

$$c_j \neq 0,\ c_j \text{ real},\ j = \mu + r + 1, \ldots, \mu + n \ .$$

Then the vector $z(t)$ of §2 is now the N-vector

$$z(t) = (c_j e^{i\tau_j t},\ -\bar{c}_j e^{-i\tau_j t} \ ,$$

$$j = 1, \ldots, \nu;\ 0, \ldots, 0;\ 0, \ldots, 0;\ c_j,\ j = \mu + r + 1, \ldots, \mu + n),$$

where the zero is repeated $(2\mu - 2\nu) + (r - \mu)$ times. In the present situation we may consider the space Ω, analogous to the one of §2, of continuous periodic N-vector functions $\varphi(t) = (\varphi_1, \ldots, \varphi_N)$ of period $T = 2\pi b_0/\omega$, $b_0 = b_1 \cdots b_\nu$, with $\varphi_{2j-1} = -\bar{\varphi}_{2j}$, $j = 1, \ldots, \mu$, φ_j real, $j = j = 2\mu + 1, \ldots, \mu + n$, and

$$\mathfrak{M}\left\{e^{-i\tau_j t}\varphi_{2j-1}(t)\right\} = c_j,\ \mathfrak{M}\left\{e^{i\tau_j t}\varphi_{2j}(t)\right\} = -\bar{c}_j, \qquad j = 1, \ldots, \nu \ ,$$

$$m\{\varphi_j(t)\} = c_j, \qquad j = \mu + r + 1, \ldots, \mu + n \ .$$

Then $z(t)\ \gamma\ \Omega$, and we may note that, for every $\varphi\ \gamma\ \Omega$, the N expressions $(2i\gamma_j)^{-1}(\varphi_{2j-1} + \varphi_{2j})$, $(2i\gamma_j)^{-1}(\rho_{j1}\varphi_{2j-1} + \rho_{j2}\varphi_{2j})$, $j = 1, \ldots, \mu$; φ_j, $j = 2\mu + 1, \ldots, \mu + n$, are all real, and thus the functions $q_j[\varphi(u), u, \varepsilon]$, $j = 1, \ldots, N$, are defined, provided the absolute values of the same expressions are all below some prescribed constant $R' > 0$. This certainly occurs if $|\varphi_j| \leq R$, $j = 1, \ldots, N$, for some $R > 0$. The transformation $\mathfrak{T}$ analogous to the one of §2 is then defined for all $\varphi \in \Omega_R$, $\Omega_R = [\varphi \in \Omega,\ |\varphi_j| \leq R,\ j = 1, \ldots, n + \mu]$ and $\mathfrak{T}\,\Omega_R \subset \Omega$ as in §2. As in §2, there is a sphere Ω_0 about $z(t)$ in Ω with $\mathfrak{T}(\Omega_0) \subset \Omega_0$. The transformation $\psi = \mathfrak{T}\varphi$ is defined by the following relations:

$$\psi_{2j-1} = c_j e^{i\tau_j t} + \varepsilon e^{i\tau_j t} \int e^{-i\tau_j u}\left\{q_{2j-1}[\varphi(u), u, \varepsilon] - d_{2j-1}\varphi_{2j-1}(u)\right\} du \ ,$$

(5.5)

$$\psi_{2j} = -\bar{c}_j e^{-i\tau_j t} + \varepsilon e^{-i\tau_j t} \int e^{i\tau_j u}\left\{q_{2j}[\varphi(u), u, \varepsilon] - d_{2j}\varphi_{2j}(u)\right\} du \ ,$$

$$j = 1, \ldots, \nu \ ,$$

$$\psi_{2j-1} = \varepsilon e^{\rho_{j1}t} \int e^{-\rho_{j1}u} q_{2j-1}[\varphi(u), u, \varepsilon]\, du \quad ,$$

(5.6)

$$\psi_{2j} = \varepsilon e^{\rho_{j2}t} \int e^{-\rho_{j2}u} q_{2j}[\varphi(u), u, \varepsilon]\, du, \qquad j = \nu + 1, \ldots, \mu ,$$

$$(5.7) \quad \psi_j = e^{-\beta_j t} \int e^{\beta_j u} q_j[\varphi(u), u, \varepsilon]\, du, \qquad j = 2\mu + 1, \ldots, \mu + r ,$$

$$(5.8) \quad \psi_j = c_j + \varepsilon \int \left\{ q_j[\varphi(u), u, \varepsilon] - d_j \varphi_j(u) \right\} du \ , \ j = \mu + r + 1, \ldots, \mu + n,$$

where $D = D[\varphi] = (d_1, \ldots, d_{2\nu}, 0, \ldots, 0, d_{\mu+r+1}, \ldots, d_{\mu+n})$, and

$$c_j d_{2j-1} = \mathfrak{M} \left\{ e^{-i\tau_j u} q_{2j-1}[\varphi(u), u, \varepsilon] \right\} ,$$

$$(5.9) \quad -\bar{c}_j d_{2j} = \mathfrak{M} \left\{ e^{i\tau_j u} q_{2j}[\varphi(u), u, \varepsilon] \right\}, \qquad j = 1, \ldots, \nu ,$$

$$c_j d_j = \mathfrak{M}\, q_j[\varphi(u), u, \varepsilon] \ , \qquad j = \mu + r + 1, \ldots, \mu + n .$$

(5.i). LEMMA. For every $\varphi \in \Omega_R$ we have $\psi \in \Omega$, and $d_{2j-1} = \bar{d}_{2j}$, $j = 1, \ldots, \nu$, d_j real, $j = \mu + r + 1, \ldots, \mu + n$.

PROOF. We have $q_{2j-1} = f_j$, $q_{2j} = - f_j$ real, $j = 1, \ldots, \nu$, and hence $q_{2j-1} = - q_{2j}$, while $e^{i\tau_j u}$, $e^{-i\tau_j u}$ are complex conjugate. By (5.9) we conclude that $d_{2j-1} = \bar{d}_{2j}$, and then the integrands in (5.5) are complex conjugate and of opposite signs. Thus, also their unique primitives of mean value zero have the same property, and (5.5) implies $\psi_{2j-1} = - \bar{\psi}_{2j}$, $j = 1, \ldots, \nu$.

Since $\rho_{j2} = \bar{\rho}_{j1}$, $j = \nu + 1, \ldots, \mu$, we can repeat on relations (5.6) the same reasoning above and thus $\psi_{2j-1} = - \bar{\psi}_{2j}$, $j = \nu + 1, \ldots, \mu$. Since β_j is real, q_j is real, $j = 2\mu + 1, \ldots, \mu + r$, by (5.7) we conclude that ψ_j is real, $j = 2\mu + 1, \ldots, \mu + r$. Finally, for $j = \mu + r + 1, \ldots, \mu + n$, relations (5.9) and (5.8) assure that d_j and ψ_j are real, $j = \mu + r + 1, \ldots, \mu + n$. Thereby, (5.i) is proved.

By (5.i) we conclude that $\mathfrak{T}\,\Omega_R \subset \Omega$ and, as in §§2, 3, there is a sphere Ω_0 about $z(t)$ in Ω_R with $\mathfrak{T}\,\Omega_0 \subset \Omega_0$, for $\varepsilon > 0$ sufficiently small. A fixed element $y(t) = (y_1, \ldots, y_N)$ exists in Ω_0, $\mathfrak{T}\, y = y$, and

$D[y]$ verifies the same relations of (5.i).

We shall consider now the determining equations of §§1, 2, 3, which are in number of $2\nu + (n - r)$, namely

$$\rho_{j1} = i\tau_j - \varepsilon d_{2j-1}, \quad \rho_{j2}(\varepsilon) = - i\tau_j - \varepsilon d_{2j}, \qquad j = 1, \dots, \nu ,$$

$$\beta_j = \varepsilon d_{\mu+j} , \qquad j = r + 1, \dots, n .$$

By the remark above we conclude that for every $j = 1, \dots, \nu$, the two equations above are equivalent, and thus we have actually only $\nu + (n - r)$ determinihg equations

$$\text{(5.10)} \qquad \begin{aligned} \rho_{j1} &= i\tau_j - \varepsilon d_{2j-1} , && j = 1, \dots, \nu , \\ \beta_j &= \varepsilon d_{\mu+j} , && j = r + 1, \dots, n . \end{aligned}$$

We shall write $c_j = \lambda_j e^{i\theta_j}$, $j = 1, \dots, \nu$, $\eta_j = c_{\mu+r+j}$, $j = 1, \dots, n - r$, $\lambda_j \neq 0$, $\eta_j \neq 0$, θ_j all real, $\theta_j (\text{mod } 2\pi)$, $\lambda = (\lambda_1, \dots, \lambda_\nu)$, $\theta = (\theta_1, \dots, \theta_\nu)$, $\eta = (\eta_1, \dots, \eta_{n-r})$,

$$d_{2j-1} = P_j + iQ_j, \ P_j, \ Q_j \text{ real}, \qquad j = 1, \dots, \nu ,$$

$$d_{\mu+r+j} = R_j, \ R_j \text{ real} , \qquad j = 1, \dots, n - r ,$$

where $P_j = P_j(a, b, \lambda, \theta, \eta, \omega, \varepsilon)$, $Q_j = Q_j(\dots)$, $R_j = R_j(\dots)$. Then equations (5.10) reduce to the following $2\nu + (n - r)$ real equations

$$\text{(5.11)} \qquad \begin{aligned} \varepsilon P_j &= \alpha_j(\varepsilon) , && j = 1, \dots, \nu , \\ \varepsilon Q_j &= a_j b_j^{-1}\omega - \gamma_j(\varepsilon) , && j = 1, \dots, \nu , \\ \varepsilon R_j &= \beta_{r+j}(\varepsilon) , && j = 1, \dots, n - r, \end{aligned}$$

where $\alpha_j(0) = 0$, $a_j b_j^{-1}\omega - \gamma_j(0) = 0$, $j = 1, \dots, \nu$, $\beta_{r+j}(0) = 0$, $j = 1, \dots, n - r$. Equations (5.11) are the $2\nu + (n - r)$ determining equations in real form. By (5.9) we have

$$P_j + iQ_j = \lambda_j^{-1} e^{-i\theta_j} \mathfrak{M}\left\{e^{-i\tau_j u} q_{2j-1}\right\} , \quad j = 1, \dots, \nu ,$$

$$R_j = \eta_j^{-1} \mathfrak{M}\{q_{\mu+r+j}\} , \qquad j = 1, \dots, n - r ,$$

and, by obvious manipulations, also

$$P_j = \lambda_j^{-1}[\cos\theta_j \mathfrak{M}\{f_j \cos a_j b_j^{-1}\omega u\} - \sin\theta_j \mathfrak{M}\{f_j \sin a_j b_j^{-1}\omega u\}] \quad ,$$

$$Q_j = \lambda_j^{-1}[-\sin\theta_j \mathfrak{M}\{f_j \cos a_j b_j^{-1}\omega u\} - \cos\theta_j \mathfrak{M}\{f_j \sin a_j b_j^{-1}\omega u\}] \; ,$$

$$j = 1, \ldots, \nu \quad ,$$

$$R_j = \eta_j^{-1}\mathfrak{M}\{f_{r+j}\}, \qquad j = 1, \ldots, n - r \quad ,$$

where the $n + \mu + 2$ arguments of f_j are

$$x_j(u) = (2i\gamma_j)^{-1}(y_{2j-1} + y_{2j}), \; j = 1, \ldots, \mu; \; x_j(u) = y_{\mu+j} \; , \quad j = \mu + 1, \ldots, n \; ;$$

(5.12)

$$x_j'(u) = (2i\gamma_j)^{-1}(\rho_{j1}y_{2j-1} + \rho_{j2}y_{2j}), \quad j = 1, \ldots, \mu; \; u, \; \varepsilon \quad .$$

By the continuity, with respect to ε proved in (4.ii), we can take the limit of both conditions (5.11) and solution (5.12) as $\varepsilon \longrightarrow 0$. By using the notation $x_j(t, \varepsilon)$, $P_j(\varepsilon)$, etc., we have

$$x_j(t, 0) = \lambda_j \tau_j^{-1}\sin(\tau_j t + \theta_j), \qquad j = 1, \ldots, \nu \quad ,$$

(5.13) $$x_j(t, 0) = 0, \; j = \nu + 1, \ldots, \mu, \; \text{ and } \; j = \mu + 1, \ldots, r \quad ,$$

$$x_j(t, 0) = \eta_{j-r} \; , \qquad j = r + 1, \ldots, n \quad ,$$

and

$$x_j'(t, 0) = \lambda_j \cos(\tau_j t + \theta_j), \qquad j = 1, \ldots, \nu \quad ,$$

(5.14)

$$x_j'(t, 0) = 0, \qquad j = \nu + 1, \ldots, \mu \quad .$$

Also,

$$P_{jo} = P_j(0) = (T\lambda_j)^{-1}\Big[\cos\theta_j \int_0^T f_j(\ldots, u, 0)\cos a_j b_j^{-1}\omega u \, du - \sin\theta_j \int_0^T f_j (\ldots, u, 0)\sin a_j b_j^{-1}\omega u \, du\Big] \quad ,$$

(5.15)

$$Q_{jo} = Q_j(o) = (T\lambda_j)^{-1}\Big[-\sin\theta_j \int_0^T f_j(\ldots, u, o)\cos a_j b_j^{-1}\omega u \, du - \cos\theta_j \int_0^T f_j (\ldots, u, o)\sin a_j b_j^{-1}\omega u\Big] \, du, \; j = 1, \ldots, \nu,$$

$$R_{jo} = R_j(o) = (T\eta_j)^{-1} \int_0^T f_{r+j}(\ldots, u, o)\, du, \qquad j = 1, \ldots, n - r,$$

where the missing $n + \mu$ arguments in the f's are the ones listed in (5.13) and (5.14). We shall use the notation $\alpha = (\alpha_1, \ldots, \alpha_\nu)$, $\gamma = (\gamma_1, \ldots, \gamma_\nu)$, $\beta = (\beta_{r+1}, \ldots, \beta_n)$. The previous considerations prove the following

(5.ii). THEOREM. If for some a, b, $\lambda(\varepsilon)$, $\theta(\varepsilon)$, $\eta(\varepsilon)$, $\alpha(\varepsilon)$, $\gamma(\varepsilon)$, $\beta(\varepsilon)$, all real and for $\varepsilon > 0$ sufficiently small, equations (5.11) hold, then system (5.1) has a real periodic solution of the form

$$
\begin{aligned}
& x_j(t, \varepsilon) = \lambda_j \tau_j^{-1} \sin(\tau_j t + \theta_j) + O(\varepsilon), && j = 1, \ldots, \nu, \\
(5.16)\qquad & x_j(t, \varepsilon) = O(\varepsilon), && j = \nu + 1, \ldots, r, \\
& x_j(t, \varepsilon) = \eta_{j-r} + O(\varepsilon), && j = r + 1, \ldots, n.
\end{aligned}
$$

In particular, for $\varepsilon \geq 0$ sufficiently small, given a, b, $\lambda(\varepsilon)$, $\theta(\varepsilon)$, $\eta(\varepsilon)$, there are always $\alpha(\varepsilon)$, $\gamma(\varepsilon)$, $\beta(\varepsilon)$ such that equations (5.11) are satisfied and system (5.1) has a real periodic solution of the form (5.16).

We shall now consider the more usual question as to whether given α, γ, β it is possible to define $\lambda = (\lambda_1, \ldots, \lambda_\nu)$, $\theta = (\theta_1, \ldots, \theta_\nu)$, $\eta = (\eta_1, \ldots, \eta_{n-r})$ in such a way that equations (5.11) are satisfied and system (5.1) has a periodic solution of the form (5.16). Let us suppose that the derivatives $\alpha_j'(0)$, $\gamma_j'(0)$, $\beta_{r+j}'(0)$ exist and are finite. By dividing relations (5.11) by ε and taking the limit as $\varepsilon \longrightarrow 0$ we have

$$
\begin{aligned}
& P_{jo}(a, b, \lambda, \theta, \eta, \omega, 0) = \alpha_j'(0), && j = 1, \ldots, \nu, \\
(5.17)\qquad & Q_{jo}(a, b, \lambda, \theta, \eta, \omega, 0) = -\gamma_j'(0), && j = 1, \ldots, \nu, \\
& R_{jo}(a, b, \lambda, \theta, \eta, \omega, 0) = \beta_{r+j}'(0), && j = 1, \ldots, n - r.
\end{aligned}
$$

We shall suppose that these relations are satisfied by some $\lambda_o = (\lambda_{jo}, j = 1, \ldots, \nu)$, $\theta_o = (\theta_{jo}, j = 1, \ldots, \nu)$, $\eta_o = (\eta_{jo}, j = 1, \ldots, n - r)$.

If we suppose that the functions f_j have continuous first partial derivatives with respect to the variables $x_1, \ldots, x_n, x_1', \ldots, x_\mu'$. Then by (§4, Note 1) the functions y_j have continuous partial derivatives with respect to all λ_j, θ_j, η_j, and the same holds for the P_j, Q_j, R_j. If equations (5.17) hold for λ_o, θ_o, η_o, and

$$(5.18) \qquad \partial(P, Q, R)/\partial(\lambda, \theta, \eta) \neq 0$$

for the same λ_o, θ_o, η_o, then equations (5.11) have a solution $\lambda(\varepsilon)$, $\theta(\varepsilon)$, $\eta(\varepsilon)$ in a neighborhood of λ_o, θ_o, η_o for all $\varepsilon \geq 0$ sufficiently small. This proves the following theorem which is essentially of the type usually proved by Poincaré periodicity conditions.

(5.iii). THEOREM. If real λ_o, θ_o, η_o exist such that equations (5.17) and (5.18) hold, then for $\varepsilon \geq 0$ sufficiently small system (5.1) has a real periodic solution of period $T = 2\pi b_o/\omega$ of the type

$$x_j(t, \varepsilon) = \lambda_j\tau_j^{-1}\sin(\tau_j t + \theta_j) + O(\varepsilon), \qquad j = 1, \dots, \nu ,$$

$$(5.19) \qquad x_j(t, \varepsilon) = O(\varepsilon), \qquad j = \nu + 1, \dots, r ,$$

$$x_j(t, \varepsilon) = \eta_{j-r} + O(\varepsilon), \qquad j = r + 1, \dots, n .$$

REMARK. The differentiability condition concerning the functions f_j used above to prove (5.iii) requires more than needed as we shall see below (cf. (5.iv), (5.vi), etc.). The present first remark may be of interest. For instance, let us require that only $f(x, t, 0)$ has continuous first partial derivatives with respect to x (and $f(x, t, \varepsilon)$ is continuous in x and ε). Then the functions P_{jo}, Q_{jo}, R_{jo} have continuous first partial derivatives with respect to λ, θ, η. If the Jacobian of the P_{jo}, Q_{jo}, R_{jo} with respect to λ, θ, η is $\neq 0$ at $(\lambda_o, \theta_o, \eta_o)$, then, for a conveniently small solid sphere σ of center $(\lambda_o, \theta_o, \eta_o)$, the mapping $u_j = P_{jo} - \alpha_j'(0)$, $v_j = Q_{jo} + \gamma_j'(0)$, $w_j = R_{jo} - \beta_{r+j}'(0)$, $(\lambda, \theta, \eta) \in \sigma$, has topological index one with respect to the origin. By force of continuity only, this holds true also for the (perturbed) continuous mapping $U_j = P_j(\lambda, \theta, \eta, \varepsilon) - \varepsilon^{-1}\alpha_j(\varepsilon)$, $V_j = Q_j + \varepsilon^{-1}(\gamma_j(\varepsilon) - a_j b_j^{-1}\omega)$, $W_j = R_j - \varepsilon^{-1}\beta_{r+j}(\varepsilon)$, $(\lambda, \theta, \eta) \in \sigma$, provided $\varepsilon \geq 0$ is sufficiently small. We shall discuss this point in more detail in the papers mentioned in the Introduction. For the present argument, see, e.g., J. Leray and J. Schauder, Topologie et équations functionnelles, Ann. Ec. Norm. (3) 51, 45-78, 1934.

Let us consider now real systems (5.1) where the functions f_j do not depend on t, i.e., autonomous systems,

$$x_j'' + 2\alpha_j x_j' + \sigma_j^2 x_j = \varepsilon f_j(x, x', \varepsilon), \qquad j = 1, \dots, \mu ,$$

$$x_j' + \beta_j x_j = \varepsilon f_j(x, x', \varepsilon), \qquad j = \mu + 1, \dots, n ,$$

satisfying the hypotheses listed at the beginning of §5, where now ω in condition (K) is actually arbitrary. As mentioned in §2 and §4 we shall suppose $\omega \in U$, where U is a sufficiently small neighborhood of any number ω_0 for which $\sigma_j(0) = a_j\omega_0/b_j$, $j = 1, \ldots, \nu$, $\rho_{j1}(0)$, $\rho_{j2}(0) \neq im\omega/b_0$, $j + \nu + 1, \ldots, n$, $m = 0, \pm 1, \ldots,$ and we may expect to satisfy (5.11) by a convenient function $\omega = \omega(\varepsilon)$ with $\omega(0) = \omega_0$. Thus

(5.iv). The same as (5.ii) for system (5.20) with $\omega(\varepsilon)$ added to the list of parameters.

For every solution $x(t)$ of (5.20) also $x(t + \theta)$ is a solution for every arbitrary constant phase θ. Thus we may expect that (at least) one of the phases θ_j, $j = 1, \ldots, \nu$, considered above, say θ_1, remains arbitrary and we put $\theta_1 = 0$. Note that now the $2\nu + (n - r)$ expressions P, Q, R are still functions of $2\nu + (n - r)$ arbitrary parameters, namely λ_j, $j = 1, \ldots, \nu$, θ_j, $j = 2, \ldots, \nu$, η_j, $j = 1, \ldots, n - r$, and $\omega \in U$. We may now denote by θ the $(\nu-1)$-vector $(\theta_2, \ldots, \theta_\nu)$. Again, as above, let us suppose that the functions $f_j(x, x', \varepsilon)$ have continuous first partial derivatives with respect to the variables $x_1, \ldots, x_n, x_1', \ldots, x_\mu'$. If equations (5.17) hold for some $\lambda_0, \theta_0, \eta_0$ and $\omega = \omega_0$, and the Jacobian

$$\partial(P, Q, R)/\partial(\lambda, \theta, \mu, \omega) \neq 0 \tag{5.21}$$

for the same $\lambda_0, \theta_0, \eta_0, \omega_0$, and $\varepsilon = 0$, then equations (5.11) have a solution $\lambda(\varepsilon)$, $\theta(\varepsilon)$, $\eta(\varepsilon)$, $\omega(\varepsilon)$, in some neighborhood of $\lambda_0, \theta_0, \eta_0$ and $\omega \in U$ for all $\varepsilon \geq 0$ sufficiently small. In analogy with (5.iii) we have now the Poincaré type theorem for autonomous systems:

(5.v). THEOREM. If real $\lambda_0, \theta_0, \mu_0, \omega_0$ exist such that equations (5.17), (5.21) hold, then for $\varepsilon \geq 0$ sufficiently small system (5.20) has a real periodic solution of period $T = 2\pi b_0/\omega$ of the type (5.19) with $\theta_1 = 0$ (and then also all other solutions with t replaced by $t + \theta$, θ arbitrary).

We proceed to prove a theorem of a more general type (others will be given in §6). For the sake of simplicity we shall consider a system (5.20) with $\nu = 1$, $r = n$, $\alpha_1 \equiv 0$, i.e., an autonomous system of the form

$$\begin{aligned} &x_1'' + \sigma_1^2 x_1 = \varepsilon f_1(x, x', \varepsilon) , \\ &x_j'' + 2\alpha_j x_j + \sigma_j^2 x_j^2 = \varepsilon f_j(x, x', \varepsilon), \qquad j = 2, \ldots, \mu , \\ &x_j' + \beta_j x_j = \varepsilon f_j(x, x', \varepsilon), \qquad j = \mu + 1, \ldots, n , \end{aligned} \tag{5.22}$$

where $1 = \nu \leq \mu \leq n$, $x = (x_1, \ldots, x_n)$, $x' = (x_1', \ldots, x_\mu')$, $\sigma_j(\varepsilon) > 0$, $\alpha_j(\varepsilon)$, $\beta_j(\varepsilon)$ real functions of ε, $\beta_j(\varepsilon) \neq 0$, (or constants), $0 \leq \varepsilon \leq \varepsilon_0$. Also, we assume that $a = b = 1$, and hence $\sigma_1(0) = \omega_0$, $\rho_{j1}(0)$, $\rho_{j2}(0) \neq im\omega$, $j = 2, \ldots, \mu$, $m = 0, \pm 1, \pm 2, \ldots$. Thus $\tau_1 = \omega_0 = \sigma_1(0)$, and there will be only one c. The functions f_j are assumed to satisfy a condition (L). There is only one equation (5.10), namely, $i\omega - \varepsilon d_1(c, \omega, \varepsilon) = i\sigma_1(\varepsilon)$, or, in real form, two equations (5.11), $P_1 = 0$, $\omega - \varepsilon Q = \sigma_1(\varepsilon)$. Since there is only one phase θ_1 it is not restrictive to take it equal to zero, i.e., $c = \lambda$, $0 < r_1 \leq \lambda \leq r_2 < R$, λ real and positive. Thus the two determining equations are actually

$$P_1(\lambda, \omega, \varepsilon) = 0, \quad \omega - \varepsilon Q_1(\lambda, \omega, \varepsilon) = \sigma_1(\varepsilon) \,. \tag{5.23}$$

Note that

$$P_1(\lambda, \omega, 0) = (T\lambda)^{-1} \int_0^T f_1[\lambda\omega^{-1}\sin \omega t, 0, \ldots, 0, \lambda \cos \omega t, 0, \ldots, 0; 0]\cos \omega t \, dt \,,$$

(5.24)

$$Q_1(\lambda, \omega, 0) = -(T\lambda)^{-1} \int_0^T f_1[\lambda\omega^{-1}\sin \omega t, 0, \ldots, 0, \lambda \cos \omega t, 0, \ldots, 0; 0]\sin \omega t \, dt \,,$$

where $T = 2\pi/\omega$. We shall suppose that for some λ_0 we have $P_1(\lambda, \omega_0, 0) = 0$ and that for two λ', λ'', $r_1 \leq \lambda' < \lambda_0 < \lambda'' \leq r_2$, the two numbers $P_1(\lambda', \omega_0, 0)$, $P_1(\lambda'', \omega_0, 0)$ have opposite signs.

(5.vi). Consider system (5.22) where the functions f_j are independent of t, Lipschitzian in x, x' and continuous in ε, for $|x_j| \leq R$, $j = 1, \ldots, n$, $|x_j'| \leq R$, $j = 1, \ldots, \mu$, $0 \leq \varepsilon \leq \varepsilon_0$. Suppose that σ_j, α_j, β_j are continuous functions of ε [or constants] and that $\sigma_j(0) > 0$, $j = 1, \ldots, \mu$, $\beta_j(0) \neq 0$, $j = \mu + 1, \ldots, n$, and either $\alpha_j(0) \neq 0$, or $\alpha_j(0) = 0$, $\sigma_j(0) \not\equiv 0$, mod $\sigma_1(0)$, $j = 2, \ldots, \mu$. Let $\omega_0 = \sigma_1(0)$, $\sigma \in [\omega_{01}, \omega_{02}]$, $\lambda \in [r_1, r_2]$, with $0 < \omega_{01} < \omega_0 < \omega_{02}$, $0 < r_1 < r_2 < R$. If there exists λ', λ_0, λ'', $r_1 \leq \lambda' < \lambda_0 < \lambda'' \leq r_2$ such that $P_1(\lambda_0, \omega_0, 0) = 0$, and $P_1(\lambda', \omega_0, 0)$, $P_1(\lambda'', \omega_0, 0)$ have opposite signs, then there is an $\varepsilon_1 > 0$ such that, for every ε, $0 \leq \varepsilon \leq \varepsilon_1$, system (5.22) has at least one periodic solution of the form

$$\begin{aligned} x_1(t, \varepsilon) &= \lambda(\varepsilon)\omega^{-1}(\varepsilon)\sin \omega(\varepsilon)t + O(\varepsilon) \,, \\ x_j(t, \varepsilon) &= O(\varepsilon), \qquad j = 2, \ldots, n \,, \end{aligned} \tag{5.25}$$

for conveniently chosen $\omega(\varepsilon) \in [\omega_{01}, \omega_{02}]$, $\lambda(\varepsilon) \in [\lambda', \lambda'']$. System (5.22) has also any other solution we obtain by replacing t by $t + \theta$, θ an arbitrary constant.

PROOF. Let us use statement (1.vi). Take $F_1(\lambda, \omega, \varepsilon) \doteq P_1(\lambda, \omega, \varepsilon)$, where $\omega \in U$, say $U = [\omega', \omega'']$, $\omega' < \omega_0 < \omega''$, $\lambda \in V = [\lambda', \lambda'']$, $\varepsilon \in [0, \varepsilon_0]$. Thus $F_1(\lambda', \omega_0, 0)$, $F_1(\lambda'', \omega_0, 0)$ have opposite signs, and, if we suppose $U = [\omega', \omega'']$ sufficiently small, also $F_1(\lambda', \omega, 0)$, $F_1(\lambda'', \omega, 0)$, $\omega \in [\omega', \omega'']$, have opposite constant signs. Take $F_2(\lambda, \omega, \varepsilon) = \omega - \sigma_1(\varepsilon) - \varepsilon Q_1(\lambda, \omega, \varepsilon)$, and note that $F_2(\omega_0, \lambda, 0) = \omega_0 - \sigma_1(0) = 0$, $F_2(\omega', \lambda, 0) < 0$, $F_2(\omega'', \lambda, 0) > 0$ for all $\lambda \in [\lambda', \lambda'']$. By (1.vi), with $M = [0, \varepsilon_2]$, $K = [\lambda', \lambda''] \times [\omega', \omega'']$, we conclude that there are a neighborhood $[0, \varepsilon_2]$ of $\varepsilon = 0$, and $\lambda(\varepsilon) \in [\lambda', \lambda'']$, $\omega(\varepsilon) \in [\omega', \omega'']$ for every $0 \leq \varepsilon \leq \varepsilon_2$, such that $F_1 = F_2 = 0$. Then (5.vi) is a consequence of (5.iv).

REMARK 1. Theorem (5.vi) is more general than the usual theorems which are proved by Poincaré periodicity condition, since no differentiability property is involved (other theorems will be given in §6).

Note that the condition of (5.vi) is certainly satisfied if $P_1(\lambda, \omega_0, 0)$ happens to have derivative with respect to λ at $\lambda = \lambda_0$ and $\partial P_1/\partial\lambda \neq 0$ at that point.

Finally, under usual conditions where the functions f_j have continuous first partial derivatives with respect to $x_1, \ldots, x_n, x_1', \ldots, x_\mu'$, then F_1, F_2 have also continuous first partial derivatives with respect to λ and ω, and $\partial(F_1, F_2)/\partial(\lambda, \omega) = F_{1\lambda}F_{2\omega} - F_{1\omega}F_{2\lambda}$, where $F_{1\lambda} = P_{1\lambda}$, $F_{2\omega} = 1 + O(\varepsilon)$, $F_{1\omega} = O(1)$, $F_{2\lambda} = O(\varepsilon)$. Thus we have

$$\partial(F_1, F_2)/\partial(\lambda, \omega) = \partial P_1/\partial\lambda + O(\varepsilon) \quad .$$

The condition just mentioned assures that this Jacobian is $\neq 0$ for $\lambda = \lambda_0$, $\omega = \omega_0$, $\varepsilon = 0$, and finally assures the uniqueness and continuity of $\omega(\varepsilon)$, $\lambda(\varepsilon)$ for $0 \leq \varepsilon \leq \varepsilon_1$, and some $\varepsilon_1 > 0$ sufficiently small. Note that, under the same conditions, the inequality

$$\int_0^T f_{1x_1'}[\lambda\omega^{-1}\sin \omega t, 0, \ldots, 0, \lambda \cos \omega t, 0, \ldots, 0; 0]\, dt < 0 \quad ,$$

or the equivalent Krylov-Bogolyubov condition $\partial P_1/\partial\lambda < 0$ [in both $\lambda = \lambda_0$, $\omega = \omega_0$, $\varepsilon = 0$], together with $\alpha_j < 0$, $j = 2, \ldots, \mu$, $\beta_j < 0$, $j = \mu + 1, \ldots, n$, assure that the same periodic solution is asymptotically orbitally stable. [For these and more general stability conditions see J. K. Hale, 9i, and E. W. Thompson, 11].

REMARK 2. The following considerations may help to understand the generality of (5.vi). The real numbers (or functions of ε) $\beta_j(\varepsilon)$, $j = \mu + 1, \ldots, n$, $\alpha_j(\varepsilon)$, $\sigma_j(\varepsilon)$, $j = 2, \ldots, \mu$, have no bearing on $P_1(\lambda, \omega, 0)$. We require for them only $\beta_j \neq 0$, $j = \mu + 1, \ldots, n$, $\sigma_j(0) > 0$, $j = 1, \ldots, n$, and either $\alpha_j(0) \neq 0$, or $\alpha_j(0) = 0$, $\sigma_j(0) \not\equiv 0 (\mathrm{mod}\ \sigma_1(0))$, $j = 2, \ldots, \mu$, $\sigma_1(0) = \omega_0$. The functions $f_j(x, x', \varepsilon)$, $j = 1, \ldots, n$, are supposed to be continuous in x, x', ε, and $f_2, \ldots, f_n$ have no bearing on $P_1(\lambda, \omega, 0)$. If we put

$$Z_1(x_1, x_1') = f_1(x_1, 0, \ldots, 0, x_1', 0, \ldots, 0; 0) ,$$

$$g_1(x, x', \varepsilon) = f_1(x, x', \varepsilon) - Z_1(x_1, x_1') ,$$

then $f_1 = Z_1 + g_1$, $g_1(x_1, 0, \ldots, 0, x_1', 0, \ldots, 0; 0) = 0$ and g_1 has no bearing on $P_1(\lambda, \omega, 0)$. Finally, if we put

$$Z_{11} = 4^{-1}[Z_1(x_1, x_1') + Z_1(-x_1, x_1') - Z_1(x_1, -x_1') - Z_1(-x_1, x_1')] ,$$

and we define Z_{12}, Z_{13}, Z_{14} analogously by changing the signs $(+, -, -)$ into $(+, +, +)$, $(-, -, +)$, $(-, +, -)$, then $Z_{11}[Z_{12}]$ is even in x_1 and odd [even] in x_1', $Z_{13}[Z_{14}]$ is odd in x_1 and odd [even] in x_1'. We have $f_1 = Z_1 + g_1 = Z_{11} + Z_{12} + Z_{13} + Z_{14} + g_1$, and Z_{12}, Z_{13}, Z_{14}, g_1 have no bearing on $P_1(\lambda, \omega, 0)$. We have

$$P_1(\lambda, \omega, 0) = (T\lambda)^{-1} \int_0^T Z_{11}(\lambda\omega^{-1} \sin \omega t, \cos \omega t)\, dt ,$$

where $T = 2\pi/\omega$, and $\omega_0 = \sigma_1(0)$. In the particular case where Z_{11} is a polynomial in x_1, x_1', hence necessarily of the form

$$Z_{11} = \sum a_{hk} x_1^{2h} x_1'^{2k-1} , \qquad h \geq 0,\ k \geq 1 ,$$

then

$$P_1(\lambda, \omega_0, 0) = 2^{-2} \sum a_{hk} \frac{(2h)!(2k)!}{h!k!(h+k)!} \left(\frac{\lambda}{2\sigma_1} \right)^{2h+2k-2}$$

where $\sigma_1 = \sigma_1(0)$. If, for instance, a_{01} and the coefficient of the maximal power of λ are ($\neq 0$ and) of opposite signs, then certainly (5.22) has a periodic solution [cf. J. K. Hale, 9c, for the analytic case].

EXAMPLE 1. The real system (5.22), without restricting its generality, can be written as follows:

$$x_1'' + \sigma_1^2 x_1 = \varepsilon[Z_{11}(x_1, x_1') + Z_{12}(x_1, x_1') + Z_{13}(x_1, x_1') + Z_{14}(x_1, x_1') + g_1(x, x', \varepsilon)] ,$$

(5.24) $$x_j'' + 2\alpha_j x_j' + \sigma_j^2 x_j = \varepsilon f_j(x, x', \varepsilon) , \qquad j = 2, \ldots, \mu ,$$

$$x_j' + \beta_j x_j = \varepsilon f_j(x, x', \varepsilon), \qquad j = \mu + 1, \ldots, n ,$$

where $Z_{11}, \ldots, Z_{14}, g_1, f_2, \ldots, f_n$ are Lipschitzian in x, x', ε where $Z_{11}[Z_{12}]$ is even in x_1 and odd [even] in x_1', $Z_{13}[Z_{14}]$ is odd in x_1 and odd [even] in x_1', where $g_1(x_1, 0, \ldots, 0, x_1', 0, \ldots, 0; 0) = 0$, where $\alpha_j(\varepsilon), \sigma_j(\varepsilon), \beta_j(\varepsilon)$ are continuous functions of ε [or constants] with $\sigma_j(0) > 0$, $j = 1, \ldots, \mu$, $\beta_j(0) \neq 0$, $j = \mu + 1, \ldots, n$, and either $\alpha_j(0) \neq 0$, or $\alpha_j(0) = 0$, $\sigma_j(0) \not\equiv 0$, mod $\sigma_1(0)$, $\sigma_1(0) = \omega_0$, $j = 2, \ldots, n$.

If we take $Z_{11} = (1 - x_1^2)x_1'$, then $P_1 = (1/2)(1 - \lambda^2/4\sigma_1^2)$. Thus for every ε, $0 \leq \varepsilon \leq \varepsilon_1$, and some $\varepsilon_1 > 0$, system (5.24) has a real periodic solution of the form

(5.25) $$x_1 = \lambda\omega^{-1}\sin(\omega t + \theta) + O(\varepsilon), \; x_j = O(\varepsilon), \qquad j = 2, \ldots, n ,$$

with $\lambda = 2\sigma_1(0) + O(\varepsilon)$. For $Z_{12} = Z_{13} = Z_{14} = 0$, $g_1 = 0$, $\mu = n = 1$, we have, as a particular case, the well known Van der Pol equation $x_1'' + \sigma_1^2 x_1 = \varepsilon(1 - x_1^2)x_1'$. By (9.i) we can see that the solution (5.25) is asymptotically orbitally stable.

If we take $Z_{11} = (1 - x_1^2 - x_1'^2)x_1'$, then $P_1 = (1/2)(1 - \lambda^2/\sigma_1^2)$ and (5.24) has a real periodic solution of the form above with $\lambda = \sigma_1(0) + O(\varepsilon)$, $\omega = \sigma_1(0) + O(\varepsilon)$.

EXAMPLE 2. As in Example 1 with $Z_{11} = (1 - |x_1|)x_1'$. Then

$$P_1 = (T\lambda)^{-1} \int_0^T (1 - \lambda\omega^{-1} |\sin \omega t|) \cos^2 \omega t \, dt ,$$

and $P_1(\lambda, \omega_0, 0) = (1/2)(1 - \lambda/\pi\sigma_1)$. Hence, system (5.24) has a periodic solution of the form (5.25) with $\lambda = \pi\sigma_1 + O(\varepsilon)$, $\omega = \sigma_1 + O(\varepsilon)$.

EXAMPLE 3. The real system

$$x'' + x - \varepsilon(1 - x^2 - y^2)x' = \varepsilon f_1(x, y, y') + \varepsilon g_1(x, x', y, y')y ,$$

$$y'' + 2y - \varepsilon(1 - x^2 - y^2)y' = \varepsilon f_2(x, x', y) + \varepsilon g_2(x, x', y, y')x ,$$

where $f_1(-x, y, y') = -f_1(x, y, y')$, $f_2(x, x', -y) = -f_2(x, x', y)$.

There are two periodic solutions of this system given by

$$x = \lambda \sin (\omega t + \theta) + O(\varepsilon),\ y = O(\varepsilon),\ \omega = 1 + O(\varepsilon),\ \lambda = 2 + O(\varepsilon)\ ;$$

$$x = O(\varepsilon),\ y = \lambda \sin (\omega t + \theta),\ \omega = 2^{1/2} + O(\varepsilon),\ \lambda = 2 + O(\varepsilon)\ .$$

[See J. K. Hale, 9c, for the analytic case. In 9i it is proved that both solutions are asymptotically orbitally stable.]

The following further examples have been studied among others, in the lines above, by R. A. Gambill and J. K. Hale [8] with emphasis on the existence of harmonic and subharmonic solutions. In [1], [3], and [9i] the stability of some of these solutions has been discussed. In all examples ε is a small parameter.

4. $$x'' + \sigma^2 x = B \cos 2\omega t + \varepsilon\alpha \cos 2\omega t \cdot x + \varepsilon b x^3\ ;$$

5. $$x'' + \sigma^2 y = B \cos \omega t + \varepsilon x^3\ ;$$

6. $$x'' + x = \varepsilon(1 - x^{2n})x' + \varepsilon p\omega \cos (\omega t + \alpha)\ ;$$

7. $$\begin{cases} x'' + \sigma_1^2 x = \varepsilon\alpha x + \varepsilon A \cos t \cdot x + \varepsilon\beta x^3 + \varepsilon\gamma x y^2 \\ y'' + \sigma_2^2 y = \varepsilon\delta y + B \cos \omega t \cdot y + \varepsilon\mu y^3 + \varepsilon\nu x^2 y\ ; \end{cases}$$

8. $$\begin{cases} x'' + 4x = \cos t + \varepsilon[\cos t \cdot x + k y^3 + b y]\ , \\ y'' + y = \varepsilon[\cos t \cdot y + c x^3]\ ; \end{cases}$$

9. $$\begin{cases} x'' + \sigma_1^2 x_1 = \varepsilon(\alpha - \gamma y^2)x' + \beta \cos t\ , \\ y'' + \sigma_2^2 y = \varepsilon(\delta - \rho x^2)y' + r \cos 2t\ ; \end{cases}$$

Example 4 is a nonlinear Mathieu equation with a large forcing term, 5 is a Duffing equation, 6 is a generalization of the Van der Pol equation, 7 is a system of nonlinear Mathieu equations, 8 is a system of nonlinear Mathieu equations of which one has a large forcing term, 9 is a system of Van der Pol equations.

§6. REAL SYSTEMS PRESENTING SYMMETRIES

We shall consider now real systems (5.1) presenting symmetries as defined below. In addition to the hypotheses listed at the beginning of §5, we shall suppose

$$\alpha_j \equiv 0, \qquad j = 1, \ldots, \mu ,$$

$$0 \leq \nu \leq \mu = r \leq n, \quad \beta_j \equiv 0, \qquad j = \mu + 1, \ldots, n .$$

Then we have $\gamma_j = \sigma_j$, $\rho_{j1} = i\sigma_j$, $\rho_{j2} = -i\sigma_j$, $j = 1, \ldots, \mu$, and equations (5.2), (5.3) become

$$(6.2) \qquad \begin{array}{ll} y_{2j-1} = i\sigma_j x_j + x_j', \ y_{2j} = i\sigma_j x_j - x_j', & j = 1, \ldots, \mu, \ y_{\mu+j} = x_j \\ & j = \mu + 1, \ldots, n , \end{array}$$

$$(6.3) \qquad \begin{array}{l} x_j = (2i\sigma_j)^{-1}(y_{2j-1} + y_{2j}), \ x_j' = 2^{-1}(y_{2j-1} - y_{2j}) , \\ \qquad\qquad j = 1, \ldots, \mu. \end{array}$$

Also, if $u = (x_1, \ldots, x_m)$, $v = (x_{m+1}, \ldots, x_\mu)$, $w = (x_{\mu+1}, \ldots, x_n)$, hence $x = (u, v, w)$, $x' = (u', v')$, for some $0 \leq m \leq \nu$, we have

$$(6.4) \qquad \begin{array}{ll} f_j(u,-v,w,-u',v',-t,\varepsilon) = f_j(u,v,w,u',v',t,\varepsilon), & j = 1, \ldots, m , \\ f_j(u,-v,w,-u',v',-t,\varepsilon) = - f_j(u,v,w,u',v',t,\varepsilon), & j = m + 1, \ldots, \mu, \\ f_j(u,-v,w,-u',v',-t,\varepsilon) = - f_j(u,v,w,u',v',t,\varepsilon), & j = \mu + 1, \ldots, n . \end{array}$$

For $m = 0$ all f_j are odd with respect to (v, u', t). For $m > 0$ the first m of the f_j are even and the remaining ones are odd with respect to (v, u', t).

We shall take $\theta_j = \pi/2$, $j = 1, \ldots, m$, $\theta_j = 0$, $j = m + 1, \ldots, \nu$, and hence $c_j = i\lambda_j$, $j = 1, \ldots, m$, $c_j = \lambda_j$, $j = m + 1, \ldots, \nu$, λ_j real; we shall take $c_j = \eta_{j-\mu-r}$, $j = \mu + r + 1, \ldots, \mu + n$. Then we have

$$(6.5) \qquad \begin{array}{ll} P_j = -(T\lambda_j)^{-1}\int_0^T f_j \sin \tau_j t \, dt, \ Q_j = -(T\lambda_j)^{-1}\int_0^T f_j \cos \tau_j t \, dt , & \\ & j = 1, \ldots, m , \\ P_j = (T\lambda_j)^{-1} \int_0^T f_j \cos \tau_j t \, dt, \ Q_j = -(T\lambda_j)^{-1}\int_0^T f_j \sin \tau_j t \, dt , & \\ & j = m + 1, \ldots, \nu , \\ R_j = (T\eta_j)^{-1} \int_0^T f_{r+j} dt, & j = 1, \ldots, n - r , \end{array}$$

where the arguments of the f's, besides t and ε, are given by the (5.12) of §5, thus, in the present case, by relations analogous to the (6.3).

We shall restrict the space Ω to all $\varphi = (\varphi_1, \dots, \varphi_{\mu+n})$ as in §5 which besides verifying $\varphi_{2j-1}(t) = -\bar{\varphi}_{2j}(t)$, $j = 1, \dots, \mu$, φ_j real, $j = 2\mu + 1, \dots, \mu + n$, have the following properties:

$$\varphi_{2j-1}(-t) = -\bar{\varphi}_{2j-1}(t),\ \varphi_{2j}(-t) = -\bar{\varphi}_{2j}(t), \qquad j = 1, \dots, m,$$

$$\varphi_{2j-1}(-t) = \bar{\varphi}_{2j-1}(t),\ \varphi_{2j}(-t) = \bar{\varphi}_{2j}(t), \qquad j = m + 1, \dots, \mu,$$

(6.6)

$$\varphi_j(-t) = \varphi_j(t), \qquad j = 2\mu + 1, \dots, \mu + n.$$

This is the same as requesting that for $j = 1, \dots, m$, all $\varphi_{2j-1}, \varphi_{2j}$ have odd real parts (r.p.) and even imaginary parts (i.p.), that for $j = m + 1, \dots, \mu$, all $\varphi_{2j-1}, \varphi_{2j}$ have even r.p. and odd i.p., and that for $j = 2\mu + 1, \dots, \mu + n$, all φ_j are real and even. Also the first two lines of (6.6) are equivalent to requesting that for $j = 1, \dots, m$, the expressions $(2i\sigma_j)^{-1}(\varphi_{2j-1} + \varphi_{2j})$ are real and even, and the expression $2^{-1}(\varphi_{2j-1} - \varphi_{2j})$ are real and odd, while for $j = m + 1, \dots, \mu$, the former are real and odd and the latter are real and even.

Because of (6.4) the integrands for the R_j and P_j are all real and odd, hence $P_j \equiv 0$, $j = 1, \dots, \nu$, $R_j \equiv 0$, $j = 1, \dots, n - \mu$. As a consequence, for $j = 2\mu + 1, \dots, \mu + n$, we have $d_j = 0$ and, by (5.8), since q_j is real and odd, we conclude that ψ_j is real and even.

Relations (5.7) are not present here since $r = \mu$.

For $j = \nu + 1, \dots, \mu$, by (5.6), and since q_{2j-1} is real and odd, and ρ_{j1} is purely imaginary, we see that the integrand has odd r.p. and even i.p., hence ψ_{2j-1} has even r.p. and odd i.p. By §5 the same holds for ψ_{2j}.

For $j = 1, \dots, m$, we have $c_j = i\lambda_j$, c_j is imaginary, q_{2j-1} is real and even, and by (6.6), the expression under { } in (5.9) has even r.p. and odd i.p., hence $c_j d_{2j-1}$ is real and d_{2j-1} is imaginary. Then the expression under { } in (5.5) has even r.p. and odd i.p. and the same holds for the integrand. Hence, the integral has odd r.p. and even i.p. and the same holds for ψ_{2j-1}. By §5 the same is true for ψ_{2j}.

For $j = m + 1, \dots, \nu$, we have $c_j = \lambda_j$, c_j is real, q_{2j-1} is real and odd, and by (6.6) the expression under { } in (5.9) has odd r.p. and even i.p., hence $c_j d_{2j-1}$ is imaginary and d_{2j-1} is imaginary. Then the expression under { } in (5.5) has odd r.p. and even i.p and the same holds for the integrand. Hence, the integral has even r.p. and odd i.p. and the same holds for ψ_{2j-1}, and finally for ψ_{2j}.

In connection with §5 we conclude that $\psi = \mathfrak{I}\varphi \in \Omega$ for every $\varphi \in \Omega$, i.e., $\mathfrak{I}\,\Omega_R \subset \Omega$.

Also, the first ν and the last $n - r$ equations (5.11) are identically satisfied and the second ν equations become

$$(6.7)\qquad a_j b_j^{-1}\omega - \varepsilon Q_j(a, b, \lambda, \eta, \omega, \varepsilon) = \sigma_j(\varepsilon), \qquad j = 1, \dots, \nu,$$

where, for $\varepsilon = 0$, we have

$$(6.8)\qquad \begin{aligned} Q_{jo} &= -(T\lambda_j)^{-1} \int_0^T f_j(\dots, t, 0)\cos a_j b_j^{-1}\omega t \, dt, \qquad j = 1, \dots, m, \\ Q_{jo} &= -(T\lambda_j)^{-1} \int_0^T f_j(\dots, t, 0)\sin a_j b_j^{-1}\omega t \, dt, \qquad j = m + 1, \dots, \nu, \end{aligned}$$

and the missing arguments are

$$(6.9)\qquad \begin{aligned} &x_j(t, 0) = \lambda_j \tau_j^{-1}\cos \tau_j t, \quad x_j'(t, 0) = -\lambda_j \sin \tau_j t, && j = 1, \dots, m, \\ &x_j(t, 0) = \lambda_j \tau_j^{-1}\sin \tau_j t, \quad x_j'(t, 0) = \lambda_j \cos \tau_j t, && j = m+1, \dots, \nu, \\ &x_j(t, 0) = 0, \qquad\qquad\qquad x_j'(t, 0) = 0, && j = \nu+1, \dots, \mu, \mu = r, \\ &x_j(t, 0) = \eta_{j-r}, && j = r+1, \dots, n. \end{aligned}$$

In connection with §5 we may summarize the preceding results as follows.

Consider the real system

$$(6.10)\qquad \begin{aligned} x_j'' + \sigma_j^2(\varepsilon)x_j &= \varepsilon f_j(x, x', t, \varepsilon), && j = 1, \dots, \mu, \\ x_j' &= \varepsilon f_j(x, x', t, \varepsilon), && j = \mu + 1, \dots, n, \end{aligned}$$

where $x = (x_1, \dots, x_n)$, $x' = (x_1', \dots, x_\mu')$, $\sigma_j(\varepsilon) > 0$ are continuous functions of ε for $0 \le \varepsilon \le \varepsilon_0$, and for some integers $0 \le m \le \nu \le \mu \le n$ the following occurs:

$\sigma_j(0) = a_j\omega/b_j$, $j = 1, \dots, \nu$, a_j, $b_j > 0$ integers;

$\sigma_j(0) \neq im\omega/b_0$, $j = \nu + 1, \dots, \mu$, $b_0 = b_1 \cdots b_\nu$, $m = 0, 1, 2, \dots$.

The functions f_j, $j = 1, \ldots, n$, satisfy condition (L) and the conditions of symmetry (6.4). As usual we shall put $a = (a_1, \ldots, a_\nu)$, $b = (b_1, \ldots, b_\nu)$, $\sigma = (\sigma_1, \ldots, \sigma_\nu)$, $\lambda = (\lambda_1, \ldots, \lambda_\nu)$, $\eta = (\eta_1, \ldots, \eta_{n-\mu})$.

(6.i). If for some a, b, λ, η, all real, and ε sufficiently small the determining equations (6.7) are satisfied, then (6.10) has a real periodic solution of the form

$$
(6.11)\quad
\begin{aligned}
x_j(t, \varepsilon) &= \lambda_j \tau_j^{-1} \cos \tau_j t + O(\varepsilon), & j &= 1, \ldots, m, \\
x_j(t, \varepsilon) &= \lambda_j \tau_j^{-1} \sin \tau_j t + O(\varepsilon), & j &= m + 1, \ldots, \nu, \\
x_j(t, \varepsilon) &= O(\varepsilon), & j &= \nu + 1, \ldots, m, \\
x_j(t, \varepsilon) &= \eta_{j-\mu} + O(\varepsilon), & j &= \mu + 1, \ldots, n,
\end{aligned}
$$

and $x_j(-t, \varepsilon) = x_j(t, \varepsilon)$, $j = 1, \ldots, m$, $j = \mu + 1, \ldots, n$, and $x_j(-t, \varepsilon) = -x_j(t, \varepsilon)$, $j = m + 1, \ldots, \mu$. In particular, given a, b, μ, λ there is always a $\sigma(\varepsilon)$ such that equations (6.7) are satisfied and system (6.10) has a real periodic solution of the form (6.11). If system (6.10) is autonomous, the parameter ω has to be added.

(6.ii). If the functions f_j have continuous first order partial derivatives and, given a, b, λ, η, there are ν numbers $\lambda_0 = (\lambda_{10}, \ldots, \lambda_{\nu 0})$ such that $Q_j(a, b, \lambda_0, \mu, \omega, 0) = 0$, $j = 1, \ldots, \nu$, and the Jacobian $\partial Q/\partial\lambda \neq 0$ at $\lambda = \lambda_0$, $\varepsilon = 0$, then for all ε sufficiently small system (6.10) has a periodic solution of the form (6.11) with convenient $\lambda(\varepsilon)$, $\lambda(0) = \lambda_0$.

Obviously here the Jacobian of §5 of order $2\nu + (n - r)$ may be identically zero while the Jacobian above $\partial Q/\partial\lambda$ of order ν may well be $\neq 0$. In the case (6.10) is autonomous, also the parameter ω can be taken into consideration as in §5.

The following result is also of interest since it is an explicit existence theorem of a $(n-\mu+2)$-parameter family of periodic solutions. It concerns autonomous systems (6.10) with $\nu = 1$, $m = 0$, or $m = 1$, $a_1 = b_1 = 1$.

(6.iii). Consider an autonomous system

$$
(6.12)\quad
\begin{aligned}
x_j'' + \sigma_j^2 x_j &= \varepsilon f_j(x, x', \varepsilon), & j &= 1, \ldots, \mu, \\
x_j' &= \varepsilon f_j(x, x', \varepsilon), & j &= \mu + 1, \ldots, n,
\end{aligned}
$$

where the functions f_j, $j = 1, \ldots, n$, are Lipschitzian in the $n + \mu$ variables $x_1, \ldots, x_n, x_1', \ldots, x_\mu'$, and continuous in ε, for $|x_s| \leq R$, $s = 1, \ldots, n$, $|x_s'| \leq R$, $s = 1, \ldots, \mu$, and $0 \leq \varepsilon \leq \varepsilon_0$. In addition suppose, for f_1 only, that $f_1(0, x_2, \ldots, x_n, 0, x_2', \ldots, x_\mu', \varepsilon) = 0$. Suppose that either all f_j, $j = 1, \ldots, n$, are odd in $(x_1, \ldots, x_\mu)$, or f_1 is even and $f_2, \ldots, f_n$ are odd in $(x_2, \ldots, x_\mu, x_1')$. Suppose $\sigma_j(\varepsilon) > 0$, $j = 1, \ldots, \mu$, are continuous functions of ε [or constants] and $\sigma_j(0) \not\equiv 0$, mod $\sigma_1(0)$, $j = 2, \ldots, \mu$. Put $\omega_0 = \sigma_1(0)$, and let r_2 be any number, $0 < r_2 < R$. Then there exists an ε_1, $0 < \varepsilon_1 \leq \varepsilon_0$, such that, for all real $\varepsilon, \lambda_1, \eta_1, \ldots, \eta_{n-\mu}$, $0 \leq \lambda_1, \eta_1, \ldots, \eta_{n-\mu} \leq r_2$, $0 \leq \varepsilon \leq \varepsilon_1$, system (6.12) has a real periodic solution of the form

$$x_1(t, \varepsilon) = \lambda_1\omega^{-1}\sin \omega t + O(\varepsilon), \quad \text{or} \quad x_1(t, \varepsilon) = \lambda_1\omega^{-1}\cos \omega t + O(\varepsilon) ,$$

$$(6.13) \qquad x_j(t, \varepsilon) = O(\varepsilon), \qquad j = 2, \ldots, \mu ,$$

$$x_j(t, \varepsilon) = \eta_{j-\mu} + O(\varepsilon) , \qquad j = \mu + 1, \ldots, n ,$$

where x_1 is odd or even in t, $x_2, \ldots, x_\mu$ are odd, $x_{\mu+1}, \ldots, x_n$ are even, where $\omega = \omega(\varepsilon, \lambda_1, \eta_1, \ldots, \eta_{n-\mu})$ is a continuous function of the same parameters, and $\omega = \omega_0$ for $\varepsilon = 0$. Also, t can be replaced by by $t + \theta$ in (6.13), θ an arbitrary constant. [See J. K. Hale, 9cf, for the analytic case, and W. R. Fuller, 10, for the present situation with $\mu = n$.]

PROOF. By (4.iii) we know that the unique $Q_1 = Q_1(\lambda_1, \mu, \omega, \varepsilon)$ where $\eta = (\eta_1, \ldots, \eta_{n-r})$, is a uniformly continuous and Lipschitzian function of λ_1, η, ω, and uniformly continuous functions of ε. The only determining equation (6.7) is

$$\omega - \varepsilon Q_1(\lambda_1, \eta, \omega, \varepsilon) = \sigma_1(\varepsilon) .$$

If $\omega \in U = [\omega_{01}, \omega_{02}]$ with $\omega_{01} < \omega_0 < \omega_{02}$, and $\omega_{02} - \omega_{01}$ sufficiently small, if $0 \leq \varepsilon \leq \varepsilon_1$ and $0 < \varepsilon_1 \leq \varepsilon_0$ sufficiently small, we put

$$F_1(\omega, \varepsilon) = \omega - \varepsilon Q_1(\omega, \varepsilon) - \sigma_1(\varepsilon) .$$

For all $\omega_{01} \leq \omega^1 \leq \omega^2 \leq \omega_{02}$, $0 \leq \varepsilon \leq \varepsilon_1$, we have

$$|Q_1(\omega^2, \varepsilon) - Q_1(\omega^1, \varepsilon)| \leq M(\omega^2 - \omega^1) ,$$

$$F_1(\omega^2, \varepsilon) - F_1(\omega^1, \varepsilon) \geq (1 - \varepsilon M)(\omega^2 - \omega^1) ,$$

where M is a constant, independent of $\lambda_1, \eta, \varepsilon, \omega$. For $0 \leq \varepsilon \leq \varepsilon'$, $\varepsilon' = \min[\varepsilon_1, 1/2M]$, the last expression above is positive. Thus F_1 is an increasing continuous function of ω with $F_1(\omega_0, 0) = 0$. Hence,

$F_1(\omega_{01}, 0) < 0$, $F(\omega_{02}, 0) > 0$, and finally $F_1(\omega_{01}, \varepsilon) < 0$, $F(\omega_{02}, \varepsilon) > 0$ for all λ_1, η as above, and $0 \leq \varepsilon \leq \varepsilon''$, ε'' sufficiently small. Thus, for the same λ_1, η, $0 \leq \varepsilon \leq \varepsilon''$, there is a unique ω with $F_1(\omega, \varepsilon) = 0$. By the usual argument of implicit function theorem we know that ω is a continuous function of the same parameters. Now (6.iii) follows from (6.i) and the previous considerations.

NOTE. If system (6.12) verifies the conditions of (6.iii) when any one of the variables x_j takes the place of x_1 and correspondingly the jth equation replaces the first one, $j = 1, \ldots, \mu$, then (6.iii) assures the existence of μ families of periodic solutions, each a $(n-\mu+2)$-parameter family.

EXAMPLE 1. The equation $x'' + x = \varepsilon f(x, x')$ with $f(0, 0) = 0$, and either $f(x, -x') = f(x, x')$, or $f(-x, x') = -f(x, x')$, has a family of periodic solutions of the form $x = \lambda\omega^{-1}\cos(\omega t + \theta) + O(\varepsilon)$, or $x = \lambda\omega^{-1}\sin(\omega t + \theta) + O(\varepsilon)$, with $\omega = \omega(\lambda, \varepsilon) = 1 + O(\varepsilon)$, λ, θ arbitrary, $\varepsilon > 0$ sufficiently small. We may take, for instance, $f = x + x^2 + x'^2$, or $f = |x| + |x'|$, or $f = x|x'|$. Also we may take $f = (2 + x)(|x'| + x)$.

EXAMPLE 2. The system $x'' + x = \varepsilon(1 - |y|)|x'|x$, $y'' + 2y = \varepsilon(1 - |x|)|y'|y$, has two families of periodic solutions of the forms

$$x = \lambda\omega^{-1}\cos(\omega t + \theta) + O(\varepsilon),\ y = O(\varepsilon),\ \omega = \omega(\lambda, \varepsilon) = 1 + O(\varepsilon),$$

$$x = O(\varepsilon),\ y = \lambda\omega^{-1}\cos(\omega t + \theta) + O(\varepsilon),\ \omega = \omega(\lambda, \varepsilon) = 2^{1/2} + O(\varepsilon).$$

EXAMPLE 3. The system $x'' + x = \varepsilon f_1(x, y, z, x')$, $y' = \varepsilon f_2(x, y, z, x')$, $z' = \varepsilon f_3(x, y, z, x')$, with $f_1(0, y, z, 0) = 0$ and either

$$f_j(-x, y, z, x') = -f_j(x, y, z, x'),\ j = 1, 2, 3,$$

or

$$f_1(x, y, z, -x') = f_1(x, y, z, x'),\ f_j(x, y, z, -x') = -f_j(x, y, z, x'),\ j = 2, 3,$$

has a four parameter family of periodic solutions, for every ε sufficiently small, of the form $x = \lambda\omega^{-1}\sin(\omega t + \theta) + O(\varepsilon)$, or

$$x = \lambda\omega^{-1}\cos(\omega t + \theta) + O(\varepsilon),\ y = \eta_1 + O(\varepsilon),\ z = \eta_2 + O(\varepsilon),$$

with $\omega = \omega(\lambda, \eta_1, \eta_2, \varepsilon)$, θ, λ, η_1, η_2 arbitrary.

EXAMPLE 4. A third order equation $x''' + a_1x'' + a_2x' + a_3x = \varepsilon f(x, x', x'', t, \varepsilon)$, with f periodic could be studied in the lines of §§5, 6. For instance, if $f(x, -x', x'', -t, \varepsilon) = - f(x, x', x'', t, \varepsilon)$, then equation $x''' + \sigma^2x' = \varepsilon f$, $\sigma > 0$, has periodic solutions of the form $x(-t, \varepsilon) = x(t, \varepsilon)$, $x(t, 0) = - c_1 a\omega\sigma^{-2}b^{-1} \cos ab^{-1}\omega t + c_2$, a, b integers, c_1, $c_2 \neq 0$ real, provided $ab^{-1}\omega - \varepsilon H = \sigma$, where $H = H(a, b, c_1, c_2, \omega, \sigma, \varepsilon)$ and

$$H = (c_1 T)^{-1} \int_0^T f[x(t,0),x'(t,0),x''(t,0),t,0]\sin ab^{-1}\omega t \, dt + O(\varepsilon) .$$

If $\sigma = a\omega/b$, and there is a $c_{10} = c_{10}(c_2)$ such that $H = 0$ and $\partial H/\partial c_1 \neq 0$ for $c_1 = c_{10}$, then the equation has a non-zero solution for every c_2; that is, a one-parameter family of periodic solutions [J. K. Hale, 9f].

BIBLIOGRAPHY

[1] Bailey, H. R., Harmonics and subharmonics for weakly nonlinear Mathieu type differential equations. To appear.

[2] Bailey, H. R., and Cesari, L., Boundedness of solutions of linear differential systems with periodic coefficients. Archive Rat. Mech. Anal. 1, 1958, 246-271.

[3] Bailey, H. R., and Gambill, R. A., On stability of periodic solutions of weakly nonlinear differential systems. J. Math. Mech. 6, 1957, 655-668.

[4] Cesari, L., (a) Un nuovo criterior di stabilità per le soluzioni delle equazioni differenziali lineari. Annali scuola Norm. Sup. Pisa (2) 9, 1940, 163-186. (b) Sulla stabilità delle soluzioni dei sistemi di equazioni differenziali lineari a coefficienti periodici. Mem. Accad. Italia (6) 11, 1941, 633-695.

[5] Cesari, L., Asymptotic Behavior and Stability Problems in Ordinary Differential Equations. Ergbn. d. Mathematik und ih. Grenzgebiete, Heft, 16, 1959, vii, 271.

[6] Cesari, L., and Hale, J. K., (a) Second order linear differential systems with periodic L-integrable coefficients. Riv. Mat. Univ. Parma, 5, 1954, 55-61; 6, 1955, p. 159. (b) A new sufficient condition for periodic solutions of weakly nonlinear differential systems. Proc. Amer. Math. Soc. 8, 1957, 757-764.

[7] Gambill, R. A., (a) Stability criteria for linear differential systems with periodic coefficients. Riv. Mat. Univ. Parma, 5, 1954, 169-181. (b) Criteria for parametric instability for linear differential systems with periodic coefficients. Ibid. 6, 1955, 37-43. (c) A fundamental system of real solutions for linear differential systems with periodic coefficients. Ibid. 7, 1956, 311-319.

[8] Gambill, R. A., and Hale, J. K., Subharmonic and ultraharmonic solutions for weakly nonlinear systems. J. Rat. Mech. Anal. 5, 1956, 353-394.

[9] Hale, J. K., (a) Evaluations concerning products of exponential and periodic functions. Riv. Mat. Univ. Parma, 5, 1954, 63-81. (b) On boundedness of the solutions of linear differential systems with periodic coefficients. Ibid. 5, 1954, 137-167. (c) Periodic solutions of nonlinear systems of differential equations. Ibid. 5, 1954, 281-311. (d) On a class of linear differential equations with periodic coefficients. Illinois J. Math. 1, 1957, 98-104. (e) Linear systems of first and second order differential equations with periodic coefficients. Ibid. 2, 1958, 586-592. (f) Sufficient conditions for the existence of periodic solutions of systems of weakly nonlinear first and second order differential equations. Journ. Math. Mech. 7, 1958, 163-172. (g) A short proof of a boundedness theorem for linear differential systems with periodic coefficients. Archive Rat. Mech. Anal. 2, 1959, 429-434. (h) On the behavior of the solutions of linear periodic differential systems near resonance points. These Contributions, 5, 1959, (i) On the stability of periodic solutions of weakly nonlinear periodic and autonomous differential systems. These contributions, 5, 1959.

[10] Fuller, W. R., Existence theorems for periodic solutions of systems of differential and differential-difference equations. Ph.D. Thesis, Purdue University, Lafayette, Indiana, 1957.

[11] Thompson, E. W., On stability of periodic solutions of autonomous differential systems. (To appear).

VII. THE APPLICATIONS OF A FIXED POINT THEOREM TO A VARIETY OF NON-LINEAR STABILITY PROBLEMS

Arnold Stokes

§1. INTRODUCTION

We use a fixed point theorem for locally convex linear spaces, due to Tychonoff[1] to reduce the study of the boundedness and stability of certain n-dimensional vector differential equations to the study of the corresponding properties of related first-order equations. In this manner, we very simply obtain a variety of known results, and in certain cases, we are able to clarify some of these results.

Fixed point theorems have been used to study the stability of systems of differential equations before, notably by Hukuwara[2], and Bellman[3], but the present approach, by using Tychonoff's theorem rather than Schauder's result[4] gives many more results in a greatly simplified fashion.

§2. DEFINITIONS

We will let E be the real vector space of all continuous functions from the non-negative reals into R^n, the n-dimensional vector space

* The results presented here are based on material contained in the author's thesis, written as partial fulfillment of the requirements for the Ph.D. degree at the University of Notre Dame. The author wishes to thank Dr. J. P. LaSalle for his valuable suggestions and constant encouragement during the development of this study. The author also is grateful to the Office of Naval Research for their assistance while at the University of Notre Dame.

1 A. Tychonoff, Ein Fixpunktsatz, Math. Ann. 111, 767-776, 1935.

2 M. Hukuwara, Sur les Points Singuliers des Equations Differentielles Lineaires, J. Faculty of Sci., Hokkaido Imp. Univ. Ser. I, Math. 2, 13-88, 1934-36.

3 R. Bellman, On the Boundedness of Solutions of Non-Linear Difference and Differential Equations, Trans. Amer. Math. Soc., 62, 357-386, 1947.

4 J. Schauder, Der Fixpunktsatz in Funktionalraum, Studia Math. 2, 171-181, 1930.

over the real field. The topology on E shall be that induced by the family of pseudo-norms $\{p_n\}_{n=1}^{\infty}$, where for $x \in E$,

$$p_n(x) = \sup_{0 \leq t \leq n} \|x(t)\|$$

where $\|x(t)\|$ is any vector norm. A fundamental system of neighborhoods is then given by $\{V_n\}_{n=1}^{\infty}$, where $V_n = \{x \in E \mid p_n(x) \leq 1\}$. Under this topology, E becomes a complete, locally convex linear space. For such a space, we have a theorem of Tychonoff's:

THEOREM 1. Let E be a complete, locally convex linear space. Let $T : E \longrightarrow E$ be continuous, and let A be a closed convex subset of E. If $T(A) \subset A$ and $\overline{T(A)}$ is compact, then there exists a fixed point of T in A.[5]

If T is a compact (completely continuous) operator and if A is bounded, then $\overline{T(A)}$ is always compact. We then obtain:

COROLLARY 1. Let E be as in Theorem 1. Let $T : E \longrightarrow E$ be continuous and compact, and let A be a closed, convex, bounded subset of E. If $T(A) \subset A$, then there exists a fixed point of T in A.

§3. REDUCTION TO A FIRST-ORDER EQUATION

We shall consider theee types of n-dimensional systems here:

(1) $\dot{x} = f(t, x)$

(2) $\dot{x} = A(t)x + f(t, x)$, where $\|X(t)\| \leq K$,
$\|X(t)X^{-1}(s)\| \leq K, K > 0$

(3) $\dot{x} = A(t)x + f(t, x)$, where $\|X(t)\| \leq Ke^{-\sigma t}$
$\|X(t)X^{-1}(s)\| \leq Ke^{-\sigma(t-s)}, K, \sigma > 0$,

---

[5] This theorem, while not explicitly stated by Tychonoff, follows from a theorem given in No. 1 in the same manner that the second fixed point theorem of Schauder's in No. 4 follows from his first theorem.

where by $X(t)$ is meant the principal matrix solution of the system $\dot{x} = A(t)x$, where $A(t)$ is a continuous function of t for $t \geq 0$. If $A(t)$ is a constant matrix, or periodic in t, examples of systems satisfying the inequalities in (2) or (3) are well-known.

In addition we shall assume that f is continuous on R^{n+1} into R^n, and that

$$(4) \qquad \|f(t, x)\| \leq G(t, \|x\|), \quad \text{for} \quad t \geq 0, \ x \in D \subset R^n$$

where $G(t, r)$ is piecewise continuous on R^2, positive for $t, r \geq 0$, and nondecreasing in r for fixed t; and D is some subset of R^n.

With (1) above, we shall associate the integral operator

$$(5) \qquad T_b(x)(t) = b + \int_0^t f(s, x(s))ds \quad ,$$

and with (2) and (3) above we shall associate the operator

$$(6) \qquad T_b(x)(t) = X(t)b + \int_0^t X(t)X^{-1}(s)f(s, x(s))ds \quad ,$$

where b is a vector in R^n.

Clearly fixed points of these operators correspond to solutions of the associated equations, the solution passing through b at $t = 0$. Also it is evident that both operators are compact in the topology of our function space E.

To apply Corollary 1, we shall proceeds as follows: Let B be a bounded subset of R^n, and let A be a subset of E defined by a positive real-valued function $g(t)$, continuous for $t \geq 0$, that is,

$$(7) \qquad A = \{x \in E \mid \|x(t)\| \leq g(t)\} \quad .$$

For such a g, A is closed, convex, and bounded in the topology given above on E. We will further assume that for $x \in A$, $x(t) \in D$ for all $t \geq 0$, so that (4) may be used.

Now take $b \in B$, $x \in A$. To apply Corollary 1 to either of the operators in (5) or (6), we must show $T_b(A) \subset A$, or, by (7), $\|T_b(x)(t)\| \leq g(t)$. Using the operator in (5), we obtain

$$\|T_b(x)(t)\| \leq \|b\| + \int_0^t \|f(s, x(s))\|ds$$

$$\leq \|b\| + \int_0^t G(s, \|x(s)\|)ds, \quad \text{by (4)} \quad ,$$

$$\leq \|b\| + \int_0^t G(s, g(s))ds, \quad \text{by the definition of } A, \text{ as } G \text{ is non-decreasing.}$$

Thus, for (5), we have $T_b(A) \subset A$ if g satisfies

$$\|b\| + \int_0^t G(s, g(s))ds \leq g(t) \quad \text{for } b \in B, \ t \geq 0 \ . \tag{8}$$

Now consider (6), and let $A(t)$ be such that $X(t)$ satisfies the inequalities in (2). In precisely the same fashion as before, we have $T_b(A) \subset A$ if g satisfies

$$K \|b\| + \int_0^t KG(s, g(s))ds \leq g(t), \quad \text{for } b \in B, \ t \geq 0 \ . \tag{9}$$

If $X(t)$ satisfies the inequalities in (3), we have $T_b(A) \subset A$ if g satisfies

$$K \|b\| e^{-\sigma t} + \int_0^t Ke^{-\sigma(t-s)}G(s, g(s))ds \leq g(t), \quad \text{for } b \in B, \ t \geq 0 \ . \tag{10}$$

Equations (8) and (9) may be combined into

$$K_1 \|b\| + \int_0^t K_1G(s, g(s))ds \leq g(t), \quad \text{for } b \in B, \ t \geq 0 \ . \tag{11}$$

where $K_1 = 1$ if equation (1) is being considered, or K, if (2) is being discussed.

Now g will clearly satisfy (11) if it satisfies the differential inequality:

(12) $\dot{g}(t) \geq K_1 G(t, g(t)),\ g(0) \geq K_1 \|b\|,$ for $b \in B,\ t \geq 0$,

and g will satisfy (10) if it satisfies the differential inequality:

(13) $\dot{g}(t) \geq -\sigma g(t) + KG(t, g(t)),\ g(0) \geq K \|b\|,$ for $b \in B,\ t \geq 0$,

where by a solution is meant a differentiable function g satisfying (12) or (13) wherever G is continuous.

Thus, to demonstrate the existence of a solution to any of the equations (1), (2), or (3) which satisfies $\|x(t)\| \leq g(t)$, we must choose $B \subset R^n$ and a g satisfying (12) or (13), (and, of course, g must possess the property that $\|x(t)\| \leq g(t)$ implies $x(t) \in D$).

§4. STATEMENT OF THE THEOREMS

(i) On existence in the large. We will consider equation (1) and assume that f is dominated by G everywhere, that is, $D = R^n$.

THEOREM 2. If the equation $\dot{r} = G(t, r)$ possesses the property that for any $r_0 > 0$ there exists a solution defined on $[0, \infty)$, passing through r_0 at $t = 0$, then for an arbitrary vector $b \in R^n$, there exists a solution of (1) defined on $[0, \infty)$ which passes through b at $t = 0$.

PROOF. Take $b \in R^n$, let $B = \{b\}$. Choose g to be a solution of $\dot{r} = G(t, r)$ with $g(0) \geq \|b\|$ which is defined on $[0, \infty)$. Clearly such a g satisfies (12), so the operator in (5) maps A into A, where A is defined as in (7). So by Corollary 1, T_b has a fixed point in A; that is, there exists a solution of (1) defined on $[0, \infty)$ passing through b at $t = 0$.

COROLLARY 2. Assume $\|f(t, x)\| \leq M(t)L(\|x\|)$, for $t \geq 0,\ x \in R^n$, where M, L are piecewise continuous, positive, and L is non-decreasing. If

$$\int_{r_0}^{\infty} \frac{1}{L(s)}\, ds = \infty \quad ,$$

for positive r_0, then for any $b \in R^n$, there exists a solution of (1) passing through b at $t = 0$.

PROOF. Observe that the equation $\dot{r} = M(t)L(r)$, $r(0) = r_0$ may be solved for any $r_0 > 0$, for if we write

$$\int_{r_0}^{r} \frac{1}{L(s)}\, ds = \int_{0}^{t} M(s) ds \quad ,$$

note that the function of r on the left is strictly increasing, so the inverse function exists, and by our assumption concerning L, the domain of the inverse functions is $[0, \infty)$, so the solution through r_0 at $t = 0$ is defined on $[0, \infty)$ for all $r_0 > 0$. The result follows from Theorem 2.

Corollary 2 is a theorem due to Wintner[6], which was later improved by him by removing the requirement that L be non-decreasing[7]. Later Conti[8] generalized this result still further, by replacing the norm function by a more general function. The above proof is considerably simpler than Wintner's original proof, however.

For an example of a function dominated as in Corollary 2, consider any function $f(t, x)$ such that $\|f(t, x)\| = o(\|x\|)$ as $\|x\| \longrightarrow \infty$ for fixed t.

(ii) On Boundedness. We will consider Equations (1) or (2), and again assume that f is dominated by G everywhere, or $D = R^n$.

THEOREM 3. If the equation $\dot{r} = G(t, r)$ possesses the property that for any $r_0 > 0$ there exists a solution defined and bounded on $[0, \infty)$, passing through r_0 at $t = 0$, then for an arbitrary vector $b \in R^n$ there exists a bounded solution of (1) or (2) defined on $[0, \infty)$ which passes through b at $t = 0$.

PROOF. The proof is precisely the same as for Theorem 1, where $B = \{b\}$, for $b \in R^n$, and here g is taken as any bounded solution to $\dot{r} = G(t, r)$, with $g(0) \geq K_1 \|b\|$.

---

6 A. Wintner, The Non-Local Existence Problem of Ordinary Differential Equations, Amer. J. of Math., 67, 277-284, 1945.

7 A. Wintner, The Infinities in the Non-Local Existence Problem of Ordinary Differential Equations, Amer. J. of Math., 68, 173-178, 1946.

8 R. Conti, Limitazioni 'in Ampiezza' della Soluzioni di un Sistema di Equazioni Differenziali e Applicazioni, Bolletino della Unione Mat. Ital., Ser. 3, 11, 344-350, 1956.

COROLLARY 3. Assume $\|f(t, x)\| < M(t)L(\|x\|)$ for $t \geq 0$, $x \in R^n$, where M, L are piecewise continuous, positive, and L is non-decreasing. If

$$\int_{r_0}^{\infty} \frac{1}{L(s)} ds = \infty ,$$

for positive r_0, and

$$\int_0^{\infty} M(s)ds < \infty ,$$

then for any $b \in R^n$, there exists a bounded solution of (1) or (2) defined on $[0, \infty)$ passing through b at $t = 0$.

PROOF. The proof follows from Theorem 3, if we observe that, as in Corollary 2, our assumption on L implies that for any $r_0 > 0$, there exists a solution to $\dot{r} = M(t)L(r)$ defined on $[0, \infty)$, passing through r_0 at $t = 0$, and further that

$$\int_0^{\infty} M(s)ds < \infty$$

implies that this solution is bounded.

With reference to the following theorem, observe that the above bounds in general depend upon the initial conditions.

For an example of a function dominated as in Corollary 3, let $f(t, x) = B(t)x$, where $B(t)$ is an $n \times n$ matrix, continuous in t, such that

$$\int_0^{\infty} \|B(t)\|dt < \infty .$$

Such a theorem appears in Bellman[9], where the proof uses Gronwall's inequality. The above corollary shows that the linearity of L is not essential, but rather that it suffices that

$$\int_{r_0}^{r} \frac{1}{L(s)} ds$$

[9] R. Bellman, Stability Theory of Differential Equations, (McGraw-Hill Inc., New York, 1953).

diverges as $r \longrightarrow \infty$.

(iii) On Ultimate Boundedness. Here we use ultimate boundedness in the sense of Yoshizawa[10]; that is, through every point there passes a solution which is not only bounded but as $t \longrightarrow \infty$, enters a bounded region, which is independent of the initial conditions. We will now consider Equation (3), and again we assume f is dominated by G everywhere.

THEOREM 4. If G is such that for every positive initial condition r_0, there exists a solution to the inequality $\dot{r} \geq -\sigma r + KG(t, r)$ passing through r_0 at $t = 0$ which is ultimately bounded, then for any $b \in R^n$, there exists a solution of (3) passing through b at $t = 0$ which is ultimately bounded.

PROOF. The proof is the same as for the above two theorems, where $B = \{b\}$, for $b \in R^n$, and here g is taken as any ultimately bounded solution to $\dot{r} \geq -\sigma r + KG(t, r)$, with $g(0) \geq K\,\|b\|$.

COROLLARY 4. For any $\epsilon > 0$, assume there exists $R(\epsilon) > 0$ such that $\|f(t, x)\| \leq N$, $\|x\| \leq R$, $t \geq 0$, and $\|f(t, x)\| \leq \epsilon\|x\|$, $\|x\| \geq R$, $t \geq 0$. Then all solutions to (3) are ultimately bounded.

PROOF. Here let

$$G(t, r) = \begin{cases} N, & r \leq R \\ \epsilon r, & r \geq R \end{cases} .$$

By Theorem 4, all we must show is that for ϵ sufficiently small, there exist ultimately bounded solutions to $\dot{r} \geq -\sigma r + KG(t, r)$ passing through r_0 at $t = 0$, where $r_0 \geq K\,\|b\|$, for b an arbitrary vector in R^n.

Now choose $\epsilon > 0$ such that $\epsilon < \sigma/K$, and choose β such that $0 < \beta < 1$. Let $g(t) = K\,\|b\|\, e^{\beta(\epsilon K-\sigma)t} + NK/\sigma$. By our choice of ϵ, $g \longrightarrow NK/\sigma$ as $t \longrightarrow \infty$, and so g is ultimately bounded. And $g(0) \geq K\,\|b\|$, so all that remains is to show that g satisfies the differential inequality:

$$\dot{g}(t) \geq -\sigma g(t) + K \begin{cases} N, & g(t) \leq R \\ \epsilon g(t), & g(t) \geq R \end{cases} .$$

[10] T. Yoshizawa, Note on the Boundedness and the Ultimate Boundedness of Solutions of $\dot{x} = F(t, x)$, Mem. of the Coll. of Sci., Univ. of Kyoto, Ser. A, 29, Math. No. 3, 275-291, 1955.

Now $\dot{g}(t) = \beta(\epsilon K - \sigma)K \|b\| e^{\beta(\epsilon K-\sigma)t}$, and

$$(14) \qquad \dot{g}(t) + \sigma g(t) = K \|b\| e^{\beta(\epsilon K-\sigma)t}[\beta\epsilon K + \sigma(1 - \beta)] + NK \ .$$

By our choice of β, this is clearly $\geq NK$ for all $t \geq 0$, and so half the inequality is satisfied. Now $Kg(t) = K^2 \|b\| e^{\beta(\epsilon K-\sigma)t} + NK^2/\sigma$, and upon comparison with (14), we see that the inequality

$$\dot{g}(t) \geq - \sigma g(t) + \epsilon K g(t)$$

reduces to the two inequalities

$$\beta \epsilon K + \sigma(1 - \beta) \geq \epsilon K$$

and

$$NK \geq \epsilon K \ NK/\sigma \ .$$

But the first of these is equivalent to $\sigma(1 - \beta) \geq \epsilon K(1 - \beta)$, or $\epsilon \leq \sigma/K$, and the second reduces to $\epsilon K/\sigma \leq 1$, or $\epsilon \leq \sigma/K$. So by choosing $\epsilon < \sigma/K$, we see that the conditions of Theorem 4 are satisfied, and so all solutions to (3) are ultimately bounded.

For an example of a function $f(t, x)$ dominated as in Corollary 4, it suffices to choose f so that $\|f(t, x)\| = o(\|x\|)$ as $\|x\| \longrightarrow \infty$, uniformly in t.

(iv) On Stability of a Critical Point. Here we will suppose (1) or (2) has a critical point at the origin, that is, $f(t, 0) = 0$ for all $t \geq 0$, and we wish to examine the stability of this point. We will say the origin is stable if given any $\epsilon > 0$, there exists a $\delta > 0$ such that for any $b \in R^n$ such that $\|b\| \leq \delta$, there exists a solution through b at $t = 0$ which is bounded in norm by ϵ for $t \geq 0$. As this is a local problem, we will assume f to be dominated by G in some neighborhood of the origin, that is $D = \{x \in R^n \mid \|x\| \leq \eta\}$, where $\eta > 0$ is given.

THEOREM 5. If the equation $\dot{r} = K_1G(t, r)$ possesses a stable critical point at the origin, then (1) or (2) also possesses a stable critical point at the origin.

PROOF. Take $\epsilon > 0$, and let $\delta > 0$ be such that if $0 < r_0 \leq \delta$, there exists a solution to $\dot{r} = K_1G(t, r)$ which passes through r_0 at $t = 0$ and is bounded by $\min(\epsilon, \eta)$ for $t \geq 0$. To apply Corollary 1, let $B = \{b \in R^n \mid \|b\| \leq \delta\}$, and let g be a solution to $\dot{r} = K_1G(t, r)$

which is bounded by $\min(\epsilon, \eta)$ for $t \geq 0$, and $g(0) = \delta$. Thus for $b \in B$, $T_b(A) \subset A$, or there exists a solution to (1) or (2) through b at $t = 0$ which lies in A. The theorem follows.

COROLLARY 5. Let $\|f(t, x)\| \leq M(t)L(\|x\|)$, for $t \geq 0$, $\|x\| \leq \eta$, where M and L are continuous, and for $r > 0$, L is positive and non-decreasing. If $L(0) = 0$, and

$$\lim_{r_0 \to 0^+} \int_{r_0}^{r} \frac{1}{L(s)} ds = \infty$$

for r small and positive, and

$$\int_0^\infty M(s)ds < \infty ,$$

then (1) or (2) possesses a stable critical point at the origin.

PROOF. By Theorem 5, if the equation $\dot{r} = K_1M(t)L(r)$ possesses a stable critical point at the origin, the result follows. But by considering the equation

$$\int_{r_0}^{r} \frac{1}{L(s)} ds = \int_0^t M(s)ds ,$$

and by the above assumptions on L and M, clearly given any $\epsilon > 0$, there exists a $\delta > 0$ such that

$$\int_{\delta}^{\epsilon} \frac{1}{L(s)} ds > \int_0^\infty M(s)ds ,$$

or, there exists a solution through r_0 at $t = 0$ which, for $r_0 \leq \delta$, is bounded by ϵ for $t \geq 0$. The corollary follows.

A result of this type, where $L(r) = r$, is given by Bellman[9], where again Gronwall's inequality is applied. As before, Corollary 5 demonstrates that the linearity of L is not important.

For an example of a function of this type, consider $f(t, x)$

---

[9] R. Bellman, Stability Theory of Differential Equations, (McGraw-Hill Inc., New York, 1953).

where the i-th component is the product of a polynomial in the components of x, without a constant term, with a function $\varphi_i(t)$, where

$$\int_0^\infty |\varphi_i(s)|ds < \infty \quad .$$

(v). On Asymptotic Stability of a Critical Point. Here we shall consider (3), and assume that (3) has a critical point at the origin, that is, $f(t, 0) = 0$. By asymptotic stability we mean, of course, that the origin is stable, and in addition, there exists a $\delta > 0$ such that through any b, $\|b\| \leq \delta$, there passes a solution which approaches 0 as $t \longrightarrow \infty$. As this is again a local problem, we will assume f to be dominated in G in some neighborhood of the origin, that is, $D = \{x \in R^n \mid \|x\| \leq \eta\}$. For our purposes, here we wish to take η as a function of ϵ, where ϵ is a small parameter, and we will also assume that $G(t, r) = G(t, r, \epsilon)$ is also a function of this parameter.

THEOREM 6. If the inequality $\dot{r} \geq - \sigma r + KG(t, r, \epsilon)$ possesses an asymptotically stable point at the origin for ϵ sufficiently small, then (3) possesses an asymptotically stable critical point at the origin.

PROOF. Take $\epsilon > 0$ so small that the differential inequality above is asymptotically stable. Then there exists a $\delta > 0$ such that for $0 < r_0 \leq \delta$, there exists a solution to the inequality which passes through r_0 at $t = 0$, tends to 0 as $t \longrightarrow \infty$, and for all $t \geq 0$, is bounded by $\eta(\epsilon)$. Then, to apply Corollary 1, let $B = \{b \in R^n \mid \|b\| \leq \delta\}$, and choose g to be a solution to the inequality having the above properties, with $g(0) = \delta$. The theorem follows.

COROLLARY 6. Assume that for any $\epsilon > 0$, we have $\|f(t, x)\| \leq \epsilon \|x\|$, for $t \geq 0$, $\|x\| \leq \eta(\epsilon)$. Then in (3), the origin is asymptotically stable.

PROOF. By Theorem 6, we must show only that for sufficiently small ϵ a solution $g(t)$ can be found to the differential inequality $\dot{g}(t) \geq - \sigma g(t) + \epsilon K g(t)$ which is bounded by $\eta(\epsilon)$ and tends to 0 as $t \longrightarrow \infty$. But for $\epsilon < \sigma/K$, we may choose $g(t) = \eta(\epsilon)e^{(\epsilon K - \sigma)t}$, and the corollary is proved.

As an example, we may take any $f(t, x)$ for which $\|f(t, x)\| = o(\|x\|)$ as $\|x\| \longrightarrow 0$ uniformly in t. This result, by now classical, is originally due to Perron.

COROLLARY 7. Assume that for any $\epsilon > 0$, we have $T(\epsilon) > 0$ such that

$$\|f(t, x)\| \leq \begin{cases} k\ \|x\| + t^b\ \|x\|^{1+a}, & 0 \leq t \leq T(\epsilon) , \\ \epsilon\ \|x\| + t^b\ \|x\|^{1+a}, & t \geq T(\epsilon) , \end{cases}$$

for $\|x\| \leq \eta(\epsilon)$, where $k, b, \geq 0$, $a > 0$. Then in (3), the origin is asymptotically stable.

PROOF. Again by Theorem 6, we must show only that for sufficiently small ϵ, a solution $g(t)$ can be found to the inequality

$$\dot{g}(t) \geq -\sigma g(t) + K \begin{cases} kg(t) + t^b g(t)^{1+a}, & 0 \leq t \leq T_1 , \\ \epsilon g(t) + t^b g(t)^{1+a}, & t \geq T_1 , \end{cases}$$

where $T_1 \geq T(\epsilon)$, which is bounded by $\eta(\epsilon)$ and tends to 0 as $t \longrightarrow \infty$.

Now, for $0 \leq r \leq 1$, $0 \leq t \leq T_1$, $-\sigma r + K(kr + t^b r^{1+a}) \leq Cr$, for some constant C, so clearly we can find a solution to the inequality over $[0, T_1]$, for any T_1, with an arbitrarily small bound on $[0, T_1]$, simply by choosing $g(0)$ sufficiently small. So we may assume $g(t) \leq \eta(\epsilon)$ on $[0, T_1]$. Let $g_0 = g(T_1)$. We now wish to define g for $t \geq T_1$.

Take $\epsilon > 0$ so that $\epsilon < \sigma/K$, and choose β so that $0 < \beta < 1$. Take $T_1 \geq T(\epsilon)$, and let $g(t) = g_0 e^{\beta(\epsilon K - \sigma)(t - T_1)}$. Clearly, by our choice of ϵ, $g \longrightarrow 0$ as $t \longrightarrow \infty$, and as $g_0 \leq \eta(\epsilon)$, $g(t) \leq \eta(\epsilon)$ for all $t \geq 0$. We now wish to show that for T_1 sufficiently large, and g_0 sufficiently small, g satisfies the differential inequality $\dot{g}(t) \geq -\sigma g(t) + \epsilon K g(t) + K t^b g(t)^{1+a}$ for $t \geq T_1$. Now

$$\dot{g}(t) - (\epsilon K - \sigma) g(t) = (\beta - 1)(\epsilon K - \sigma) g_0 e^{\beta(\epsilon K - \sigma)(t - T_1)} = \alpha g(t) ,$$

where $\alpha = (\beta - 1)(\epsilon K - \sigma) > 0$ by our choice of β and ϵ. So we must show that $\alpha g(t) \geq K t^b g(t)^{1+a}$, or, as g is positive, $\alpha \geq K t^b g(t)^a$.

But

$$K t^b g(t)^a = g_0^a K e^{-a\beta(\epsilon K - \sigma) T_1} \cdot t^b e^{a\beta(\epsilon K - \sigma) t} .$$

Now choose T_1 so large that $t^b e^{a\beta(\epsilon K - \sigma) t} < \sqrt{\alpha}$ for $t \geq T_1$, and choose g_0 so small that $g_0^a K e^{-a\beta(\epsilon K - \sigma) T_1} \leq \sqrt{\alpha}$. Clearly both these choices can be made, and the corollary is proved.

Corollary 7 appears in Coddington and Levinson[11]. There we also have an example of a system to which Corollary 7 applies.

---

[11] E. A. Coddington and N. Levinson, Theory of Ordinary Differential Equations, (McGraw-Hill Inc., New York, 1950).

VIII. QUADRATIC DIFFERENTIAL EQUATIONS AND NON-ASSOCIATIVE ALGEBRAS

Lawrence Markus

§1. STATEMENT OF THE PROBLEM AND THE RESULTS

In order to obtain a thorough knowledge of the qualitative behavior of the solutions for a class of non-linear differential equations, we classify and analyse differential systems which have quadratic polynomials as coefficients. We find all such differential systems in the plane, and an interesting collection of such systems in higher dimensional spaces.

The method of classification is based on an algebraic technique and thus differs from the geometric methods which are customarily used in the qualitative theory of differential equations. While the algebraic method is somewhat more intricate than the geometric analysis in dimension two, it is available in higher dimensions where the geometry is more difficult to manage.

To each quadratic differential system we attach a certain non-associative, but commutative, real linear algebra. The problem of affine equivalence of differential systems is shown to be the same as the isomorphism problem for the corresponding algebras. This is analogous to the classification of linear differential systems by canonical forms for the coefficient matrix, that is, by means of certain linear endomorphisms of a vector space.

Thus we first find all real, commutative, two-dimensional, linear algebras, see Theorems 6, 7, and 8. This yields all quadratic differential systems, up to affine equivalence, in the affine plane. It is easy to show, Theorem 9, that there are just six geometrical types of quadratic differential systems with isolated critical points in the plane. Representation for these six geometric types are pictured in Figures I through VI. Finally there is a brief discussion of quadratic differential systems in the real projective plane and in higher dimensional spaces. The standard summation notation is used throughout.

§2. THE ALGEBRA CORRESPONDING TO A QUADRATIC DIFFERENTIAL SYSTEM

DEFINITION. A quadratic differential system is

$$(\mathcal{Q}) \qquad \frac{dx^i}{dt} = a^i_{jk}x^jx^k \ , \qquad i = 1, 2, \ldots, n \ ,$$

where the n^3 real constants a^i_{jk} are normalized so that $a^i_{jk} = a^i_{kj}$.

DEFINITION. The related real linear algebra $\mathfrak{A}$ of the quadratic differential system $\mathcal{Q}$ is defined by the multiplication table for a basis $u_1, u_2, \ldots, u_n$ as $u_j \cdot u_k = a^i_{jk}u_i$. Clearly the n-dimensional algebra $\mathfrak{A}$ is commutative but it may not be associative.

THEOREM 1. Two quadratic differential systems

$$(\mathcal{Q}) \qquad \dot{x}^i = a^i_{jk}x^jx^k$$

and

$$(\hat{\mathcal{Q}}) \qquad \dot{y}^i = \hat{a}^i_{jk}y^jy^k \qquad i, j, k = 1, 2, \ldots, n$$

are equivalent under a non-singular linear transformation

$$x^i = b^i_\ell y^\ell$$

if and only if their related algebras $\mathfrak{A}$ and $\hat{\mathfrak{A}}$, respectively, are isomorphic.

PROOF. Suppose $x^i = b^i_\ell y^\ell$ carries the system $\mathcal{Q}$ into $\hat{\mathcal{Q}}$. Write $B^\ell_i b^i_k = b^\ell_i B^i_k = \delta^\ell_k$, the Kronecker symbol, so $y^\ell = B^\ell_i x^i$. Then

$$\dot{y}^\ell = B^\ell_i \dot{x}^i = B^\ell_i a^i_{rs} b^r_j b^s_k y^j y^k = \hat{a}^\ell_{jk} y^j y^k \ .$$

Thus

$$\hat{a}^\ell_{jk} = B^\ell_i b^r_j b^s_k a^i_{rs} \quad \text{or} \quad \hat{a}^\ell_{jk} b^i_\ell = b^r_j b^s_k a^i_{rs} \ .$$

Now consider the related algebras with multiplication tables

$$(\mathfrak{A}) \qquad u_j \cdot u_k = a^i_{jk}u_i$$

and

$$(\hat{\mathfrak{A}}) \qquad \hat{u}_j \cdot \hat{u}_k = \hat{a}^i_{jk}\hat{u}_i \ .$$

The linear transformation of $\hat{\mathfrak{A}}$ onto $\mathfrak{A}$ defined by $\hat{u}_\ell \longrightarrow b^i_\ell u_i$ is now shown to be an isomorphism. For

$$\hat{u}_j \cdot \hat{u}_k = \hat{a}^i_{jk}\hat{u}_i \longrightarrow \hat{a}^i_{jk}b^\ell_i u_\ell$$

and

$$b^r_j u_r \cdot b^s_k u_s = b^r_j b^s_k a^\ell_{rs} u_\ell \quad .$$

But

$$\hat{a}^i_{jk} b^\ell_i = b^r_j b^s_k a^\ell_{rs} \quad .$$

Conversely, if there is an isomorphism between $\mathfrak{A}$ and $\hat{\mathfrak{A}}$, which is expressed in terms of the given bases by $\hat{u}_\ell \longrightarrow b^i_\ell u_i$, then the linear transformation $x^i = b^i_\ell y^\ell$ carries $\mathcal{Q}$ into $\hat{\mathcal{Q}}$, as required.

Here we say that $\mathcal{Q}$ and $\hat{\mathcal{Q}}$ are affinely (or linearly) equivalent.

It is of interest to relate the algebraic properties of $\mathfrak{A}$ to the behavior of the solution curves of $\mathcal{Q}$. The next three theorems give certain general results in this direction.

THEOREM 2. The quadratic differential system

$$(\mathcal{Q}) \qquad \dot{x}^i = a^i_{jk}x^j x^k \qquad i, j, k = 1, 2, \ldots, n$$

has an isolated critical point at the origin if and only if the related algebra $\mathfrak{A}$ has no nilpotent element $e \neq 0$ for which $e \cdot e = 0$.

PROOF. If $\mathcal{Q}$ has another critical point P than the origin 0, then the line OP consists of critical points since $\mathcal{Q}$ is homogeneous. Take OP to be the x^1-axis and then the coefficients of $\mathcal{Q}$ (in the new coordinates) satisfy $a^i_{11} = 0$ for $i = 1, 2, \ldots, n$. But then the first basis vector u_1, for the related algebra $\mathfrak{A}$, satisfies $u_1 \cdot u_1 = 0$.

Conversely if $e \cdot e = 0$, take a linear transformation of the x-space so that $e = u_1$, the first basis vector of $\mathfrak{A}$. Then $a^i_{11} = 0$ and the x^1-axis consists of critical points of $\mathcal{Q}$. Q.E.D.

THEOREM 3. The quadratic differential system

$$(\mathcal{Q}) \qquad \dot{x}^i = a^i_{jk}x^j x^k \qquad i, j, k = 1, 2, \ldots, n$$

has a solution which is a ray to or from the origin

if and only if the related algebra $\mathfrak{A}$ has a non-zero idempotent (i.e., an element $e \neq 0$ for which $e \cdot e = e$).

PROOF. If $\mathcal{Q}$ has a ray solution, say the x^1-axis, then $a^i_{11} = 0$ for $i = 2, 3, \ldots, n$. Thus the basis vector u_1 in $\mathfrak{A}$ satisfies $u_1 \cdot u_1 = \lambda u_1$ for some $\lambda \neq 0$. Take $e = 1/\lambda\ u_1$. Conversely if $\mathfrak{A}$ has an idempotent, say $u_1 \cdot u_1 = u_1$, then $a^i_{11} = 0$ for $i = 2, 3, \ldots, n$ and $a^1_{11} = 1$ so the x^1-axis is a ray solution of $\mathcal{Q}$. Q.E.D.

COROLLARY. If $n = 1, 3, 5, 7, \ldots$ is odd, and if $\mathcal{Q}$ has an isolated critical point at the origin ($\mathfrak{A}$ has no non-zero element $e \cdot e = 0$), then $\mathcal{Q}$ has a solution ray($\mathfrak{A}$ has an idempotent element).

PROOF. This is an immediate consequence of the topological result that the sphere S^{n-1} cannot support a continuous, non-vanishing, tangent vector field. Q.E.D.

THEOREM 4. The quadratic differential system

$$\mathcal{Q}) \qquad \dot{x}^i = a^i_{jk}x^j x^k \qquad i, j, k = 1, 2, \ldots, n$$

has an invariant r-plane, $1 \leq r < n$, through the origin if and only if the related algebra $\mathfrak{A}$ has an r-dimensional subalgebra.

PROOF. Say the r-plane spanned by the coordinate axes of $x^1, x^2, \ldots, x^r$ is invariant under the flow of $\mathcal{Q}$. Then $a^i_{jk} = 0$ for $r + 1 \leq i \leq n$ and $j, k \leq r$. Thus the basis vectors $u_1, u_2, \ldots, u_r$ of $\mathfrak{A}$ span an r-dimensional algebra. The converse also follows from the same observation. Q.E.D.

If $\mathfrak{A}$ has an r-dimensional ideal then, in a certain sense, $\mathcal{Q}$ has a projection on an (n-r)-dimensional quadratic differential system. However if $\mathfrak{A}$ splits as a direct product, then so does $\mathcal{Q}$, as is indicated in the next theorem.

THEOREM 5. Let $\mathcal{Q}$ be a quadratic differential system with related algebra $\mathfrak{A}$. If there exist ideals $\mathfrak{B}$ and $\mathfrak{C}$, intersecting only in zero and spanning $\mathfrak{A}$, then $\mathcal{Q}$ is affinely equivalent to a product differential system of the form

$$\dot{x}^i = a^i_{jk} x^j x^k \qquad i, j, k = 1, 2, \ldots, r$$

and

$$\dot{x}^i = a^i_{jk} x^j x^k \qquad i, j, k = r + 1, \ldots, n \ .$$

PROOF. We require that $a^i_{jk} = 0$ for $i \leq r$ with $j > r$ or $k > r$; and for $i > r$ with $j < r$ or $k < r$. Choose a basis $u_1, u_2, \ldots, u_r, u_{r+1}, \ldots, u_n$ of $\mathfrak{A}$ such that $u_1, \ldots, u_r$ is a basis for the ideal $\mathfrak{B}$ and $u_{r+1}, \ldots, u_n$ is a basis for the ideal $\mathfrak{C}$. Then the multiplication tensor a^i_{jk} has the desired properties. Q.E.D.

§3. EXAMPLES AND APPLICATION OF QUADRATIC DIFFERENTIAL SYSTEMS

In the study of the potential flow from a doublet we are lead to the differential equation

$$\frac{dy}{dx} = \frac{y^2 - x^2}{2xy}$$

whose solution curves describe the streamlines. The equi-potential lines are the orthogonal trajectories which thereby satisfy

$$\frac{dy}{dx} = \frac{-2xy}{y^2 - x^2} \ .$$

The quadratic differential system

$$\dot{x} = 2xy, \quad \dot{y} = y^2 - x^2$$

corresponds to the algebra 4) of Theorem 8, which is the complex numbers.

Next consider the quadratic differential system

$$\dot{x} = yz - x^2$$
$$\dot{y} = xz - y^2$$
$$\dot{z} = xy - z^2$$

which occurs in a solution of the Einstein gravitational field equations, cf. [9]. The corresponding algebra has a basis of idempotents e_1, e_2, e_3 with

$$e_1 \cdot e_1 = e_1 \, , \qquad e_2 \cdot e_2 = e_2 \, , \qquad e_3 \cdot e_3 = e_3$$

$$e_1 \cdot e_2 = -\frac{1}{2} e_3, \quad e_1 \cdot e_3 = -\frac{1}{2} e_2, \quad e_2 \cdot e_3 = -\frac{1}{2} e_1 \quad .$$

There is just one (independent) nilpotent of index two, namely, $e_1 + e_2 + e_3$.

Next consider the Jacobi elliptic functions[1] $u = \text{sn } t$, $v = \text{cn } t$, $w = \text{dn } t$ for a fixed parameter value of k^2 on $0 < k < 1$. Then

$$\frac{du}{dt} = vw \, , \quad \frac{dv}{dt} = -uw \, , \quad \frac{dw}{dt} = -k^2 uv \quad .$$

This corresponds to an algebra with basis e_1, e_2, e_3 of nilpotents such that

$$e_1 \cdot e_1 = 0 \, , \qquad e_2 \cdot e_2 = 0 \, , \qquad e_3 \cdot e_3 = 0$$

$$e_1 \cdot e_2 = -\frac{1}{2} k^2 e_3 \, , \quad e_1 \cdot e_3 = -\frac{1}{2} e_2 \, , \quad e_2 \cdot e_3 = \frac{1}{2} e_1 \quad .$$

Let

$$E_1 = \frac{1}{k} e_1, \quad E_2 = \frac{1}{k} e_2, \quad E_3 = e_3$$

and then

$$E_1 \cdot E_1 = 0, \quad E_2 \cdot E_2 = 0, \quad E_3 \cdot E_3 = 0$$

and

$$E_1 \cdot E_2 = -\frac{1}{2} E_3, \quad E_1 \cdot E_3 = -\frac{1}{2} E_2, \quad E_2 \cdot E_3 = \frac{1}{2} E_1 \quad .$$

Thus for all k, the corresponding quadratic differential systems for the elliptic functions are linearly equivalent.

One of the major applications of quadratic differential systems occurs in interaction processes which depend on collisions of entities. For example, second order chemical processes, or biological interactions are often of this nature.

Following Volterra [18], we consider $n \geq 2$ species σ_1, σ_2, $\ldots$, σ_n of creatures with populations N_1, N_2, $\ldots$, N_n respectively. Say that the members of σ_i each have mass β_i and also assume that each species

---

[1]   The author wishes to thank Professor G. Birkhoff for calling attention to this example and for discussing with him the topic of quadratic differential equations.

has a natural net birth-death rate ε_i. Assume that when individuals of species σ_i and σ_j meet, there is a certain probability that the member of σ_i will destroy and eat the member of σ_j. Then the fluctuations of the populations with time t can be studied by the differential system

$$\frac{dN_i}{dt} = \left(\varepsilon_i + \sum_{j=1}^{n} \frac{\beta_j}{\beta_i} a_{ij} N_j \right) N_i$$

for $i = 1, 2, \ldots, n$. Here the a_{ij} are real biological constants and the conservation of mass requires that

$$\frac{a_{ij}}{\beta_i} = - \frac{a_{ji}}{\beta_j} \quad .$$

One computes easily $\beta_1 \dot{N}_1 + \beta_2 \dot{N}_2 + \cdots + \beta_n \dot{N}_n = \varepsilon_1 \beta_1 N_1 + \cdots + \varepsilon_n \beta_n N_n$, which leads to a first integral in certain important cases. For example, if $\varepsilon_1 = \varepsilon_2 = \cdots = \varepsilon_n = 0$ we have a quadratic differential system with an integral $\beta_1 N_1 + \cdots + \beta_n N_n = \text{const}$, which is a hyperplane.

To be more definite consider the case of two interacting species,

$$\dot{N}_1 = \left(\varepsilon_1 + \frac{\beta_2 a_{12}}{\beta_1} N_2 \right) N_1$$

$$\dot{N}_2 = \left(\varepsilon_2 + \frac{\beta_1 a_{21}}{\beta_2} N_1 \right) N_2 \quad .$$

Then an integral is

$$N_1^{\varepsilon_2} \exp \left\{ \frac{\beta_1 a_{21} N_1}{\beta_2} \right\} N_2^{-\varepsilon_1} \exp \left\{ \frac{-\beta_2 a_{12} N_2}{\beta_1} \right\} = \text{const.}$$

In case $\varepsilon_1 = \varepsilon_2 = 0$ we have the quadratic differential system

$$\dot{N}_1 = \beta_2 \gamma N_1 N_2, \quad \dot{N}_2 = - \beta_1 \gamma N_1 N_2$$

for

$$\beta_1 > 0, \ \beta_2 > 0, \ \gamma = \frac{a_{12}}{\beta_1} > 0 \quad .$$

This corresponds to the algebra 3) of Theorem 6.

As a second example of Volterra's theory consider the problem of the three fish. There are three species σ_1, σ_2, σ_3 with populations N_1, N_2, N_3, respectively. We assume that the net birth-death rates $\varepsilon_1 = \varepsilon_2 = \varepsilon_3 = 0$, that is, each species is in equilibrium with the environment whenever the other two species are removed. Assume that the first type of fish eats the second type, which eats the third, which eats the first. Then $a_{12} > 0$, $a_{23} > 0$, $a_{31} > 0$. The resulting quadratic differential system corresponds to an algebra with a basis of nilpotents e_1, e_2, e_3 such that

$$e_1 \cdot e_1 = 0, \qquad e_2 \cdot e_2 = 0, \qquad e_3 \cdot e_3 = 0$$
$$e_1 \cdot e_2 = Ae_1 - Ce_2, \; e_1 \cdot e_3 = - Be_1 + Ee_3, \; e_2 \cdot e_3 = De_2 - Fe_3 \quad .$$

Here the positive constants A, B, C, D, E, F are given by

$$2A = \frac{\beta_2 a_{12}}{\beta_1}, \qquad - 2B = \frac{\beta_3 a_{13}}{\beta_1}, \qquad - 2C = \frac{\beta_1 a_{21}}{\beta_2}$$
$$2D = \frac{\beta_3 a_{21}}{\beta_2}, \qquad 2E = \frac{\beta_1 a_{31}}{\beta_3}, \qquad - 2F = \frac{\beta_2 a_{32}}{\beta_3} \quad .$$

Volterra's theory can be modified to apply to military situations, where it is termed Lanchester's law [5], but here the conservation of mass is no longer required since civilized warriors do not (as yet) eat one another.

Finally we point out that by increasing the dimension of the space of dependent variables we can make a differential system with non-homogeneous quadratic polynomial coefficients into a quadratic differential system. For example, consider the damped harmonic oscillator with the equations of motion $x + h\dot{x} + \omega^2 x = 0$, for positive constants h and ω. In the affine phase space this is written

$$\dot{x} = y, \quad \dot{y} = - \omega^2 x - hy \quad .$$

Consider the differential system in the real projective plane, with homogeneous coordinates (x, y, z) and write the differential equations in an affine plane "at infinity" where $x \neq 0$. In this affine plane use the coordinates

$$v = \frac{y}{x}, \quad z = \frac{1}{x}$$

and then

$$\dot{z} = -zv, \quad \dot{v} = -\omega^2 - 2hv - v^2 \ .$$

Now introduce a new coordinate w and consider the quadratic differential system in R^3,

$$\dot{z} = -zv, \quad \dot{v} = -\omega^2 w^2 - 2hvw - v^2, \quad \dot{w} = 0 \ .$$

The solutions of this enlarged differential system, in the plane $w = 1$, are solutions for the harmonic oscillator.

§4. REAL COMMUTATIVE ALGEBRAS OF DIMENSION TWO

LEMMA. Let $\mathfrak{A}$ be a real 2-dimensional commutative linear algebra. If $\mathfrak{A}$ does not contain a (non-zero) nilpotent element with square zero, then $\mathfrak{A}$ contains an idempotent e with $e \cdot e = e \neq 0$.

PROOF. For some basis u_1, u_2 of $\mathfrak{A}$ we have

$$u_1 \cdot u_1 = a_{11}^1 u_1 + a_{11}^2 u_2, \; u_1 \cdot u_2 = a_{12}^1 u_1 + a_{12}^2 u_2, \; u_2 \cdot u_2 = a_{22}^1 u_1 + a_{22}^2 u_2.$$

If $a_{11}^2 = 0$, then $a_{11}^1 \neq 0$ and take

$$e = \frac{1}{a_{11}^1} u_1$$

which is idempotent. If $a_{22}^1 = 0$, then $a_{22}^2 \neq 0$ and one could take

$$e = \frac{1}{a_{22}^2} u_2$$

which is idempotent.

Thus suppose $a_{11}^2 \neq 0$ and $a_{22}^1 \neq 0$. Consider the real cubic polynomial $a_{22}^1 x^3 + (2a_{12}^1 - a_{22}^2)x^2 + (a_{11}^1 - 2a_{12}^2)x - a_{11}^2$. This has a real non-zero root x_0. Thus

$$a_{22}^2 x_0^2 + 2a_{12}^2 x_0 + a_{11}^2 = x_0[a_{22}^1 x_0^2 + 2a_{12}^1 x_0 + a_{11}^1] \ .$$

If $a_{22}^2 x_0^2 + 2a_{12}^2 x_0 + a_{11}^2 = a_{22}^1 x_0^2 + 2a_{12}^1 x_0 + a_{11}^1 = 0$, then $E = u_1 + x_0 u_2$ is nilpotent which is impossible. Then $E \cdot E = \lambda E$ where $\lambda = a_{11}^1 + 2a_{12}^1 x_0 + a_{22}^1 x_0^2 \neq 0$. Define the idempotent $e = 1/\lambda \ E$. Q.E.D.

We first classify the quadratic differential systems in the plane where the origin is not an isolated critical point. Here the related algebra contains a non-zero nilpotent element with zero square, that is, a nilpotent of index two.

THEOREM 6. There are ten (types of) real commutative 2-dimensional algebras which contain a nilpotent of index two.

1) $e_1 \cdot e_1 = 0,\ e_2 \cdot e_2 = 0,\ e_1 \cdot e_2 = 0$

2) $e_1 \cdot e_1 = 0,\ e_2 \cdot e_2 = 0,\ e_1 \cdot e_2 = e_2$

3) $e_1 \cdot e_1 = 0,\ e_2 \cdot e_2 = 0,\ e_1 \cdot e_2 = e_1 + e_2$

4) $e_1 \cdot e_1 = 0,\ e_1 \cdot e_2 = 0,\ e_2 \cdot e_2 = e_2$

5) $e_1 \cdot e_1 = 0,\ e_1 \cdot e_2 = 0,\ e_2 \cdot e_2 = e_1$

6) $e_1 \cdot e_1 = 0,\ e_1 \cdot e_2 = e_1,\ e_2 \cdot e_2 = ke_2,$ real $k \neq 0$

7) $e_1 \cdot e_1 = 0,\ e_1 \cdot e_2 = e_1,\ e_2 \cdot e_2 = e_1 + 2e_2$

8) $e_1 \cdot e_1 = 0,\ e_1 \cdot e_2 = e_2,\ e_2 \cdot e_2 = e_1$

9) $e_1 \cdot e_1 = 0,\ e_1 \cdot e_2 = e_2,\ e_2 \cdot e_2 = -e_1$

10) $e_1 \cdot e_1 = 0,\ e_1 \cdot e_2 = e_2,\ e_2 \cdot e_2 = ke_1 + e_2,$ real $k \neq 0.$

The first three algebras each have a basis of nilpotent elements of index two whereas the last seven algebras each have just one independent nilpotent of index two. In Cases 6) and 10) the algebras are distinct for distinct values of the parameter k.

PROOF. First assume that the algebra has a basis of nilpotents of index two, say e_1 and e_2. Then we need only determine the real constants c, d in

$$e_1 \cdot e_2 = ce_1 + de_2 \quad .$$

If $c = 0,\ d = 0$ we have the first Case 1). If $c = 0,\ d \neq 0,$ then a scalar change in e_1 yields the algebra 2). If $c \neq 0,\ d = 0$ we again obtain an algebra isomorphic with 2). If $c \neq 0,\ d \neq 0$ let

$$E_1 = \frac{1}{d}\, e_1, \quad E_2 = \frac{1}{c}\, e_2$$

and compute $E_1 \cdot E_2 = E_1 + E_2$, which yields the Case 3).

The Case 1) has a zero multiplication and hence is different from 2) or 3). In the algebras 2) and 3) the only nilpotents of index 2 are scalar multiples of e_1 and e_2. The algebra 3) is distinguished from 3) by the property that the product of two independent nilpotents is a nilpotent of index 2.

From now on assume that the algebra has just one independent nilpotent e_1 which we take as the first vector of a basis, e_1 and e_2. Write $e_1 \cdot e_2 = ce_1 + de_2$. If $c = d = 0$, then $e_1 \cdot e_2 = 0$. If $c = 0$, $d \neq 0$, replace e_1 by $1/d\ e_1$ and then $e_1 \cdot e_2 = e_2$. If $c \neq 0$, $d = 0$ replace e_2 by $1/c\ e_2$ to obtain $e_1 \cdot e_2 = e_1$. If $c \neq 0$ and $d \neq 0$ let $E_1 = 1/d\ e_1$, $E_2 = c/d\ e_1 + e_2$ to obtain $E_1 \cdot E_1 = 0$, $E_1 \cdot E_2 = E_2$. Thus one can always select a basis e_1, e_2 for which $e_1 \cdot e_1 = 0$ and $e_1 \cdot e_2 = 0$, or $e_1 \cdot e_2 = e_1$, or $e_1 \cdot e_2 = e_2$. Exactly one of these alternatives must hold. For only in the first case does e_1 (which is the unique nilpotent up to a scalar multiple) annihilate the algebra upon left multiplication; in the second case e_1 generates a 1-dimensional ideal and this is not so in the third alternative.

Now take a basis in the algebra $\mathfrak{A}$ with $e_1 \cdot e_1 = 0$, $e_1 \cdot e_2 = 0$. Write $e_2 \cdot e_2 = fe_1 + ge_2$ for real constants f, g. By a change to a new basis, still preserving the properties $e_1 \cdot e_1 = 0$, $e_1 \cdot e_2 = 0$ we can always obtain $e_2 \cdot e_2 = e_2$ or $e_2 \cdot e_2 = e_1$. These yield the algebras 4) and 5) above. They differ in that 4) contains an idempotent whereas 5) does not.

Now take a basis in the algebra $\mathfrak{A}$ with $e_1 \cdot e_1 = 0$, $e_1 \cdot e_2 = e_1$. Take a new basis $E_1 = \alpha e_1$, $\alpha \neq 0$, and $E_2 = \lambda e_1 + e_2$ so that $E_1 \cdot E_1 = 0$ and $E_1 \cdot E_2 = E_1$. Write $e_2 \cdot e_2 = fe_1 + ge_2$ and then

$$E_2 \cdot E_2 = \frac{(2-g)\lambda + f}{\alpha} E_1 + gE_2 \quad .$$

If $g = 0$, choose $\alpha = 1$, $\lambda = 1/2\ f$ which makes E_2 nilpotent and this is not allowed here. Thus $g \neq 0$. If $g \neq 2$, we can obtain $E_2 \cdot E_2 = gE_2$. If $g = 2$, one obtains either $E_2 \cdot E_2 = E_1 + 2E_2$ or $E_2 \cdot E_2 = 2E_2$. These yield the algebras 6) and 7) above. The algebra 7) has no 1-dimensional subalgebra, other than that generated by a nilpotent element, and so 7) is different from every algebra in 6). If $k = 2$ in 6) then every element generates a 1-dimensional algebra whereas if $k \neq 2$ there is only one (up to scalar multiples) non-nilpotent element which generates a 1-dimensional subalgebra. Now consider the Case 6) with $k \neq 2$. Here the 1-dimensional space generated by e_2 is distinguished. But then $e_1 \cdot e_2 = e_1$ normalizes e_2 so that k is an invariant of the algebra and distinct values of k determine non-isomorphic algebras.

Finally take a basis in $\mathfrak{A}$ corresponding to the third alternative:

$$e_1 \cdot e_1 = 0, \quad e_1 \cdot e_2 = e_2 \quad .$$

Write $e_2 \cdot e_2 = fe_1 + ge_2$. Replace e_2 by $E_2 = \mu e_2$, $\mu \neq 0$. Then $E_2 \cdot E_2 = \mu^2 fe_1 + \mu g E_2$. If $g = 0$, we can obtain algebras 8) or 9). If $g \neq 0$, we obtain algebra 10). In this last case we exclude $k = 0$ since this yields an algebra with a basis of nilpotents.

In the algebras 8), 9) and 10) the idempotents αe_1 each define a linear transformation upon multiplication of $\mathfrak{A}$. Moreover only e_1 produces an eigenvalue of $+1$ and that for the eigenvectors be_2, $b \neq 0$. Thus if there is an isomorphism of algebra 8) onto 9) or 10) it must carry e_1 to e_1 and e_2 to be_2. This distinguishes between algebras 8), 9), and 10). Also for distinct values of k in the algebras 10) we obtain non-isomorphic algebras. Q.E.D.

The only algebras occurring in this Theorem 6 which are associative are 1), 4), 5), and 6), for $k = 1$. The algebra 6) for $k = 2$ is power associative and all the remaining algebras of Theorem 6 are not power associative. The necessary and sufficient condition for the power associativity of real commutative algebras is the identity $x^2 \cdot x^2 = (x^2 \cdot x) \cdot x$, cf. [1], and a direct computation yields the desired result.

THEOREM 7. Every real commutative 2-dimensional algebra $\mathfrak{A}$, containing no nilpotents of index two, and with a basis of idempotents, is isomorphic with exactly one of the following:

1) $e_1 \cdot e_1 = e_1$, $e_2 \cdot e_2 = e_2$, $e_1 \cdot e_2 = 0$

2a) $e_1 \cdot e_1 = e_1$, $e_2 \cdot e_2 = e_2$, $e_1 \cdot e_2 = e_2$

b) $e_1 \cdot e_1 = 2e_1$, $e_2 \cdot e_2 = e_2$, $e_1 \cdot e_2 = e_2$

c) $e_1 \cdot e_1 = \alpha e_1$, $e_2 \cdot e_2 = e_2$, $e_1 \cdot e_2 = e_2$, for $0 < \alpha < 1$ and $\alpha > 2$

3a) $e_1 \cdot e_1 = 2e_1$, $e_2 \cdot e_2 = \beta e_2$, $e_1 \cdot e_2 = e_1 + e_2$, with $\beta \neq 0$, $\beta \neq 2$

b) $e_1 \cdot e_1 = \alpha e_1$, $e_2 \cdot e_2 = \beta e_2$, $e_1 \cdot e_2 = e_1 + e_2$ with $\alpha \neq 0$, $\alpha \neq 2$, $\beta \neq 0$, $\beta \neq 2$, $\alpha\beta \neq 4$, $\alpha + \beta \neq \alpha\beta$, and $\dfrac{2}{\alpha - 1} \leq \beta \leq \dfrac{2 + \alpha}{\alpha}$ on $\alpha < -1$ and on $\alpha > 2$.

Distinct values of the parameters yield non-isomorphic algebras.

PROOF. Take a basis E_1 and E_2 of idempotents in $\mathfrak{A}$.
Then $E_1 \cdot E_2 = cE_1 + dE_2$.
If $c = d = 0$, we have Case 1) of the theorem.
If $c = 0$ and $d \neq 0$ write

$$e_1 = \frac{1}{d} E_1, \; e_2 = E_2$$

to obtain an algebra of the form, Case 2, $e_1 \cdot e_1 = \alpha e_1$, $e_2 \cdot e_2 = e_2$, $e_1 \cdot e_2 = e_2$. The parameter restrictions on α will be explained later. If $c \neq 0$ and $d = 0$, the algebras are isomorphic to those of Case 2.

If $c \neq 0$ and $d \neq 0$, let

$$e_1 = \frac{1}{d} E_1, \quad e_2 = \frac{1}{c} E_2$$

to obtain an algebra of the form, Case 3,

$$e_1 \cdot e_1 = \alpha e_1, \quad e_2 \cdot e_2 = \beta e_2, \quad e_1 \cdot e_2 = e_1 + e_2 \; .$$

The condition that there are no nilpotents imposes no restriction on Case 1, but requires $\alpha \neq 0$ in Case 2, and $\alpha \neq 0$, $\beta \neq 0$, $\alpha\beta \neq 4$ in Case 3.

Next we show that no algebra of one of the above three types 1), 2), or 3) is isomorphic with an algebra of a different type.

In Case 1) the only idempotents are e_1 and e_2 and their product is zero. This characterizes Case 1). In Case 2) the only idempotents are

$$\frac{1}{\alpha} e_1, \; e_2,$$

and

$$\frac{1}{\alpha} e_1 + (1 - \frac{2}{\alpha}) e_2 \; .$$

Here the idempotent e_2 generates a 1-dimensional ideal, yet it does not annihilate the algebra upon multiplication. Thuse Case 2) differs from Case 1). In Case 2) we distinguish the subcases 2a) where $\alpha = 1$, for only in this case does the corresponding idempotent $e_1 - e_2$ also generate an ideal. Also the subcase 2b), where $\alpha = 2$, is the only algebra of 2) which contains just two idempotents.

In Case 3), the idempotents are

$$\frac{1}{\alpha} e_1, \; \frac{1}{\beta} e_2 \; ,$$

and

$$\frac{2-\beta}{4-\alpha\beta}\, e_1 + \frac{2-\alpha}{4-\alpha\beta}\, e_2 \quad .$$

If $\alpha + \beta \neq \alpha\beta$, it is easy to see that no idempotent generates a 1-dimensional ideal. Thus Case 3) differs from 1) and 2). Finally we show that $e_1 \cdot e_1 = \alpha e_1$, $e_2 \cdot e_2 = \beta e_2$, $e_1 \cdot e_2 = e_1 + e_2$ with $\alpha \neq 0$, $\beta \neq 0$, $\alpha\beta \neq 4$ is an algebra of Case 2) in case $\alpha + \beta = \alpha\beta$. Here there are idempotents

$$\bar{E}_1 = \frac{1}{\alpha}\, e_1, \quad \bar{E}_2 = \frac{2-\beta}{4-\alpha\beta}\, e_1 + \frac{2-\alpha}{4-\alpha\beta}\, e_2 \quad .$$

If $\alpha = 2$ and $\alpha + \beta = \alpha\beta$, then $\beta = 2$ so $\alpha\beta = 4$ which is impossible. Thus $\alpha \neq 2$ and also $\beta \neq 2$. Compute $\bar{E}_1 \cdot \bar{E}_2 \neq 0$. But $\bar{E}_2$ generates a 1-dimensional ideal and thus we have an algebra of Case 2). The subcase 3a), $\alpha = 2$, is distinguished in that there are just two idempotents. If $\beta = 2$ (so $\alpha \neq 2$) we again have Case 3a).

We next study the restrictions on the parameter α in Case 2) and (α, β) in Case 3), so as to have a unique form for each possible algebra.

Thus consider $e_1 \cdot e_1 = \alpha e_1$, $e_2 \cdot e_2 = e_2$, $e_1 \cdot e_2 = e_2$ for $\alpha \neq 0$, $\alpha \neq 1$, $\alpha \neq 2$. Here e_2 is the unique idempotent which generates an ideal. Using

$$\frac{1}{\alpha}\, e_1 + \left(1 - \frac{2}{\alpha}\right) e_2$$

and e_2 as a base for the algebra we find that the algebras with values α and

$$\frac{\alpha}{\alpha - 1}$$

are isomorphic and that no other value of the parameter yields an isomorphic algebra. Thus we pick out a single representative for each algebra of Case 2c) by demanding

$$\alpha > \frac{\alpha}{\alpha - 1} \quad .$$

Thus $0 < \alpha < 1$ and $\alpha > 2$ enforces a one-to-one correspondence between parameter values α and isomorphism classes of algebras in 2c).

Consider Case 3a) with $\alpha = 2$ (so $\beta \neq 2$). There are just two idempotents

$$\frac{1}{2}\, e_1 \quad \text{and} \quad \frac{1}{\beta}\, e_2 \quad .$$

The product of these idempotents is

$$\frac{1}{\beta}\left(\frac{1}{2} e_1\right) + \frac{1}{2}\left(\frac{1}{\beta} e_2\right) .$$

Thus isomorphic algebras in 3a) must yield the same value for

$$\frac{1}{\beta} + \frac{1}{2} .$$

Thus β is different for different algebras. The case $\beta = 2$ (so $\alpha \neq 2$) is isomorphic to an algebra with $\alpha = 2$ and $\beta \neq 2$.

Now turn to Case 3b):

$$e_1 \cdot e_1 = \alpha e_1, \; e_2 \cdot e_2 = \beta e_2, \; e_1 \cdot e_2 = e_1 + e_2$$

with

$$\alpha \neq 0, \; \alpha \neq 2, \; \beta \neq 0, \; \beta \neq 2, \; \alpha\beta \neq 4, \; \alpha + \beta \neq \alpha\beta .$$

There are three idempotents

$$E_1 = \frac{1}{\alpha} e_1, \quad E_2 = \frac{1}{\beta} e_2 ,$$

and

$$E_3 = \frac{2 - \beta}{4 - \alpha\beta} e_1 + \frac{2 - \alpha}{4 - \alpha\beta} e_2 .$$

The table

$$E_1 \cdot E_2 = \frac{1}{\beta} E_1 + \frac{1}{\alpha} E_2 ,$$

$$E_1 \cdot E_3 = \frac{\alpha + \beta - \alpha\beta}{4 - \alpha\beta} E_1 + \frac{1}{\alpha} E_3 ,$$

$$E_2 \cdot E_3 = \frac{1}{\beta} E_3 + \frac{\alpha + \beta - \alpha\beta}{4 - \alpha\beta} E_2 ,$$

shows that the algebras of 3b) with parameter values $(\bar{\alpha}, \bar{\beta}) = (\alpha, \beta)$, or (β, α), or

$$\left(\alpha, \frac{4 - \alpha\beta}{\alpha + \beta - \alpha\beta}\right), \quad \text{or} \quad \left(\frac{4 - \alpha\beta}{\alpha + \beta - \alpha\beta}, \alpha\right)$$

or

$$\left(\frac{4 - \alpha\beta}{\alpha + \beta - \alpha\beta}, \beta\right), \quad \text{or} \quad \left(\beta, \frac{4 - \alpha\beta}{\alpha + \beta - \alpha\beta}\right)$$

are isomorphic. Also these are the only isomorphic parameter sets.

To obtain a unique parametrization of the algebras in 3b), we must find a fundamental domain, in the (α, β)-plane with the deletions $\alpha \neq 0$, $\alpha \neq 2$, $\beta \neq 0$, $\beta \neq 2$, $\alpha\beta \neq 4$, $\alpha + \beta \neq \alpha\beta$ under the group generated by the transformations

$$T : (\alpha, \beta) \longrightarrow \left(\alpha, \frac{4 - \alpha\beta}{\alpha + \beta - \alpha\beta} \right)$$

and

$$R : (\alpha, \beta) \longrightarrow (\beta, \alpha) \quad .$$

We compute $T^2 = R^2 = I$, the identity. Also $RTR = TRT$. Call $Z = RT$ (first T then R) and $Z^3 = I$, $RZ = Z^2R$. Thus the group consists of six elements I, Z, Z^2, R, ZR, and Z^2R.

Use the fact that

$$\frac{4 - \alpha\beta}{\alpha + \beta - \alpha\beta}$$

is a monotonic decreasing function of β, for each fixed α. A careful study of the geometry of the six transformations of the group I, Z, Z^2, R, ZR, Z^2R shows that every allowable point of the (α, β)-plane is equivalent to one and only one point of the fundamental domain

$$\frac{2}{\alpha - 1} \leq \beta \leq \frac{2 + \alpha}{\alpha}$$

with $\alpha > 2$, and

$$\frac{2}{\alpha - 1} \leq \beta \leq \frac{2 + \alpha}{\alpha}$$

for $\alpha < -1$. Q.E.D.

A long computation based on the identity $x^2 \cdot x^2 = (x^2 \cdot x) \cdot x$ shows that algebras 1) and 2a) of Theorem 7 are associative but that none of the other algebras of Theorem 7 are power associative.

THEOREM 8. A real commutative two-dimensional algebra $\mathfrak{A}$, with just one idempotent and no nilpotent elements of index two, is isomorphic with exactly one of the following:

1) $e_1 \cdot e_1 = e_1$, $e_1 \cdot e_2 = 0$, $e_2 \cdot e_2 = e_1 + ge_2$, with $0 \leq g < 2$

2) $e_1 \cdot e_1 = e_1$, $e_1 \cdot e_2 = e_1 + e_2$, $e_2 \cdot e_2 = fe_1$,
with $f < -1$

3a) $e_1 \cdot e_1 = e_1$, $e_1 \cdot e_2 = \frac{1}{2} e_2$, $e_2 \cdot e_2 = e_1$

b) $e_1 \cdot e_1 = e_1$, $e_1 \cdot e_2 = de_2$, $e_2 \cdot e_2 = e_1 + ge_2$,
with $d \neq 0$, $g \geq 0$, $g^2 < 4(1 - 2d)$

c) $e_1 \cdot e_1 = e_1$, $e_1 \cdot e_2 = \frac{1}{2} e_2$, $e_2 \cdot e_2 = - e_1$

d) $e_1 \cdot e_1 = e_1$, $e_1 \cdot e_2 = de_2$, $e_2 \cdot e_2 = - e_1 + ge_2$,
with $d \neq 0$, $d \neq 1$, $g^2 \neq 4d^2$, $g^2 < 4(2d - 1)$, $g \geq 0$.

4) $e_1 \cdot e_1 = e_1$, $e_1 \cdot e_2 = e_2$, $e_2 \cdot e_2 = - e_1$.

Distinct values of the parameters yield non-isomorphic algebras.

PROOF. Let e_1 be the idempotent element of $\mathfrak{A}$. Then there is a basis e_1, e_2 in $\mathfrak{A}$ such that exactly one of the following three cases holds:

CASE 1: $e_1 \cdot e_2 = 0$

CASE 2: $e_1 \cdot e_2 = e_1 + e_2$

CASE 3: $e_1 \cdot e_2 = de_2$, for $d \neq 0$.

To see this write $e_1 \cdot e_2 = ce_1 + de_2$. If $c = d = 0$ we have Case 1. If $c \neq 0$, $d = 0$ replace e_2 by $E_2 = - ce_1 + e_2$ to obtain Case 1. If $c = 0$, $d \neq 0$ we have Case 3. Finally suppose $c \neq 0$, and $d \neq 0$. Here replace e_2 by

$$E_2 = \frac{c}{1 - d} e_1 - e_2$$

(if $d \neq 1$) to obtain Case 3 again. If $d = 1$, $c \neq 0$ use

$$E_2 = \frac{1}{c} e_2$$

to obtain Case 2. Thus every $\mathfrak{A}$ lies in Case 1, 2, or 3.

Also an algebra $\mathfrak{A}$ can fall in just one of these three cases. For in $\mathfrak{A}$ the element e_1 is distinguished as the unique idempotent. The linear transformation $T_1 : \mathfrak{A} \longrightarrow \mathfrak{A}$ of multiplication by e_1 has a null space in just Case 1. In Case 2 T_1 has just one independent eigenvector, e_1. In Case 3 T_1 has two independent eigenvectors.

Now consider an algebra $\mathfrak{A}$ of Case 1. We consider three subcases designated by:

1A) $e_2 \cdot e_2 = ge_2$ for $g \neq 0$

1B) $e_2 \cdot e_2 = fe_1$ for $f > 0$

1C) $e_2 \cdot e_2 = \pm e_1 + ge_2$ for $g \neq 0$.

Suppose $e_1 \cdot e_1 = e_1$, $e_1 \cdot e_2 = 0$, $e_2 \cdot e_2 = fe_1 + ge_2$. Subcase 1A) is inadmissible since it allows a basis of idempotents in $\mathfrak{A}$. If $g = 0$, $f \neq 0$ then there are nilpotents unless $f > 0$. If $g \neq 0$, $f \neq 0$ we have Case 1C), upon replacing e_2 by $E_2 = |f|^{-1/2}e_2$.

In Case 1B) replace e_2 by $E_2 = f^{-1/2}e_2$ to obtain the result $e_1 \cdot e_1 = e_1$, $e_1 \cdot E_2 = 0$, $E_2 \cdot E_2 = e_1$ which is listed in the theorem. In Case 1C) the possibility $e_2 \cdot e_2 = - e_1 + ge_2$ is not allowed since it admits a basis of idempotents. Also in case $e_2 \cdot e_2 = e_1 + ge_2$ we must have $g^2 < 4$ to prevent a basis of idempotents. But the values $+ g$ and $- g$ yield isomorphic algebras under the automorphism $e_1 \longrightarrow e_1$, $e_2 \longrightarrow - e_2$. Therefore every algebra $\mathfrak{A}$ of Case 1 is listed under 1) in the theorem.

It is easy to check that each algebra of 1) has exactly one idempotent and no nilpotent of index two. Also in an algebra $\mathfrak{A}$ of 1) the element e_1 is distinguished by $e_1 \cdot e_1 = e_1$. Also the subspace λe_2 is the null space of T_1, multiplication by e_1. The condition $e_2 \cdot e_2 = e_1 + ge_2$ fixes the pair $\{e_2, - e_2\}$. But then the restriction $g \geq 0$ shows that no algebra in 1) corresponds to two distinct admissible values of g.

Now consider Case 2), $e_1 \cdot e_1 = e_1$, $e_1 \cdot e_2 = e_1 + e_2$, $e_2 \cdot e_2 = fe_1 + ge_2$. If $g \neq 0$ we replace e_2 by

$$E_2 = - \frac{g}{2} e_1 + e_2$$

to obtain $e_1 \cdot e_1 = e_1$, $e_1 \cdot E_2 = e_1 + E_2$, $E_2 \cdot E_2 = he_1$. Therefore we can always take $g = 0$ in Case 2. If $f \geq - 1$ there are two independent idempotents in $\mathfrak{A}$ and so we are lead to the Case 2) listed in the theorem.

Each algebra $\mathfrak{A}$ of Case 2) has no nilpotent of index two and only one idempotent. In an algebra $\mathfrak{A}$ of 2) the element e_1 is distinguished by $e_1 \cdot e_1 = e_1$. The affine space $\{e_2 + \alpha e_1\}$ is distinguished by the relation $e_1 \cdot e_2 = e_1 + e_2$. But the only member of $\{e_2 + \alpha e_1\}$ whose square is a scalar multiple of e_1 is e_2, which is thereby distinguished in $\mathfrak{A}$. Thus distinct values of f yield non-isomorphic algebras in Case 2).

Now consider Case 3, $e_1 \cdot e_1 = e_1$, $e_1 \cdot e_2 = de_2$, $e_2 \cdot e_2 = fe_1 + ge_2$, with $d \neq 0$. If $f = 0$ there is either a nilpotent of index two or two idempotents in $\mathfrak{A}$ so we must have $f \neq 0$. Replace e_2 by $E_2 = |f|^{-1/2}e_2$ and thereby we can assume that $f = + 1$ or

$f = -1$. Further we separate the case where the operator T_1, of multiplication by e_1, is the identity, that is $d = 1$. We are then lead to three subcases of 3:

I) $e_1 \cdot e_1 = e_1,\ e_1 \cdot e_2 = e_2,\ e_2 \cdot e_2 = \pm e_1 + ge_2$

II) $e_1 \cdot e_1 = e_1,\ e_1 \cdot e_2 = de_2,\ e_2 \cdot e_2 = e_1 + ge_2,$
$d \neq 0,\ d \neq 1$

III) $e_1 \cdot e_1 = e_1,\ e_1 \cdot e_2 = de_2,\ e_2 \cdot e_2 = -e_1 + ge_2,$
$d \neq 0,\ d \neq 1.$

An algebra $\mathfrak{A}$ of Case 3 has a distinguished element e_1 with $e_1 \cdot e_1 = e_1$. Thus the operator T_1 is an algebraic invariant and in 3I) it is the identity. Case 3 II) is distinguished from 3 III) in that the eigenvector e_2 has a square with $e_2 \cdot e_2 = e_1 + ge_2$, instead of $e_2 \cdot e_2 = -e_1 + ge_2$.

In 3 I) the possibility $e_2 \cdot e_2 = e_1 + ge_2$ is eliminated since such an algebra always has a basis of idempotents. Also the possibility $e_2 \cdot e_2 = -e_1 + ge_2$ is restricted since if $g^2 = 4$ there are nilpotents whereas if $g^2 > 4$ there are two idempotents.

Thus take $g^2 < 4$ and define new basis vectors $E_1 = e_1$, $E_2 = \lambda e_1 + \mu e_2$ where $\mu > 0$ and λ are defined by

$$\lambda = -\frac{\mu g}{2}, \quad \mu^2 = \frac{4}{4 - g^2} \quad .$$

Then one computes directly $E_1 \cdot E_1 = E_1,\ E_1 \cdot E_2 = E_2,\ E_2 \cdot E_2 = -E_1$, which is algebra 4) in the theorem. This algebra is the complex numbers.

We next show that in 3 II) the admissible values of the parameters are $g^2 < 4(1 - 2d)$ and $g = 0,\ d = \frac{1}{2}$. Use $(\alpha e_1 + \beta e_2)^2 = (\alpha^2 + \beta^2)e_1 + (2\alpha\beta d + \beta^2 g)e_2$ to show that there are neither nilpotents nor idempotents, with $\beta \neq 0$, for $g^2 < 4(1 - 2d)$. But if $g^2 > 4(1 - 2d)$ then there are two idempotents and so this is not admissible. If $g^2 = 4(1 - 2d)$ there are two idempotents in $\mathfrak{A}$ unless $g = 0,\ d = \frac{1}{2}$. Now the interchange of e_2 and $-e_2$ replaces g by $-g$ and so we can take $g \geq 0$. Thus algebras of 3 II) are those of Cases 3a) and 3b) in the theorem.

In Case 3b) the arithmetical condition $g^2 < 4(1 - 2d)$ shows that $d \neq 1,\ d \neq \frac{1}{2}$. Thus 3a) is different from any algebra in 3b). In an algebra $\mathfrak{A}$ of 3b) we can distinguish the idempotent e_1, then the pair $\{e_2, -e_2\}$ by $e_2 \cdot e_2 = e_1 + ge_2$. Since $g \geq 0$ there is just one algebra for each admissible parameter value.

Next consider the cases 3 III). We shall show that the admissible values are exactly those where $d \neq 0$, $d \neq 1$, $g \neq \pm 2d$, $g^2 < 4(2d - 1)$ and also $g = 0$, $d = \frac{1}{2}$. Use $(\alpha e_1 + \beta e_2)^2 = (\alpha^2 - \beta^2)e_1 + (2\alpha\beta d + \beta^2 g)e_2$. Since $g \neq \pm 2d$, there are no nilpotents of index two. Since $g^2 < 4(2d - 1)$, or $g = 0$, $d = \frac{1}{2}$, there are no idempotents other than e_1. Now suppose $d \neq 0$, $d \neq 1$ but $g^2 = 4d^2$ or $g^2 \geq 4(2d - 1)$ (excepting $g = 0$, $d = \frac{1}{2}$). If $g^2 = 4d^2$ there are nilpotents of index two. If $g^2 \geq 4(2d - 1)$ then there are idempotents other than e_2 (provided we exempt $g = 0$, $d = \frac{1}{2}$). Therefore we have the algebras of Cases 3c) and 3d) in the theorem.

Finally we must note that two parameter values in 3d) define non-isomorphic algebras. For such an algebra $\mathfrak{A}$ first distinguish the idempotent e_1, then the eigenspace of e_2, then the pair $\{e_2, - e_2\}$. Interchanging e_2 and $- e_2$ replaces g by $- g$ and since $g \geq 0$ there is a unique algebra for each admissible parameter value. Q. E. D.

Only the algebra 4) in Theorem 8 is associative. Moreover none of the other of these algebras in Theorem 8 is even power associative.

In Theorems 6, 7, and 8 we have found all real commutative algebras of dimension two. An interesting by-product of our classification is the following result.

COROLLARY. The only real commutative two-dimensional algebra which is power associative, yet not associative is:

$$e_1 \cdot e_1 = 0, \; e_1 \cdot e_2 = e_1, \; e_2 \cdot e_2 = 2e_2 \quad .$$

This is a Jordan algebra, that is, it satisfies the identity $(uv)u^2 = u(vu^2)$.

§5. QUALITATIVE BEHAVIOR OF QUADRATIC DIFFERENTIAL EQUATIONS IN THE AFFINE PLANE

We shall study the topological behavior of the solution curve familites for quadratic differential systems in the plane, with an isolated critical point at the origin. Thus we shall utilize the algebras listed in Theorems 7 and 8.

Consider the algebra 1) of Theorem 7:

$$e_1 \cdot e_1 = e_1, \; e_2 \cdot e_2 = e_2, \; e_1 \cdot e_2 = 0 \quad .$$

This corresponds to the quadratic differential system

$$\dot{x} = x^2, \quad \dot{y} = y^2 \ .$$

Corresponding to the three idempotents e_1, $e_1 + e_2$, and e_2 there are three lines through the origin which define solution curves. The computation

$$\dot{\theta} = \frac{x\dot{y} - y\dot{x}}{x^2 + y^2} = \frac{xy}{r^2}(y - x)$$

where $\theta = \arctan y/x$, displays these ray solutions along the lines $x = 0$, $y = 0$, and $y - x = 0$. We designate a ray solution by $+$ if the radial velocity is positive and $-$ if the radial velocity is negative. For each sector between the ray solutions we write an arrow to the right $\longrightarrow$ if the angular velocity is positive and an arrow to the left $\longleftarrow$ if the angular velocity is negative. Thus the differential system

$$\dot{x} = x^2, \quad \dot{y} = y^2$$

is represented by the combinatorial scheme

$$\longrightarrow + \longleftarrow + \longrightarrow + \longleftarrow - \longrightarrow - \longleftarrow - \longrightarrow \ ,$$

where the first $+$ (in the order of reading left to right) represents the positive half-axis solution.

The combinatorial scheme of each quadratic differential system, with an isolated critical point and a finite number of solution rays in the plane, is defined similarly. The combinatorial scheme is only specified up to a cyclic ordering; thus

$$\longleftarrow + \longrightarrow + \longleftarrow - \longrightarrow - \longleftarrow - \longrightarrow + \longleftarrow$$

also represents

$$\dot{x} = x^2, \quad \dot{y} = y^2 \ .$$

DEFINITION. Two quadratic differential systems, $\mathcal{Q}_1$ and $\mathcal{Q}_2$, with isolated critical points in the plane, are called geometrically equivalent in case $\mathcal{Q}_1$ has the same cyclic combinatorial scheme as $\mathcal{Q}_2$, or as $\mathcal{Q}_2$ transformed by an orientation reversing reflection.

DEFINITION. Two autonomous differential systems $\mathcal{S}_1$ and $\mathcal{S}_2$ on a differentiable manifold M^n are called topologically equivalent in case there exists a homeomorphism of the M^n onto itself carrying each critical point of $\mathcal{S}_1$ onto one of $\mathcal{S}_2$, and each directed (but non-parametrized) solution curve of $\mathcal{S}_1$ onto one of $\mathcal{S}_2$, and conversely carrying the solution curves of $\mathcal{S}_2$ onto those of $\mathcal{S}_1$. If M^n is orientable and the homeomorphism is orientation preserving, then $\mathcal{S}_1$ and $\mathcal{S}_2$ are o-topologically equivalent.

LEMMA. Let $\mathcal{Q}_1$ and $\mathcal{Q}_2$ be quadratic differential systems, with isolated critical points and finitely many ray solutions in the affine plane. Let σ_1 and σ_2 be their corresponding cyclic combinatorial schemes, as above. Then $\mathcal{Q}_1$ and $\mathcal{Q}_2$ are o-topologically equivalent if σ_1 and σ_2 are the same.

PROOF. This is an immediate consequence of the general theory of separatrix configurations developed in Theorem 7.1 of [14]. Q.E.D.

It should be noted that topologically equivalent differential systems can fail to be equivalent under a differentiable homeomorphism.

THEOREM 9. Each quadratic differential system, with an isolated critical point in the affine plane, is geometrically equivalent to exactly one of the following six differential systems:

I. $\dot{x} = x^2$, $\dot{y} = y^2$ which represent Cases 1), 2a), 2c), and 3b) for $\alpha > 2$, $0 < \beta < 2$, $4 - \alpha\beta < 0$, and 3b) for $\alpha < -1$, $-1 < \beta < 0$ in Theorem 7.

II. $\dot{x} = 3x^2 + 2xy$, $\dot{y} = y^2 + 2xy$, which represents Cases 3b) for $\alpha > 2$, $0 < \beta < 2$, $4 - \alpha\beta > 0$, and 3b) for $\alpha < -1$, $0 < \beta < 1$, in Theorem 7.

III. $\dot{x} = 2x^2 + 2xy$, $\dot{y} = -y^2 + 2xy$, which represents Cases 2b), 3a) for $\beta < 0$, and 3a) for $\beta > 2$, in Theorem 7.

IV. $\dot{x} = 2x^2 + 2xy$, $\dot{y} = y^2 + 2xy$, which represents Cases 3a) for $0 < \beta < 2$ in Theorem 7.

All of the above have four or six directions of approach to the origin, and next there are systems with just two directions of approach.

V. $\dot{x} = x^2 + y^2$, $\dot{y} = y^2$ which represents Cases 1), 3a), and 3b) of Theorem 8.

VI. $\dot{x} = x^2 - 2y^2 + 2xy$, $\dot{y} = 2xy$ which represents Cases 2), 3c), 3d) and 4) of Theorem 8.

PROOF. By examination of

$$\dot{\theta} = \frac{x\dot{y} - y\dot{x}}{x^2 + y^2}$$

we sketch the solution curves and find the schemes for the quadratic differential systems as follows.

In Theorem 7, we find

1) $\longrightarrow + \longleftarrow + \longrightarrow + \longleftarrow - \longrightarrow - \longleftarrow - \longrightarrow$

2a) $\longleftarrow + \longrightarrow + \longleftarrow - \longrightarrow - \longleftarrow - \longrightarrow + \longleftarrow$

2b) $\longrightarrow + \longrightarrow + \longleftarrow - \longleftarrow - \longrightarrow$

2c) $\longrightarrow + \longleftarrow + \longrightarrow + \longleftarrow - \longrightarrow - \longleftarrow - \longrightarrow$ for all α

3a) $\longrightarrow + \longleftarrow + \longleftarrow - \longrightarrow - \longrightarrow$ for $\beta < 0$

$\longleftarrow + \longleftarrow + \longrightarrow - \longrightarrow - \longleftarrow$ for $0 < \beta < 2$

$\longrightarrow + \longrightarrow + \longleftarrow - \longleftarrow - \longrightarrow$ for $\beta > 2$

3b) $\longleftarrow + \longrightarrow + \longleftarrow + \longrightarrow - \longleftarrow - \longrightarrow - \longleftarrow$
for $\alpha > 2,\ 0 < \beta < 2,\ 4 - \alpha\beta > 0,$ and for
$\alpha < -1,\ 0 < \beta < 1$

$\longrightarrow + \longleftarrow + \longrightarrow + \longleftarrow - \longrightarrow - \longleftarrow - \longrightarrow$
for $\alpha > 2,\ 0 < \beta < 2,\ 4 - \alpha\beta < 0$ and for
$\alpha < -1,\ -1 < \beta < 0.$

In Theorem 8, we find

1) $\longrightarrow + \longleftarrow - \longrightarrow$ for all g

2) $\longleftarrow + \longrightarrow - \longleftarrow$ for all f

3a) $\longrightarrow + \longleftarrow - \longrightarrow$

3b) $\longrightarrow + \longleftarrow - \longrightarrow$ for all d and g

3c) $\longleftarrow + \longrightarrow - \longleftarrow$

3d) $\longleftarrow + \longrightarrow - \longleftarrow$ for all d and g

4) $\longleftarrow + \longrightarrow - \longleftarrow$

An examination of these combinatorial schemes, and those corresponding to the above differential systems after an orientation reversing map $x \longrightarrow -x$, $y \longrightarrow y$, yields the geometrical classification stated in the theorem. Q.E.D.

A necessary and sufficient condition that quadratic differential equations $\mathcal{Q}_1$ and $\mathcal{Q}_2$, with isolated critical points in the plane, be topologically equivalent is that they have the same separatrix configurations, cf. [14]. An examination of Figures I through VI shows that I and

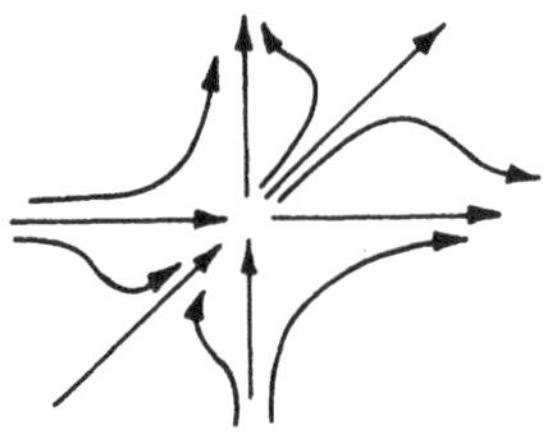

Fig. I.

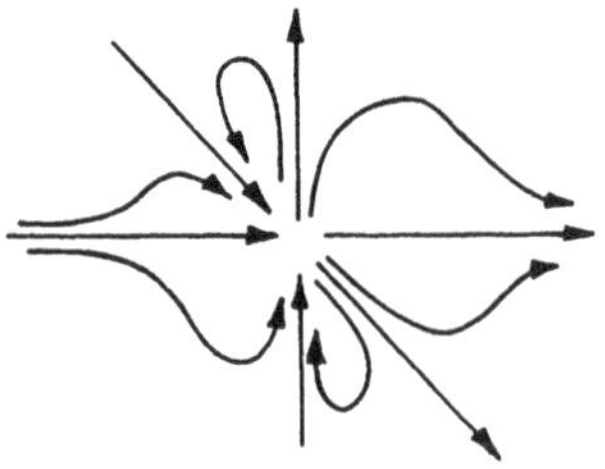

Fig. II.

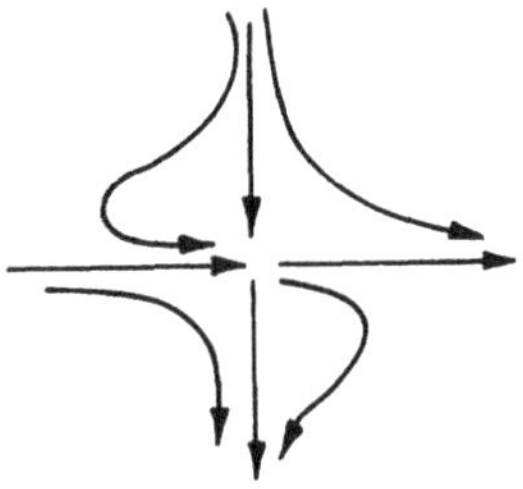

Fig. III.

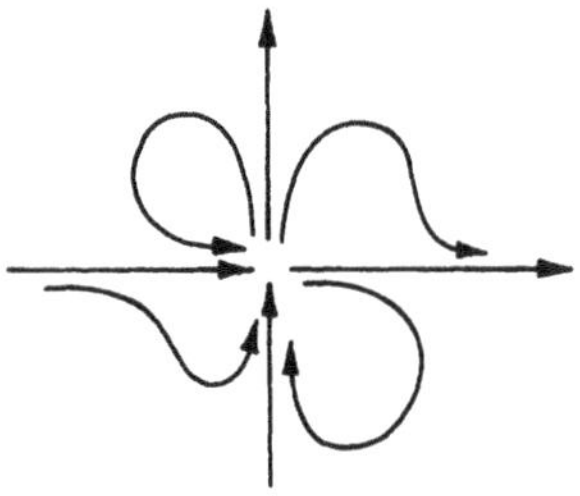

Fig. IV.

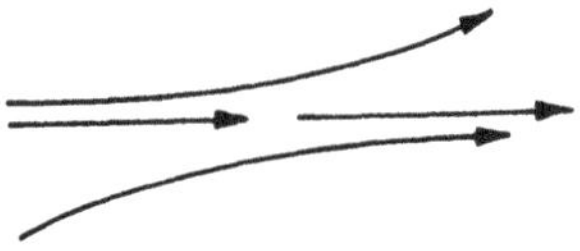

Fig. V.

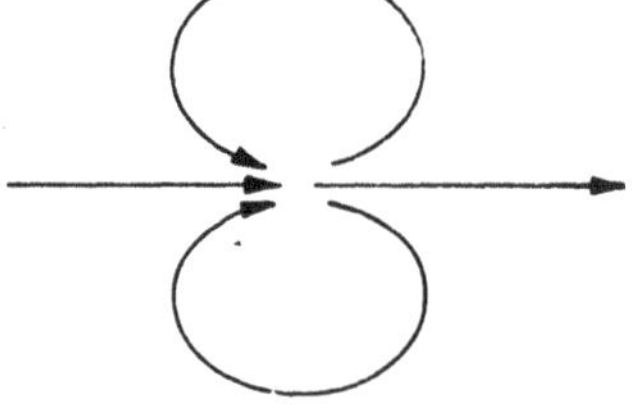

Fig. VI.

III are topologically equivalent and also II and IV are topologically equivalent. Thus there are only four topologically different planar quadratic differential systems having isolated critical points.

§6. QUADRATIC DIFFERENTIAL EQUATIONS AND LINE ELEMENT FIELDS IN REAL PROJECTIVE SPACE

Consider a quadratic differential system in the affine n-space R^n

$$\mathcal{Q}) \qquad \frac{dx^i}{dt} = \sum_{j,k=1}^{n} a^i_{jk} x^j x^k \qquad i = 1, 2, \ldots, n \ .$$

Now $\mathcal{Q}$ defines a non-directed line-element field $\mathcal{L}$ in R^n, with the critical points of $\mathcal{Q}$ deleted. We shall only consider $\mathcal{Q}$ with isolated critical points in R^n in this section.

Now $\mathcal{L}$ can be extended to a unique differentiable line element field on the real projective space P^n, with a minimal set of critical points occurring on the P^{n-1} at infinity. We define linear equivalence of two line-element fields $\mathcal{L}_1$ and $\mathcal{L}_2$ on R^n, and topological equivalence of $\mathcal{L}_1$ and $\mathcal{L}_2$ on either R^n or P^n just as was done for differential systems (except that here the solution curves have no direction or sense).

THEOREM 10. Let $\mathcal{Q}_1$ and $\mathcal{Q}_2$ be quadratic differential systems, with isolated critical points in R^n. Let the corresponding line-element fields be $\mathcal{L}_1$ and $\mathcal{L}_2$ in R^n. If $\mathcal{Q}_1$ and $\mathcal{Q}_2$ are linearly equivalent in R^n, then so are $\mathcal{L}_1$ and $\mathcal{L}_2$. Conversely if $\mathcal{L}_1$ and $\mathcal{L}_2$ are linearly equivalent in R^n, then $\mathcal{Q}_1$ is linearly equivalent to $\mathcal{Q}_2$, or to the quadratic differential system $c\,\mathcal{Q}_2$ obtained by multiplying each coefficient of $\mathcal{Q}_2$ by the fixed positive number c.

PROOF. If $\mathcal{Q}_1$ and $\mathcal{Q}_2$ are linearly equivalent in R^n, then certainly $\mathcal{L}_1$ and $\mathcal{L}_2$ are linearly equivalent in R^n.

Next suppose that after a linear transformation of R^n, $\mathcal{L}_1$ coincides with $\mathcal{L}_2$. Then there are two quadratic differential systems

$$\mathcal{Q}_2) \qquad \dot{x}^i = P^i(x), \qquad i = 1, 2, \ldots, n \ ,$$

and the transform of $\mathcal{Q}_1$, say

$$\hat{\mathcal{Q}}_1) \qquad \dot{x}^i = \hat{P}^i(x) \ , \qquad i = 1, 2, \ldots, n \ ,$$

which yield the same line-element field. Therefore $P^i(x) = \lambda(x)\hat{P}^i(x)$ where $\lambda(x) \neq 0$ in $R^n - \{0\}$, the space R^n with the origin deleted. Since

$$\lambda(x)^2 = \sum_{i=1}^{n} [P^i(x)]^2 \Big/ \sum_{i=1}^{n} [\hat{P}^i(x)]^2 \ ,$$

we see that $\lambda(x)$ is a real analytic function on $R^n - \{0\}$. By the identity theorem for power series, $\lambda(x)$ is a constant in a neighborhood of each point and is therefore everywhere constant in $R^n - \{0\}$. Thus $\lambda(x) = \lambda_0$.

Since $\mathcal{Q}_1$ and $-\mathcal{Q}_1$ are linearly equivalent, we can take $\lambda_0 = c > 0$. Q.E.D.

THEOREM 11. Let $\mathcal{Q}_1$ and $\mathcal{Q}_2$ be quadratic differential systems, with isolated critical points in R^n. Let $\mathcal{L}_1$ and $\mathcal{L}_2$ be the corresponding line-element fields in the real projective space P^n. If $\mathcal{Q}_1$ and $\mathcal{Q}_2$ are linearly equivalent in R^n, then $\mathcal{L}_1$ and $\mathcal{L}_2$ are topologically equivalent in P^n.

PROOF. A linear transformation of R^n onto itself can be extended to a projective transformation of P^n onto itself. This projective transformation is a homeomorphism of P^n onto itself which carries the solution curves of $\mathcal{L}_1$ onto those of $\mathcal{L}_2$, since $\mathcal{L}$ is a unique extention of $\mathcal{Q}$. Q.E.D.

It may happen that a topological equivalence of $\mathcal{L}_1$ and $\mathcal{L}_2$ in R^n can not be extended to a topological equivalent in P^n. We shall not classify line-element fields in P^n; not even in the projective plane P^2.

§7. QUADRATIC DIFFERENTIAL EQUATIONS IN 3-SPACE R^3

We shall only investigate real commutative 3-dimensional algebras which are the direct product of a 2-dimensional algebra and the real numbers R.

REMARK. Let $\mathfrak{A}$ be a real commutative 2-dimensional algebra

THE LINE-ELEMENT FIELD FOR $\dot{x} = x^2$, $\dot{y} = y^2$ IN P^2

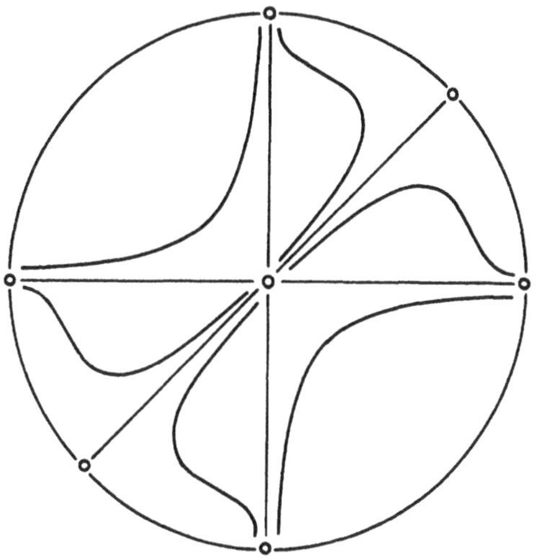

FIGURE VII

and R the real numbers. Then $\mathfrak{A} \times R$ contains a nilpotent of index two if and only if $\mathfrak{A}$ does. This is easily seen from the multiplication $(e_1, \lambda_1) \cdot (e_2, \lambda_2) = (e_1 \cdot e_2, \lambda_1\lambda_2)$ in $\mathfrak{A} \times R$.

THEOREM 12. Let $\mathfrak{A}_1$ and $\mathfrak{A}_2$ be real commutative 2-dimensional algebras containing no nilpotents of index two. Then $\mathfrak{A}_1 \times R$ is isomorphic with $\mathfrak{A}_2 \times R$ if and only if $\mathfrak{A}_1$ is isomorphic with $\mathfrak{A}_2$.

PROOF. Assume $\mathfrak{A}_1 \times R$ is isomorphic with $\mathfrak{A}_2 \times R$. Now $\mathfrak{A}_1$ (or $\mathfrak{A}_1 \times 1$) is a 2-dimensional ideal in $\mathfrak{A}_1 \times R$. We shall consider the image $\mathfrak{A}_1'$ of $\mathfrak{A}_1$ in $\mathfrak{A}_2 \times R$.

Let e_1, e_2, e_3 be a basis for $\mathfrak{A}_2 \times R$ such that e_1, e_2 are a basis for $\mathfrak{A}_2$. Take e_1, e_2 so that $\mathfrak{A}_2$ has one of the canonical forms for commutative 2-dimensional algebras found in Theorems 7 and 8, and also demand $e_1 \cdot e_3 = e_2 \cdot e_3 = 0$, $e_3 \cdot e_3 = e_3$.

If $\mathfrak{A}_1$ and $\mathfrak{A}_2$ are each isomorphic with the algebra $\mathfrak{C}$, having a basis of orthogonal idempotents (algebra 1) in Theorem 7), then the theorem holds.

Consider the 2-dimensional subalgebra $\mathfrak{A}_1'$ in $\mathfrak{A}_2 \times R$. If $\mathfrak{A}_1' = \mathfrak{A}_2 \times e_3$ the theorem holds. Otherwise choose a basis $u = \alpha e_1 + \beta e_2 + \gamma e_3$ and $v = \xi e_1 + \eta e_2 + \zeta e_3$ of $\mathfrak{A}_1'$. Since $\mathfrak{A}_1'$ is an ideal $u \cdot e_3 = \gamma e_3$ must lie in $\mathfrak{A}_1'$. Then

1) $\gamma = \zeta = 0$ in which case $\mathfrak{A}_1' = \mathfrak{A}_2$ and the theorem holds,

or

2) $\gamma \neq 0$ (or $\zeta \neq 0$) and e_3 is in $\mathfrak{A}_1'$.

In the second case choose a basis $w = \lambda e_1 + \mu e_2$ and e_3 for $\mathfrak{A}_1'$. In fact $w \cdot w = \nu w$ for some real $\nu \neq 0$, so $\mathfrak{A}_1' \cong \mathfrak{C}$. In this case $\mathfrak{A}_1 \cong \mathfrak{C}$.

Now repeat the argument for the image $\mathfrak{A}_2'$ of $\mathfrak{A}_2$ in $\mathfrak{A}_1 \times R$ to show that either $\mathfrak{A}_2' = \mathfrak{A}_1 \times 1$ or else $\mathfrak{A}_2' \cong \mathfrak{C}$. In any case $\mathfrak{A}_1 \cong \mathfrak{A}_2$. Q.E.D.

Thus we have constructed an infinite class of quadratic differential systems in R^3, with isolated critical points, such that no two of them are linearly equivalent.

BIBLIOGRAPHY

[1] ALBERT, A. A., "Power associative rings," Trans. Am. Math. Soc. 64 (1948), pp. 552-593.

[2] BROUWER, L. E. J., "Over de loodrechte trajectorien der baankrommen eener vlakke eenladige projectieve groep," Nieuw. Archief. (2) 11, (1915), pp. 265-290.

[3] COLEMAN, C., Homogeneous differential equations, Princeton University thesis, 1955.

[4] DULAC, H., Points singuliers des équations différentielles, Mém. Sci. Math. Fasc. 61, Paris (1934).

[5] ENGEL, J. H., "A verification of Lanchester's Law," J. of Op. Research Soc. of Am. (1954), pp. 163-171.

[6] FORSTER, H., "Uber das Verhalten der Integralkurven einer gewöhnlichen Differentialgleichung erster Ordnung in der Umgebung eines singularen Punktes," Math. Zeitsch. 43, (1938), pp. 271-320.

[7] FROMMER, M., "Die Integralkurven einer gewöhnlichen Differenzialgleichung in der Umgebung rationaler Unbestimmtheitsstellen," Math. Ann. 99, (1928), pp. 222-272.

[8] GAVURIN, M., "On systems of differential equations of the form $y' = Ay^2 - 2By + C$," Doklady Akad. Nauk, SSSR 84, (1952), p. 205.

[9] KASNER, E., "Solutions of the Einstein equations involving functions of only one variable," Trans. Am. Math. Soc. 27, (1925), pp. 155-162.

[10] KESTIN, J. and ZAREMBA, A., "Geometrical methods in the analysis of ordinary differential equations," Appl. Sci. Research Bull. 3, (1953), pp. 149-189.

[11] LEFSCHETZ, S., "Notes on Differential Equations," Contributions II, Princeton Annals Studies 29, (1952), pp. 61-73.

[12] ——————, "The ambiguous case in planar differential systems," Bol. de la Soc. Mat. Mexicana (1957), pp. 63-74.

[13] LIAGHINA, L., "The integral curves of the equations $y' = \frac{ax^2+bxy+cy^2}{dx^2+exy+fy^2}$," Uspeki Matem. Nauk, VI, 2(42), (1951), pp. 171-183.

[14] MARKUS, L., "Global structure of ordinary differential equations in the plane," Trans. Am. Math. Soc. 76, (1954), pp. 127-148.

[15] NEMICKII, V. and STEPANOV, V., Qualitative Theory of Differential Equations, Princeton, 1959, esp. Ch. II.

[16] SANSONE, G. and CONTI, R., Equazioni Differenziali Non Lineari, Rome, 1956, esp. Chapter II.

[17] STEBAKOV, S., "Qualitative investigation of the system $\dot{x} = P(x, y)$, $\dot{y} = Q(x, y)$ by means of isoclines," Doklady Akad. Nauk. SSSR. 82, (1952), pp. 677-680.

[18] VOLTERRA, V., Lecons sur la théorie mathematique de la lutte pour la vie, Paris, 1931.

University of Minnesota
Minneapolis, Minnesota

IX. SUR UNE PROPRIÉTÉ DE L'ENSEMBLE DES TRAJECTOIRES BORNÉES DE CERTAINS SYSTÈMES DYNAMIQUES

Georges Reeb

Nous exposerons la remarque (cf. théorème 1) qui fait l'objet de la présente note sur un cas particulier, les généralisations dont cette remarque est susceptible étant immédiates.

Considérons la variété à bord $S_n \times I$ (où I est l'intervalle $[-1+1]$ et où S_n est la sphère à n dimensions munie de la structure d'espace de Riemann usuelle) et désignons par E un champ de vecteurs sur $S_n \times I$ (de classe C_1) vérifiant les propriétés suivantes:

(i) E ne s'annule en aucun point de $S_n \times I$.

(ii) Sur $S_n \times \{-1\}$ le champ E est dirigé vers l'intérieur de $S_n \times I$, tandis que sur $S_n \times \{1\}$ il est dirigé vers l'extérieur de $S_n \times I$.

Il est clair que de tels champs E existent. Parmi les trajectoires de E nous pouvons distinguer les trajectoires entrantes (c. à d. celles qui rencontrent $S_n \times \{-1\}$) et les trajectoires sortantes (celles qui recontrent $S_n \times \{1\}$). Désignons la réunion des trajectoires entrantes par R et la réunion des trajectoires sortantes par S. Les ensembles S et R sont ouverts et homéomorphes à $S_n \times [0, 1]$ si $S \cap R = \emptyset$.

Nous désignerons par H l'hypothèse suivante:

(H) $$S \cap R = \emptyset \quad .$$

Le complémentaire compact de $S \cap R$ est noté A. L'ensemble A mérité d'être appelé l'ensemble des <u>trajectoires bornées</u> de E.

Si H est vérifiée, l'ensemble A jouit d'un certain nombre de propriétés dont voici quelques unes des plus importantes:

(1) $H(A) \cong H(S_n)$ (en homologie de Čech)

(La propriété (1) est évidente si on remarque que S et R sont homéomorphes à $S_n \times [0, 1]$ et que l'homologie de S_n/A vérifie l'excision complète).

(2) l' ensemble A admet des transformations, homotopes à l'identité, sans points fixes.

(La propritété (2) est évidente puisque A est un ensemble de trajectoires bornées).

La remarque que nous avons en vue résulte de (1) et (2):

THÉORÈME 1. Si n est pair et si H est vérifiée, alors A n'est pas homéomorphe à un polyèdre, ni localement connexe.

En effet si A était localement connexe le théorème de points fixes lui serait applicables donc $\chi(A) = 0$ et il y aurait contradiction avec $\chi(A) = \chi(S_n) = 2$.

Le théorème met en évidence une propriété topologique de A, assez intéressante sous réserve d'établir les deux points suivants:

(i) H est vérifiée par certains champs E même si n est pair.

(ii) si n est impair, l'ensemble A peut-être un polyèdre.

Nous allons maintenant construire deux exemples qui apporteront la preuve de ces deux points.

1. Exemple montrant que si n est impair il se peut que A soit un polyèdre

Un tel example est très facile à fabriquer et correspond à un phénomène bien familier. Soit u un champ de vecteurs de classe C_1 sur S_n, ne s'annulant en aucun point de S_n. On définira $E(x, t)$ par ses deux composantes dans $S_n \times I$, à savoir:

$$E(x, t) = [u(x), |t|] \quad .$$

Ici l'ensemble A se réduit à $S_n \times \{0\}$.

2. Exemple montrant que H peut-être vérifiée même si n est pair

Nous allons décrire cet exemple dans le cas particulier où $n = 2$, le cas général pouvant être calqué assez simplement sur le cas où

$n = 2$. A cet effect on considèrera $S_2 \times I$ comme l'espace obtenu en recollant convenablement (le bord $S_n \times \{1\}$ étant recollé sur $S_n \times \{-1\}$) quatre exemplaires de $S_2 \times I$; nous définirons E dans chacun de ces exemplaires, d'où le champ E total.

Mais au préalable introduisons dans S_2 un axe polaire et un système de coordonnées sphériques (θ latitude, φ longitude). Les calottes définies par

$$1 \leq \theta \leq \frac{\pi}{2} \text{ et } -\frac{\pi}{2} \leq \theta \leq -1$$

seront désignées respectivement par Γ_0 et Γ_1; la zone $-1 \leq \theta \leq 1$ est notée B. On désignera par e le champ de vecteurs de composantes $(\sigma, \cos\theta)$; ce champ est analytique et admet les parallèles de S_2 comme lignes intégrales. Une rotation de $\frac{\pi}{2}$ autour d'un diamètre équatorial de S_2 transformera les coordonnées (θ, φ) et les éléments Γ_0, Γ_1, B, e en nouveau système de coordonnées et de nouveaux éléments qui seront notés $(\bar{\theta}, \bar{\varphi})$, $\bar{\Gamma}_0$, $\bar{\Gamma}_1$, $\bar{B}$, $\bar{e}$. Nous définissons maintenant E dans les quatres exemplaires de $S_2 \times I$:

(i) <u>Premier exemplaire</u>:

Les composantes de E sont:

$$(e_0, 1 - t) \quad \text{si} \quad (\theta, \varphi) \in B$$

$$(e_0, (1 - t) + (\theta - 1)) \quad \text{si} \quad (\theta, \varphi) \in \Gamma_0$$

$$(e_0, (1 - t) - (\theta + 1)) \quad \text{si} \quad (\theta, \varphi) \in \Gamma_1 \quad .$$

On remarque que sur $S_2 \times \{-1\}$ le champ E est dirigé vers l'intérieur de $S_2 \times I$. Les trajectoires de E ne traversent pas $B \times \{1\}$ mais elles traversent $\Gamma_0 \times \{1\}$ et $\Gamma_1 \times \{1\}$.

(ii) <u>Deuxième exemplaire</u>:

Les composantes de E sont:

$$(e_0 \frac{(1-t)}{2}, \frac{t+1}{2}) \quad \text{si} \quad (\theta, \varphi) \in B \quad .$$

$$(e_0 \frac{(1-t)}{2}), ((1 + t) + (\theta - 1)) (\frac{(1-t)}{2} + \frac{1+t}{2}) \quad \text{si} \quad (\theta, \varphi) \quad \Gamma_0$$

$$(e_0 \frac{(1-t)}{2}, ((1 + t) - (\theta + 1)) \frac{1-t}{2} + \frac{1+t}{2}) \quad \text{si} \quad (\theta, \varphi) \in \Gamma_1 \quad .$$

On remarquera que $t = 1$ implique $E(x, 1) = (\sigma, 1)$.

(iii) Troisième exemplaire:

$$\left((t+1)\, e_1,\ \frac{1-t}{2}\right) \quad \text{si} \quad (\bar\theta, \bar\varphi) \in \bar B$$

$$\left((t+1)\, e_1,\ \frac{1-t}{2} + (\bar\theta - 1)(1+t)\right) \quad \text{si} \quad (\bar\theta, \bar\varphi) \in \Gamma_0$$

$$\left((t+1)\, e_1,\ \frac{1-t}{2} - (\bar\theta + 1)(1+t)\right) \quad \text{si} \quad (\bar\theta, \bar\varphi) \in \Gamma_1$$

(iv) Quatrième exemplaire:

$$\left(2e_1,\ \frac{t+1}{2}\right) \quad \text{si} \quad (\bar\theta, \bar\varphi) \in \bar B$$

$$\left(2e_1,\ 2(\bar\theta - 1) + (t+1)\right) \quad \text{si} \quad (\bar\theta, \bar\varphi) \in \bar\Gamma_0$$

$$\left(2e_1,\ -2(\bar\theta + 1) + (t+1)\right) \quad \text{si} \quad (\bar\theta, \bar\varphi) \in \bar\Gamma_1 \ .$$

Le champ E que nous venons de définir n'est pas exactement de classe C_1 (en effet sur les zones de recollement et sur $B \cap (\Gamma_0 \cup \Gamma_1) \times I$) le champ E est seulement de classe C_0). Cependant notre exemple conserve sa valeur démonstrative; il serait d'ailleurs facile, en modifiant légèrement les définitions analytiques, de donner effectivement un exemple de classe C_∞ (Voir également la dernière phase de cette note).

Pour voir que notre exemple établit effectivement la propriété voulue il suffit de remarquer que les trajectoires de E qui entrent par $B \times \{-1\}$ sont arretées dès le bord du premier exemplaire de $S_2 \times I$, tandis que les trajectoires qui entrent par $(\Gamma_0 \cup \Gamma_1) \times \{-1\}$ traversent les deux premiers exemplaires de $S_2 \times I$ mais sont arretées au bord du troisième (parce que les calottes Γ_0 et Γ_1 ont été choisies suffisamment petites).

Il reste une dernière remarque: les exemples ci-dessus sont peut-être critiquables parce qu'une légère modification de E peut entraîner la non-validité de H; on peut facilement parer à ce dernier inconvénient comme le suggère dans le cas où n est impair l'example du champ:

$$\left(u(x)t,\ (t + \tfrac{1}{2})(t - \tfrac{1}{2})\right) \ .$$

Si n est pair on peut également faire une construction analogue. Cette dernière remarque, combinée avec le théorème d'approximation de Weierstrass établit le fait suivant: le théorème 1 reste valable même si le champ E est supposé analytique.

X. ON LAGRANGE STABLE MOTIONS IN THE NEIGHBORHOOD OF CRITICAL POINTS

Pinchas Mendelson

1. We consider an autonomous system of ordinary differential equations

$$(1.1) \qquad \frac{dx_i}{dt} = f_i(x_1, \ldots, x_n), \qquad (i = 1, 2, \ldots, n),$$

defined in some region Ω in the n-dimensional Euclidean space E^n and having a unique solution there.

We shall employ the standard vector notation: the n-tuplet $(x_1, \ldots, x_n)$ will be denoted by x.

Let P be a point of Ω. The (unique) trajectory or orbit, of system (1.1), which passes through P, is denoted by Γ_P; its parametrization by (the time) t is assumed to be chosen in such a way that $\Gamma_P(0) = P$. The positive semi-orbit $\{\Gamma_P(t) \mid t \geq 0\}$ and the negative semi-orbit $\{\Gamma_P(t) \mid t \leq 0\}$ are designated by Γ_P^+ and Γ_P^-, respectively.

Let ω be an open subset of Ω and let $H(\omega)$ be the boundary of ω in Ω. A point $p \in H(\omega)$ is said to be a <u>point of egress</u> from the set ω (with respect to the system (1.1) and the set Ω) if there exists an $\eta_1 > 0$ such that $\Gamma_P(t) \in \omega$ for all $-\eta_1 < t < 0$. If, moreover, there exists an $\eta_2 > 0$ such that $\Gamma_P(t) \in \Omega - \bar{\omega}$ for all $0 < t < \eta_2$, then P is said to be a <u>point of strict egress</u> from ω.

Consider ω fixed. The set of all points of egress from ω will be called E. The set of all points of strict egress from ω will be called E^*. Clearly $E^* \subset E$.

If $P \in \omega$ and $\Gamma_P^+ \not\subset \omega$, then Γ_P^+ must intersect $E \subset H(\omega)$; we denote the first such point of intersection by $\tilde{P}$ (note that the open arc of the semi-orbit Γ_P^+ between P and $\tilde{P}$ is contained in ω).

2. The following theorem, due to T. Ważewski ([3], [4]) will be used in the sequel:

THEOREM (T. Ważewski). Let every point of egress from ω (with respect to the system (1.1) and the set Ω) be a point of strict egress, and let a certain set Z satisfy the following conditions:

(2.1)
(i) $Z \subset \omega \cup E$
(ii) $Z \cap E$ is a deformation retract of E
(iii) $Z \cap E$ is not a deformation retract of Z,

then there exists a point $P_0 \in \omega \cap Z$ such that $\Gamma^+_{P_0} \subset \omega$.

The above theorem, sometimes referred to as Ważewski's principle, was formulated by A. Pliś [2] in a somewhat more general form. His statement of the principle is as follows:

THEOREM (A. Pliś). Let $E = E^*$ and let certain sets Z, E_1 satisfy the following conditions:

(2.2)
(i) $E_1 \subset E$
(ii) $Z \subset \omega \cup E_1$, $Z \neq \Phi$
(iii) E_1 is not a quasi-isotopic deformation retract[1] of $Z \cup E_1$ in $\omega \cup E_1$,

then there exists a point $P_0 \in \omega \cap Z$ such that either $\Gamma^+_{P_0} \subset \omega$ or $\tilde{P}_0 \in E - E_1$.

The results contained in this note will be obtained by a suitable application of Ważewski's principle in the neighborhood of an isolated critical point of system (1.1).

3. Let Φ be a system such as (1.1) and let $0 \in \Omega$ be a (not necessarily isolated) critical point of Φ. Let ω be an open topological n-sphere containing 0 in its interior; we assume ω small enough so

---

[1] Let I denote the closed unit interval $[0, 1]$. A set A is a quasi-isotopic deformation retract of a set B in the topological space E if there exists a continuous mapping $\varphi : B \times I \longrightarrow E$ having the following properties:

(i) $\varphi(P, 0) = P$ for all $P \in B$.

(ii) $\varphi(P, s) = P$ for all $P \in A$ and any $0 \leq s \leq 1$.

(iii) $\varphi(P, 1) \in A$ for all $P \in B$.

(iv) $\varphi_s : B \longrightarrow E$ defined by $\varphi_s(P) = \varphi(P, s)$, $P \in B$, s fixed, is a homeomorphism for all $0 \leq s < 1$.

that $\omega \subset \Omega$. The sets $E \subset H(\omega)$ and $E^* \subset H(\omega)$ are comprised, as above, of the points of egress and the points of strict egress from ω, respectively.

> THEOREM. The equality $E = E^*$ is a sufficient condition for the existence of a semi-orbit (other than 0) wholly contained in ω.

PROOF. We always have $E^* \subset E$. If E is empty the equality $E^* = E$ holds trivially. In this case it follows from the definition of E that no motion which starts in ω can intersect $H(\omega)$ when continued in the positive direction. In other words, every orbit which passes through a point $Q \in \omega$ satisfies $\Gamma_Q^+ \subset \omega$.

Similarly, if $H(\omega) - E$ is empty and $E = E^*$, every point of $H(\omega)$ is a point of strict egress from ω; therefore no motion which starts in ω can intersect $H(\omega)$ when continued in the negative direction. Every orbit which passes through $Q \in \omega$ satisfies $\Gamma_Q^- \subset \omega$.

We may therefore assume that $E = E^*$ is a non-empty proper subset of $H(\omega)$.

The set E may be connected, or it may be disconnected; in the latter case we denote its components by E_α where α ranges over some index set $\mathfrak{A}$. We shall treat these cases separately.

Suppose first that E is disconnected. Then $E = \bigcup_{\alpha \in \mathfrak{A}} E_\alpha$ and the cardinality of $\mathfrak{A}$ is at least 2.

Let P be any point of ω other than 0. Let P_α be an arbitrary point of E_α and let π_α be a polygonal path with the following properties:

(i) π_α has P for its initial point and P_α for its terminal point.

(ii) The set π_α, except for its terminal point P_α, is contained in $\omega - 0$.

Let $Z = \bigcup_{\alpha \in \mathfrak{A}} (\pi_\alpha \cup E_\alpha)$. Clearly the set Z is connected. Furthermore, it follows from the above construction that $Z \cap E = \bigcup_{\alpha \in \mathfrak{A}} E_\alpha$ and that $0 \notin Z$.

Thus $Z \cap E = E$ is (trivially) a deformation retract of E. On the other hand, $Z \cap E$ — being disconnected — is clearly not a deformation retract of Z, which is connected. Ważewski's theorem may therefore be applied to the set Z and we conclude that there exists a point $Q \in Z$ such that the semi orbit $\Gamma_Q^+ \subset \omega$. This completes the proof in the case when E is disconnected.

Assume next that E has only one component. Let P be an arbitrary point of E. Let $H_\delta(0)$ be the surface of a sphere of radius δ and center 0; δ is assumed small enough so that $H_\delta(0) \subset \omega$. Let π be a polygonal path with the following properties:

(i) The initial point of π is on $H_\delta(0)$; its terminal point is P,

(ii) The set π, except for its terminal point P is contained in $\omega - 0$.

Let $Z = H_\delta(0) \cup \pi \cup E$. Clearly $Z \cap E = E$ and $Z \neq \Phi$. Let $\omega_1 = \omega - 0$. Then $H(\omega_1) = H(\omega) \cup 0$. Since 0 is a critical point, 0 is not a point of egress from ω_1 (with respect to the system Φ and the set Ω). Hence the set E' of points of egress from ω_1 is identical with the set E. Clearly E, as a consequence of the Jordan-Brouwer Separation Theorem, is not a quasi-isotopic deformation retract of $Z \cup E$ in $\omega_1 \cup E$. Conditions (2.2) are satisfied, with $E_1 = E$; there exists a point $Q \in \omega_1 \cap Z$ such that $\Gamma_Q^+ \subset \omega_1 \subset \omega$. This completes the proof of the theorem.

We recall that a trajectory Γ_P is said to be positively stable in the sense of Lagrange, or simply L^+-stable if $\overline{\Gamma_P^+}$ is compact; it is L^--stable if $\overline{\Gamma_P^-}$ is compact. Thus the theorem proved above furnishes a sufficient condition for the existence of an L^+ or L^--stable motion in the neighborhood of the critical point 0. It is of interest to contrast this local condition with the global one derived by the author [1], namely that the system Φ have no improper saddle points[2].

4. In the event that system (1.1) is of class C^2 we can employ the theorem proved above to derive sufficient conditions, expressed analytically, for the existence of Lagrange stable motions in the neighborhood of the critical point 0. The particular analytical expressions obtained will depend on the choice of the set ω. Two examples will be worked out by way of illustration.

(4.1) Let ω be the open sphere of radius λ and center 0.

---

[2] The system Φ is said to have an improper saddle point if there exists a sequence of points $\{P_n\} \subset \Omega$ and two sequences $\{\tau_n\}$, $\{t_n\}$ of real numbers satisfying the following conditions:

(i) $0 < \tau_n < t_n$, $(n = 1, 2, \dots)$.

(ii) $\tau_n \longrightarrow +\infty$ as $n \longrightarrow +\infty$.

(iii) $P_n \longrightarrow P \in \Omega$ as $n \longrightarrow +\infty$.

(iv) $\Gamma_{P_n}(t_n) \longrightarrow Q \in \Omega$ as $n \longrightarrow +\infty$.

(v) $\{\Gamma_{P_n}(\tau_n)\}$ has no limit point in Ω.

We assume, without loss of generality, that 0 is at the origin. Let $\rho = \{x_1^2 + \dots + x_n^2\}^{1/2}$. The condition $E = E^*$ means that every point of egress from ω is a point of strict egress; in other words, $\rho = \rho(t)$ along any orbit which intersects ω cannot have a relative maximum on the surface $\rho = \lambda$. This, in turn, is certainly the case if for every

$P \in H(\omega)$ such that:

(a) P is not a critical point; and

(b) $\left.\dfrac{d\rho}{dt}\right|_P = 0$,

we would have $\left.\dfrac{d^2\rho}{dt^2}\right|_P > 0$. In fact, this last statement is actually stronger than the statement $E = E^*$ and could be weakened somewhat if necessary.

In terms of the notation employed for system (1.1) we have:

$$(4.11) \qquad \rho^2 = \sum_{i=1}^{n} x_i^2$$

$$(4.12) \qquad \frac{d\rho}{dt} = \dot{\rho} = \frac{1}{\rho} \sum_{i=1}^{n} x_i f_i$$

$$(4.13) \qquad \frac{d^2\rho}{dt^2} = \ddot{\rho} = \frac{1}{\rho}\left(\sum_{i=1}^{n} f_i^2 + \sum_{i,j=1}^{n} \frac{\partial f_i}{\partial x_j} f_j x_i - \dot{\rho}^2 \right) .$$

Thus the sufficient condition under consideration obtains the following form:

COROLLARY 1. Let $\lambda > 0$ be arbitrary. If for every regular point $P(\bar{x}_1, \dots, \bar{x}_n) = P(\bar{x})$ such that

$$\rho_P^2 = \sum_{i=1}^{n} \bar{x}_i^2 = \lambda^2$$

and at which

$$\rho\dot{\rho} = \sum_{i=1}^{n} \bar{x}_i f_i(\bar{x}) = 0 \quad ,$$

there holds the inequality

$$\sum_{i=1}^{n} f_i^2(\bar{x}) + \sum_{i,j=1}^{n} \bar{x}_i f_j(\bar{x}) \left.\frac{\partial f_i}{\partial x_j}\right|_{\bar{x}} > 0 , \tag{4.14}$$

then there exists a semi-orbit wholly contained in the open sphere of radius λ and center 0.

(4.2) Next let ω be an open box with sides parallel to the coordinate axes; i.e., the open set defined by the inequalities $-\lambda_i < x_i < \mu_i$, $\lambda_i > 0$, $\mu_i > 0$, $(i = 1, \ldots, n)$. Following the same line of reasoning as in (4.1) we obtain

COROLLARY 2. Let $\lambda_i > 0$, $\mu_i > 0$, $(i = 1, \ldots, n)$, be arbitrary. If for every regular point $P(\bar{x}_1, \ldots, \bar{x}_n) = P(\bar{x})$ lying on the ith face of the boundary of ω (i.e., such that $\bar{x}_i = -\lambda_i$ or $\bar{x}_i = \mu_i$) at which $f_i(\bar{x}_1, \ldots, \bar{x}_n) = 0$, there holds the inequality

$$\sum_{j=1}^{n} \frac{\partial f_i}{\partial x_j} f_j > 0 , \tag{4.21}$$

then there exists a semi-orbit wholly contained within the box $-\lambda_i < x_i < \mu_i$, $(i = 1, \ldots, n)$.

Other topologically equivalent choices of ω will result in still further analytical expressions for the above sufficient condition.

Columbia University

BIBLIOGRAPHY

[1] MENDELSON, P., "On dynamical systems without improper saddle points," to appear.

[2] PLIŚ, A., "On a topological method of the study of the behavior of the integrals of ordinary differential equations," Bull. Acad. Polon. des Sciences Cl. III, Vol. II, No. 9 (1954), pp. 415-418.

[3] WAŻEWSKI, T., "Sur un principe topologique de l'examen de l'allure asymptotique des intégrales des équations différentielles," Ann. Soc. Polon. de Math., Vol. XX (1947), pp. 279-313.

[4] WAŻEWSKI, T., "Sur une méthode topologique de l'examen de l'allure asymptotique des intégrales des équations différentielles," Proceedings of the International Congress of Mathematicians (1954), Vol. 3, pp. 132-139.

XI. THE LOCAL THEORY OF PIECEWISE CONTINUOUS DIFFERENTIAL EQUATIONS

I. IDEAL SYSTEMS

J. André and P. Seibert

INTRODUCTION

The theory of automatic control, which is an important field of modern engineering, gave rise to the investigation of a class of differential equations which, until very recently, have not been studied by mathematicians; these equations involve piecewise continuous functions. Among the previous contributions to the subject we mention a paper of Solncev [10] which gives a stability theory for two-dimensional piecewise continuous systems, the monograph of Flügge-Lotz [5] who investigates various piecewise linear systems and stresses mainly the engineering point of view, two notes by the authors [1a], [1b] concerned with n-dimensional piecewise linear systems, a number of recent papers on the problem of optimal control (e.g., those of Bellman, Glicksberg, Gross [2], Boltyanskiĭ, Gamkrelidze, Pontryagin [3], [6], [9], Krasovskiĭ [8], Bushaw [4]), and a paper of Krasovskiĭ [7] on stability in the large of piecewise continuous systems.

The present paper gives a generalization of the two above-mentioned notes by the authors. The systems under consideration are piecewise of type C^2 and the set of discontinuity is assumed to consist of certain hypersurfaces of class C^3. The latter are called the "switching spaces" of the system, their points "switching points".[1] The precise hypotheses and the definition of a solution will be given in §1. In §2 we study the

[1] It will be noted that in most of the papers on optimal control ([2], [3], [6], [8], [9]) the right hand sides of the differential equations considered depend discontinuously on t and not, as in the other papers (including the present one), on x. However, since the problem treated in these papers is to find optimal switching times for a given initial point, the solution of this problem for all initial points is actually equivalent to that of finding certain switching spaces in the sense considered here (provided, the uncontrolled system is autonomous). The systems studied by Krasovskiĭ in [7] (which are generalizations of those occurring in the control problems) have right hand sides which also depend discontinuously on t.

behavior of the solutions near the switching spaces and give a classification of the switching points. The latter consist of three principal types: transition points (i.e., points at which a solution traverses the switching space), end-points (at which two solutions "end", (i.e., are not continuable beyond the point), and starting points (at which two solutions "start", i.e., are not continuable into the past). At switching points of some of the other classes (which are less frequent) solutions may fuse or fiburcate. In §5 we describe qualitatively the entire sets of solutions passing through given closed subsets of a switching space. For this purpose we introduce "local flows" (in §4) which are defined on compact subsets of the phase space and can be considered as "local dynamical systems" in analogy to the concept of local groups. The system of solutions around the switching spaces can then be conceived as a complex of local flows, connected with each other by certain neighborhood relations.

The second part of the paper will be devoted to systems with switching delay. These formally belong to a class of difference-differential equations.[1] They are distinguished from the systems considered in the present part of the paper (also called "ideal systems") by the property that the discontinuous function changes sign shortly after the trajectory has reached the switching space, rather than at the exact moment of transition. They usually represent a better approximation to physical reality than the ideal systems and, in contrast with the latter, their solutions are continuable indefinitely into the future, i.e., they have no end-points.

§1. THE CONCEPT OF PIECEWISE CONTINUOUS DIFFERENTIAL EQUATIONS AND THEIR SOLUTIONS

1. <u>The system</u> (S). Consider m real single-valued functions $s_\mu(x)$, defined in E^n (n-dimensional euclidean space) and satisfying the following conditions.

(a) The functions $s_\mu(x)$ are of class C^3 (i.e., they possess continuous partial derivatives of first, second and third order).

(b) At no point of E^n does any function $s_\mu(x)$ vanish simultaneously with its gradient

$$\left(\frac{\partial s_\mu}{\partial x_1}, \ldots, \frac{\partial s_\mu}{\partial x_n} \right) .$$

[1] Strictly speaking, this is true only for systems with "constant time lag", not for those with "threshold".

Under these conditions the sets

(1.1) $$S_\mu : s_\mu(x) = 0$$

($\mu = 1, \ldots, m$) are smooth hypersurfaces of E^n. We introduce the following notations:

$$\underset{=}{S} = \bigcup_{\mu=1}^{m} S_\mu \quad ,$$

$$s(x) = (s_1(x), \ldots, s_m(x)) \quad .$$

We now consider the domains into which the space is decomposed by the hypersurfaces S_μ. Denoting by $\operatorname{sgn}(\alpha_1, \ldots, \alpha_m)$ the vector $(\operatorname{sgn} \alpha_1, \ldots, \operatorname{sgn} \alpha_m) = (\alpha_1/|\alpha_1|, \ldots, \alpha_m/|\alpha_m|)$ (α_μ real, $\neq 0$), we can associate to every point $x \in S^n - \underset{=}{S}$ a vector $e(x)$ by setting

(1.2) $$e(x) = \operatorname{sgn} s(x) \quad .$$

Furthermore, we associate to every vector e with coordinates ± 1 an (open) domain[1]

(1.3) $$D_e = \{x \in E^n - \underset{=}{S} : e(x) = e\} \quad .$$

Apparently, every such domain is bounded by hypersurfaces S_μ.

Finally, we associate to every vector e (for which D_e is $\neq \emptyset$) an n-dimensional vector function $f(x, e)$ which is defined and of class C^2 throughout E^n [2] and which does not vanish on the boundary of D_e.

After these preparations we consider the differential equation

(1.4) $$\dot{x} = f(x) \qquad (x = (x_1, \ldots, x_n), \quad \dot{} = d/dt)$$

where the vector function $f(x)$ is given by

(1.5) $$f(x) = f(x, e) \quad \text{for} \quad x \in D_e$$

---

1 which may be empty or disconnected.

2 As long as we restrict ourselves to ideal systems, it is sufficient that $f(x, e)$ is defined in an open set containing the closure $\bar{D}_e$ of D_e.

and undefined on $\underset{=}{S}$. Using (1.2) and (1.5), we can write (1.4) in the form

$$\text{(S)} \qquad \dot{x} = f(x, \operatorname{sgn} s(x)) \ .$$

2. The Concept of Solution. Given a point p of a domain D_e, there exists exactly one solution of the differential equation

$$\text{(S, e)} \qquad \dot{x} = f(x, e)$$

which passes through p at the time $t = 0$. We denote it by $x(t, p)$. Since the system (S) coincides with (S, e) for all $x \in D_e$, the function $x(t, p)$ is also a solution of (S) if it is restricted to a t-interval I for which[1]

$$x(I, p) \subset D_e$$

holds. We now extend the concept of a solution of (S) by the following definition:

DEFINITION 1.1. We call a (single valued) vector function $x(t)$, defined in an interval I, a solution of the system (S) if it satisfies the following conditions:

1° It is continuous throughout I.

2° For every $t \in I$, such that $x(t) \in D_e$ [which is equivalent to $e(x(t)) = \operatorname{sgn} s(x(t)) = e$], the function $x(t)$ is differentiable and satisfies equation (S, e).

3° The set of values $t \in I$ for which $x(t) \in \underset{=}{S}$ holds, has no cluster point in I.

A solution according to this definition satisfies equation (S) in the ordinary sense for almost all values $t \in I$. In the second part of the paper we will extend the notion of solution in order to include also certain paths contained entirely in $\underset{=}{S}$ which are related to a phenomenon in the theory of discontinuous control (after end-point motions).

§2. SWITCHING POINTS AND THEIR CLASSIFICATION

3. Topological classification of switching points. The hypersurfaces S_μ are called the switching spaces of the system (S). Points in a switching space, or switching points, will usually be denoted by u.

---

[1] $x(I, p)$ denotes, as usual, the set $\{x(t, p)\}_{t \in I}$.

Unless the contrary is stated, it will always be assumed that u belongs to exactly one switching space S_μ, and we will therefore drop the index μ. Under this assumption u belongs to the boundaries of exactly two domains D_e which we denote by D_{e^+} and D_{e^-}, respectively. The corresponding vectors e^+ and e^- differ only in their μ-th coordinate which we assume to be $+1$ in e^+ and -1 in e^-. The solutions of the equations

$$\dot{x} = f(x, e^{\pm})$$

with the initial value u will be denoted by $x^{\pm}(t, u)$.

We first classify the switching points u with respect to each of the cruves $x^+(t, u)$ and $x^-(t, u)$ by introducing the following definitions: If the limits

$$(3.1) \qquad \gamma_1^{\pm} = \lim_{t \to -0} e_\mu(x^{\pm}(t, u)), \quad \gamma_2^{\pm} = \lim_{t \to +0} e_\mu(x^{\pm}(t, u))$$

both exist, the point u will be called

an A-point (with respect to $x^{\pm}(t, u)$) if $\gamma_1^{\pm} = \mp 1$, $\gamma_2^{\pm} = \pm 1$,

an E-point if $\gamma_1^{\pm} = \pm 1$, $\gamma_2^{\pm} = \mp 1$,

an L-point if $\gamma_1^{\pm} = \gamma_2^{\pm} = \mp 1$,

an R-point if $\gamma_1^{\pm} = \gamma_2^{\pm} = \pm 1$.

In the case of an A-point there exists a t-interval $B = (0, \beta)$ for which $e_\mu(x^{\pm}(t, u)) = +1$, and consequently $x^{\pm}(B, u) \subset D_{e^{\pm}}$. Since $x^{\pm}(t, u)$ satisfies $(S, e^{\pm})$ and this equation coincides with (S) for $x \in D_{e^{\pm}}$, the function $x^{\pm}(t, u)$, restricted to the interval B, is a solution of (S). Analogously we find that for some interval $(-\alpha, 0)$ the same function does not satisfy (S).

In the case of an E-point the situation is vice versa, i.e., $x^{\pm}(t, u)$ satisfies (S) for small negative values of t, but not for small positive values.

If u is an R-point, $x^{\pm}(t, u)$ is a solution of (S) in a certain open neighborhood of $t = 0$, while in the case of an L-point there exists an interval around 0 in which $x^{\pm}(t, u)$ does not satisfy (S).

Consider now the case where some of the limits (3.1) fail to exist. This situation occurs whenever the intersections of one or both of the curves $x^{\pm}(t, u)$ with S cluster at u. We call u

an A^*-point if $\gamma_2^{\pm}$ exists and is equal to ± 1,

an E^*-point if $\gamma_1^{\pm}$ exists and is equal to ± 1,

an L^*-point in all other cases.

It is easy to see that A^*-, E^*-, L^*-points have the above stated properties of A-, E-, L-points, respectively, except that $x^{\pm}(t, u)$ usually satisfies equation (S) in certain intervals clustering at $t = 0$ and contained (in the case of A^*- [E^*]-points) in the semi-neighborhoods which are void of solutions in the case of A- [E-] points.

We now extend the classes A, E, L by the sets A^*, E^*, L^*, respectively. The set of all X-points u (X = A, E, L, R) with respect to $x^{\pm}(t, u)$ will be denoted by $X^{\pm}$. Then the complete topological classification of the switching points is obtained by considering the intersections of the sets X^+ and Y^- (X, Y = A, E, L, R). Since the points of the sets $X^+ \cap Y^-$ and $X^- \cap Y^+$ are of the same topological character with respect to the solutions passing through them, the sets of topologically equivalent switching points are given by the expressions

$$XY = (X^+ \cap Y^-) \cup (X^- \cap Y^+) \tag{3.2}$$

(X, Y = A, E, L, R). The symmetry relation $XY = YX$ reduces the number of these classes to 10.

4. Behavior of solutions at switching points. If u is of type AE, there exists a unique solution with u as its initial value which is defined in an open interval around $t = 0$. This solution, which we denote by $x(t, u)$, is given by

$$x(t, u) = \begin{cases} x^+(t, u) & \text{for } t \geq 0 \\ x^-(t, u) & \text{for } t \leq 0 \end{cases}$$

in the case $a \in A^+ \cap E^-$, and vice versa if $u \in A^- \cap E^+$. At the point u the solution crosses the switching space; u is therefore called a transition point. In the topological behavior of the solution $x(t, u)$ there is no difference between a transition point and a point at which $f(x)$ is continuous. (Vid. Figure 1, also for the following cases.)[1]

If u is an AA - [EE-] point, the positive [negative] half

[1] Dotted lines represent half trajectories which are not solution of (S).

trajectories[1] of $x^{\pm}(t, u)$ are both solutions of (S), so that the uniqueness theorem fails to hold in this case. Apparently, these two half trajectories cannot be continued beyond u into the past [future]. An AL - [EL-] point is the initial point of exactly one solution of (S) which is defined only for $t \geq 0$ [$t \leq 0$]. We call the AA- and AL-points starting points, the EE- and EL- points end-points.

If u is an AR - [ER-] point, both $x^+(t, u)$ and $x^-(t, u)$ are solutions of (S), one being defined only for $t \leq 0$ [$t \geq 0$], the other for positive and negative t-values. This implies the existence of two solutions actually passing through u. Assume, e.g., $u \in A^+ \cap R^-$ [$u \in E^+ \cap R^-$]. (In the case $u \in A^- \cap R^+$[$u \in E^- \cap R^+$] the situation is analogous.) Then

$$x_1(t, u) = x^-(t, u)$$

and

$$x_2(t, u) = \begin{cases} x^-(t, u) & \text{for} \quad t \leq 0 \ [t \geq 0] \ , \\ x^+(t, u) & \text{for} \quad t \geq 0 \ [t \leq 0] \ , \end{cases}$$

are both solutions of (S) passing through u. Since for $t \leq 0$ [$t \geq 0$] these two solutions are identical, the situation can be described as a bifurcation [fusion] of trajectories.[2]

Through an LL- point there apparently exists no solution of (S). - If u is an LR- point, there is exactly one solution [either $x^+(t, u)$ or $x^-(t, u)$] with u as its initial value. It is defined for positive and negative values of t and does not cross the switching space. Apart from the transition (AE-) points, the LR- points are the only switching points at which the existence and uniqueness theorem holds in the strict sense. - If, finally, u belongs to the set RR, both $x^+(t, u)$ and $x^-(t, u)$ are solutions of (S) around $t = 0$. This implies the existence of altogether four solutions passing through u, the latter therefore being a point of fusion and of bifurcation simultaneously.

5. Normal and exceptional switching points. Beside the topological qualities of a switching point which we analysed in the preceding setion, it is of significance whether the curves $x^{\pm}(t, u)$ associated to the point u are tangent to S or not. We introduce the following notions:

---

[1] i.e., the restrictions of $x^{\pm}(t, u)$ to $t \geq 0$ [$t \leq 0$].

[2] The concept of solutions introduced in the paper André-Seibert [1a] was slightly less general than the one we use here. According to the former, there exist fusions of trajectories, but no bifurcations.

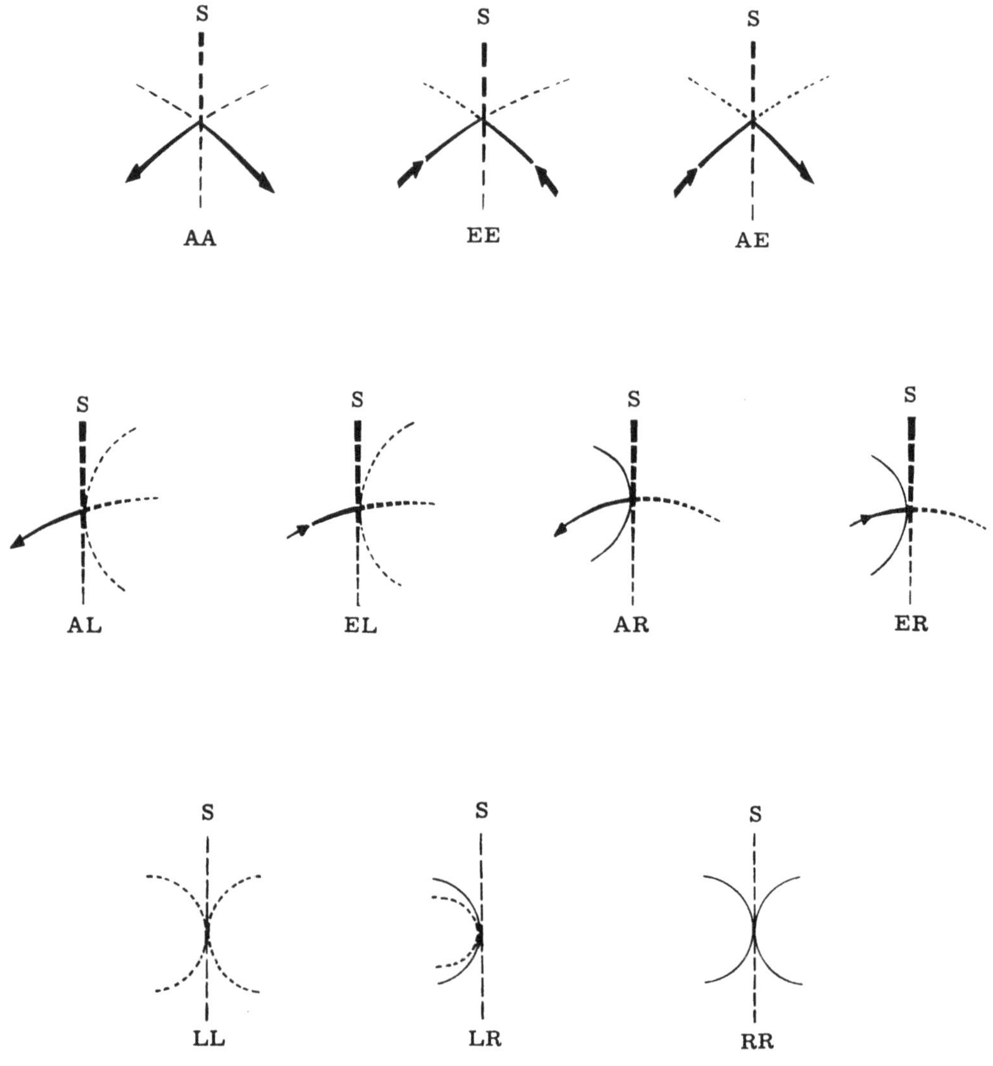

Fig. 1.

A point $u \in S$ will be called a <u>normal switching point</u> if neither of the curves $\overset{+}{x}{}^{-}(t, u)$ are tangent to the switching space S at u, otherwise an <u>exceptional point</u>. In particular, all points of the types A^*, E^*, L^* are apparently exceptional. The same is true for all points of the types LX, RX (X = A, E, L, R), since for all of these at least one of the curves $x^{\pm}(t, u)$ fails to traverse S at the point u. Denoting the set of all normal points by S^o, we therefore have the inclusion

$$(5.1) \qquad S^o \subseteq AA \cup AE \cup EE \quad .$$

The sets of normal starting, end-, and transition points will be denoted by A, E, T, respectively:

$$(5.2) \qquad A = AA \cap S^o, \quad E = EE \cap S^o, \quad T = AE \cap S^o \quad .$$

We call an exceptional point u <u>of first order</u>, if only one of the curves $\overset{+}{x}{}^{-}(t, u)$ is tangent to S at u and the contact is of 1st order; in all other cases we say u is <u>of higher order</u>. By S' we denote the set of all exceptional points, by S'' that of all exceptional points of higher order, and by $S^1 = S' - S''$ the set of exceptional points of first order. The intersections of any subset Q of S with S^o, S', S'', S^1 will be denoted by Q^o, Q', Q'', Q^1.[1] Since at points of types LL, LR, RR both curves $\overset{+}{x}{}^{-}(t, u)$ are tangent to S, we have the relation

$$(5.3) \qquad LL \cup LR \cup RR \subseteq S'' \quad .$$

6. <u>Determination of the type of a switching point</u>. In order to state the criteria for the type of a given switching point $u \in S = S_\mu$ it is sufficient to assume $s_\mu \in C^2$, $f \in C^1$ instead of the stronger conditions formulated in §1.1. Then the functions $\overset{+}{x}{}^{-}(t, u)$ which satisfy the equations $\dot{x} = f(x, \overset{+}{e}{}^{-})$ are of type C^2 and we can apply the following Taylor expansion:

$$(6.1) \qquad \begin{aligned} s_\mu(\overset{+}{x}{}^{-}(t, u)) &= \frac{d}{d\tau} s_\mu(\overset{+}{x}{}^{-}(\tau, u))\Big|_{\tau=0} t + 1/2 \frac{d^2}{d\tau^2} s_\mu(\overset{+}{x}{}^{-}(\tau, u)\Big|_{\tau=\theta t} t^2 = \\ &= \overset{+}{c}{}_1^{-}(u)t + 1/2\, \overset{+}{c}{}_2^{-}(\theta t, u)t^2 \qquad (0 < \theta < 1) \quad . \end{aligned}$$

---

[1] In the case of the sets A, E, T [vid. (5.2)] we omit the superscript o. -- It should be noted that the sets S^o, S', S'', S^1 comprise only that part of S which is not contained in any other switching space.

Here

$$c_i^{\pm}(t, u) = \frac{d^i}{dt^i} s_\mu(x^{\pm}(t, u))$$

and

$$c_i^{\pm}(u) = c_i^{\pm}(0, u) \quad (i = 1, 2) \quad .$$

For the present purpose it is sufficient to examine the signs of the coefficients $c_1^{\pm}(u)$ and $c_2^{\pm}(u)$. An obvious calculation yields the following formulas:[1]

$$(6.2) \qquad c_1^{\pm}(u) = f(u, e^{\pm}) \frac{d}{dx} s_\mu(u) \quad ,$$

$$(6.3) \qquad \begin{aligned} c_2^{\pm}(u) = {} & f(u, e^{\pm}) \left(\frac{d^2}{dx^2} s_\mu(u) \right) f(u, e^{\pm})' \\ & + f(u, e^{\pm}) \left(\frac{d}{dx} f(u, e^{\pm}) \right)' \frac{d}{dx} s_\mu(u) \quad ; \end{aligned}$$

Here $\frac{d}{dx}$ denotes the gradient when applied to a scalar and the Jacobian matrix when applied to a vector:

$$\frac{ds_\mu}{dx} = \begin{pmatrix} \frac{\partial s_\mu}{\partial x_1} \\ \vdots \\ \frac{\partial s_\mu}{\partial x_n} \end{pmatrix} , \quad \frac{df}{dx} = \frac{d}{dx}(f_1, \ldots, f_n) = \begin{pmatrix} \frac{\partial f_1}{\partial x_1}, & \ldots, & \frac{\partial f_n}{\partial x_1} \\ & \ldots & \\ \frac{\partial f_1}{\partial x_n}, & \ldots, & \frac{\partial f_n}{\partial x_n} \end{pmatrix}$$

The operator $\frac{d^2}{dx^2}$, applied to a scalar, denotes the Jacobian matrix of the gradient. From the formula (6.1) we immediately obtain the following theorem:

THEOREM 6.1. If u is a point of the switching space S (and of no other switching space), the following implications hold:[2]

---

[1] By $'$ we denote the operation of transposition.

[2] The upper [lower] signs +, - in the superscripts corresponds to the upper [lower] signs <, >.

$$c_1^{\pm}(u) \gtrless 0 \quad \text{implies} \quad u \in A^{\pm} ,$$

$$c_1^{\pm}(u) \lessgtr 0 \quad \text{implies} \quad u \in E^{\pm}$$

$$c_1^{\pm}(u) = 0 \quad \text{and} \quad c_2^{\pm}(u) \gtrless 0 \quad \text{together imply} \quad u \in R^{\pm} ,$$

$$c_1^{\pm}(u) = 0 \quad \text{and} \quad c_2^{\pm}(u) \lessgtr 0 \quad \text{together imply} \quad u \in L^{\pm} .$$

The coefficients $c_1^{\pm}(u)$, $c_2^{\pm}(u)$ are given by the formulas (6.2) and (6.3).

In particular, u is

a normal transition point if $c_1^{+}(u)$ and $c_1^{-}(u)$ are either both positive or both negative,

a normal starting point if $c_1^{+}(u) > 0$ and $c_1^{-}(u) < 0$,

a normal endpoint if $c_1^{+}(u) < 0$ and $c_1^{-}(u) > 0$.

Unconsidered in this theorem remain only those exceptional points of higher order which involve contacts of higher than first order [i.e., at least 3-point contact].

The three last implications require only the hypotheses $f \in C^0$, $s \in C^1$. If these are satisfied, the functions $c_1^{\pm}$ are continuous and we obtain the following corollary:[1]

COROLLARY 6.1. The sets A, E, T of normal switching points are open relative to S.

Since every point of S^1 involves a contact of first order, we conclude from the third and fourth implication in Theorem 6.1 and from the continuity of $c_2^{\pm}(u)$ (assuming again $f \in C^1$, $s \in C^2$):

COROLLARY 6.2.[2] The set S^1 of exceptional points of first order consists of the sets $(AL)^1$, $(EL)^1$, $(AR)^1$, $(ER)^1$, all of which are open relative to S'.

---

[1] The boundaries of the sets A, E, T may contain parts of intersections of S with other switching spaces. The latter, however, being closed sets, do not affect the validity of Corollary 6.1.

[2] All sets with superscript 1 consist of exceptional points of first order.

§3. TOPOLOGICAL PROPERTIES OF THE CLASSES OF SWITCHING POINTS

7. <u>The exceptional classes</u>. Let u be an exceptional point of S ($= S_\mu$) [vid. §2.5]. Then the conditions

$$(7.1) \qquad \begin{cases} s_\mu(u) = 0 \\ c_1^{\pm}(u) = f(u, e^{\pm}) \dfrac{d}{dx} s_\mu(u) = 0 \end{cases}$$

hold, the second either for e^+ or e^- [vid. (6.1), (6.2)]. The functions $s_\mu(u)$ and $c_1^{\pm}(u)$ are of class C^1. Therefore, if the Jacobian matrix

$$J_1^{\pm}(u) = \frac{d}{dx} (s_\mu(u), c_1^{\pm}(u))$$

$$\left[= \frac{d}{dx} (s_\mu(x), c_1^{\pm}(x)) \Big|_{x=u} \right]$$

of the system (7.1) is of rank 2, it follows from the implicit function theorem that the set of all exceptional points in a certain neighborhood of u constitute an (n-2)-dimensional differentiable submanifold of S [or, if $c_1^{\pm}(u)$ both vanish, the union of two such manifolds]. We introduce the following terminology:

If the matrices $J_1^{\pm}(u)$ are of rank 2 at all points of S [or of a part Q of S] satisfying the corresponding conditions (7.1), we say, the system (S) satisfies condition A on the switching space S [or on the subset Q].

We can then state the following theorem:

THEOREM 7.1. If condition A is satisfied on a subset Q[1] of the switching space S, the set Q' of exceptional points in Q consists of (n-2)-dimensional differentiable submanifolds of S if it is not empty. Every point $u \in Q'$ possesses a neighborhood N such that $Q' \cap N$ is contained in the union of at most two of the connected manifolds constituting Q' (namely, one (connected) component of each of

---

[1] It is assumed that Q contains no points of switching spaces other than S.

the sets $\{u \in Q' : c_1^+(u) = 0\}$ and $\{u \in Q' : c_1^-(u) = 0\}$). If, in particular, U is an exceptional point of first order (vid. §2.5), the set $Q' \cap N$ is contained in a single (connected) manifold.

We now establish a similar theorem for the set S'' of exceptional points of higher order [vid. §2.5]. Every point $u \in S''$ satisfies one of the two following sets of equations:

$$s_\mu(u) = c_1^+(u) = c_1^-(u) = 0 \quad , \tag{7.2}$$

$$s_\mu(u) = c_1^\pm(u) = c_2^\pm(u) = 0 \tag{7.3}$$

(the latter either for c_1^+ or c_1^-). Denote by $J_2(u)$, $J_3^\pm(u)$ the Jacobian matrices of these systems:

$$J_2(u) = \frac{d}{dx}\,(s_\mu(u),\ c_1^+(u),\ c_1^-(u)) \quad ,$$

$$J_3^\pm(u) = \frac{d}{dx}\,(s_\mu(u),\ c_1^\pm(u),\ c_2^\pm(u)) \quad .$$

We say, <u>condition B</u> is satisfied on S [or $Q \subset S$], if the matrices $J_2(u)$, $J_3^\pm(u)$ are of rank 3 at every point of S [or Q] satisfying the corresponding set of conditions (7.2), (7.3).

Then, in analogy to Theorem 7.1, we obtain

THEOREM 7.2. If condition B is satisfied on a subset[1] Q of the switching space S, the set Q'' of exceptional points of higher order in Q consists of $(n-3)$-dimensional differentiable submanifolds of Q' if it is not empty. Every point $u \in Q''$ possesses a neighborhod N such that $Q'' \cap N$ is contained in the union of at most three of the connected manifolds constituting Q''.

REMARK 7.1. By comparing the number of (scalar) equations entering into the conditions A and B with the number of independent variables, it is easy to see that, in general, both conditions hold in the entire switching space.

---

1 Vid. Footnote on p. 236

REMARK 7.2. Since S'' is of lower dimension than S' (provided that condition B is satisfied on S' and the latter is not empty), $S^1 [= S' - S'']$ is apparently dense in S':

$$\overline{S^1} = S' \quad . \tag{7.4}$$

8. <u>The classes of normal switching points</u>. For the following considerations we assume conditions A and B to be satisfied throughout the switching space S. Then the following lemma holds:

LEMMA 8.1. If $c_1^{\pm}(u) = 0$ and $c_2^{\pm}(u) \neq 0$ hold, the function $c_1^{\pm}$ assumes positive and negative values in every neighborhood of u.

PROOF. Under the assumptions $c_1^+(u) = 0$, $c_2^+(u) > 0$ the curve $x^+(t, u)$ is contained in the set $\bar{D}_{e^+}$ for $|t| < \delta$ and δ sufficiently small [vid. (6.1)]. Now assume that c_1^+ is non-negative in a neighborhood N of u relative to S:

$$c_1^+(v) \geq 0 \quad \text{for} \quad v \in N \quad , \tag{8.1}$$

and take N so small that

$$c_2^+(v) > 0 \quad \text{for} \quad v \in N \quad . \tag{8.2}$$

Then, due to the continuous dependence of x^+ on the initial values, there exists a neighborhood $N_1 \subset N$ of u such that, if I denotes the interval $(0, \delta)$ and δ is chosen small enough, the set $x^+(I, N_1)$ contains no point of S outside N. Then (8.1) and (8.2) imply

$$x^+(I, N_1) \subset \bar{D}_{e^+} \quad .$$

Since $x^+(I, N_1)$ is a homeomorphic image of the product set $N_1 \times I$ and N_1 is an open set on the hypersurface S, the set $x^+(I, N_1)$ apparently contains a semi-neighborhood $\mathcal{N} \cap D_{e^+}$ of u ($\mathcal{N}$: neighborhood of u). On the other hand, it follows from the hypotheses (and (6.1)) that $x^+(t,u) \subset D_{e^+}$ tends to u for $t \longrightarrow -0$, which leads to a contradiction.

If $c_1^+ < 0$ is assumed in N instead of $c_1^+ > 0$, the preceding argument needs only to be modified by considering the interval $(-\delta, 0)$ instead of $(0, \delta)$. - For the coefficients c_1^-, c_2^- the proof is exactly the same.

We now consider the boundaries relative to S of the sets A, E, T and denote them by ∂A, ∂E, ∂T.

From Theorem 7.1 and Lemma 8.1 we easily obtain the following pair of inclusions:

$$(AL)^1 \cup (AR)^1 \subseteq \partial A \cap \partial T \quad , \tag{8.3}$$

$$(EL)^1 \cup (ER)^1 \subseteq \partial E \cap \partial T \quad . \tag{8.4}$$

Consider, e.g., the case $u \in (A^+ \cap L^-)^1$. [The other cases are treated in obvious analogy.] Here $c_1^+ > 0$, $c_1^- = 0$, $c_2^- > 0$ [Theorem 6.1]. According to the lemma there exist points $v, w \in S$ arbitrarily near u with $c_1^+(v) > 0$, $c_1^-(v) > 0$; $c_1^+(w) > 0$, $c_1^-(w) < 0$. This implies $v \in T$, $w \in A$ [Theorem 6.1].

After these preparations we can prove the following theorem:

THEOREM 8.1. If the conditions A and B are satisfied on the switching space S, one of the following statements is true, provided that S intersects no other switching space:

(a) S consists entirely of normal switching points of one type, i.e., $S = A$, E, or T.

(b) The sets T and $A \cup E$ are both non-empty and their boundaries ∂T and $\partial(A \cup E)$ coincide. Thus the closure $\bar{T}$ of T separates A from E and the components of A and E from each other and, conversely, the components of T are separated from each other by $\overline{A \cup E}$.

If S intersects other switching spaces S_λ, the preceding statements hold for every component of $S - \cup_\lambda (S_\lambda \cap S)$.

PROOF. We first show that the set of exceptional points S' is empty if all normal points are of the same type. (The converse is obvious because $\partial A, \partial E, \partial T \subseteq S'$.) Corollary 6.2 yields

$$S^1 \subseteq (AL)^1 \cup (AR)^1 \cup (EL)^1 \cup (ER)^1 \quad . \tag{8.5}$$

From (8.3), (8.4) it follows that all sets on the right-hand side of (8.5) are empty, hence $S^1 = \emptyset$. Finally, (7.4) implies $S' = \emptyset$, so that we have the case (a).

Suppose now that S' and therefore also S^1 are non-empty. Then it follows immediately from (8.3), (8.4), and (8.5) that both T and $A \cup E$ are non-empty. The same relations imply

$$(8.6) \qquad (\partial A)^1 \subseteq \partial T, \ (\partial E)^1 \subseteq \partial T, \ (\partial T)^1 \subseteq \partial(A \cup E).$$

Observing that ∂A, ∂E, ∂T are closed subsets of S and using (7.4) and (8.6), we obtain

$$\partial A = \partial A \cap S' \subseteq \partial A \cap \overline{S^1} = \overline{\partial A \cap S^1} = \overline{(\partial A)^1} \subseteq \overline{\partial T} = \partial T$$

and the analogous relations

$$\partial E \subseteq \partial T \quad \text{and} \quad \partial T \subseteq \partial(A \cup E) \ .$$

Hence,

$$(8.7) \qquad \partial(A \cup E) = \partial T \ ,$$

which yields the separation property of case (b).

The extension to the case where S intersects other switching spaces is immediate.

§4. LOCAL FLOWS

9. <u>Definitions and elementary properties</u>. As we have seen in §2.4, the qualitative behavior of the solution of a system (S) differs in many respects from that of a dynamical system in the usual sense, defined, e.g., by the solutions of differential equations satisfying a Lipschitz condition. In order to study families of solutions of (S) qualitatively, we introduce the notion of "local flows".

DEFINITION 9.1. Let Y be a compact connected topological space. Denote by t_0, t_1 two continuous maps of Y into the negative and positive halves of the real line (both including 0) respectively, and by $I(y)$ the interval $[t_0(y), t_1(y)]$. Then a continuous mapping φ of the set $\Omega = \{(y, t) : y \in Y, t \in I(y)\}$ onto Y defines a local flow $\mathcal{F} = (Y, I, \varphi)$ if φ has the following properties:

(a) $y \in Y$, $t \in I(y)$ implies $I(\varphi(t, y)) = I(y) - t = [t_0(y) - t, t_1(y) - t]$.

(b) $\varphi(0, y) = y$ holds for all $y \in Y$.

(c) If $y \in Y$, $t \in I(y)$, $t + t' \in I(y)$, the relation $\varphi(t + t', y) = \varphi(t', \varphi(t, y))$ holds.

(d) $\varphi(t, y)$ is a topological mapping of the set $\{y \in Y : t \in I(y)\}$ for every fixed t.

In the case where Y is a k-dimensional manifold with boundary[1] we call F a local k-flow.

The sets $\varphi(I(y), y))$ [y fixed] are called the paths of the local flow; we denote them by $\gamma(y)$. The set $\bigcup_{p \in P} \gamma(p)$ $(P \subset Y)$ will be denoted by $\gamma(P)$. It follows from (a) and (c) that

$$q \in \gamma(p) \quad \text{implies} \quad \gamma(q) = \gamma(p) \ .$$

To every point $y \in Y$ we associate two points

$$y_0 = \varphi(t_0(y), y), \quad y_1 = \varphi(t_1(y), y)$$

and define a pair of sets Y_0, Y_1 by

$$Y_0 = \{y_0(y)\}_{y \in Y}, \quad Y_1 = \{y_1(y)\}_{y \in Y} \ .$$

We call Y_0 and Y_1 the entrance and exit set of $\mathcal{F}$. Apparently, the paths of $\mathcal{F}$ are in one-one correspondence with the points of each of the sets Y_0, Y_1; $y_0(y)$ and $y_1(y)$ are the end-points of the path through y.

PROPOSITION 9.1. The entrance and exit set Y_0, Y_1 of a local flow are connected sets.

PROOF. Since the mappings t_0, t_1, φ are continuous, so are y_0 and y_1. Therefore the sets Y_0, Y_1, which are images under y_0 and y_1 of the connected set Y, are connected.

PROPOSITION 9.2. The mapping $\varphi(t, y)$ is univalent (as a function of t for every point $y \in Y_0$).[2]

---

[1] I.e., a Hausdorff space with the property that each of its points possesses a neighborhood which is homeomorphic either to an open k-sphere or to a semi-k-sphere $x_1^2 + \dots + x_k^2 < 1$, $x_1 \geq 0$.

[2] It should be noted that Definition 9.1 excludes the case where $I(y)$ is the whole real line. Therefore, in particular, critical points [$\varphi(t, y) \equiv y$] are excluded due to condition (a).

PROOF. Suppose that $\varphi(t^1, y) = \varphi(t^2, y)$. Then obviously $\varphi(-t^1, \varphi(t^1, y)) = \varphi(-t^1, \varphi(t^2, y))$ holds. By applying (b) and (c) we obtain

$$\varphi(0, y) = y = \varphi(t^2 - t^1, y) .$$

Finally (a) yields $I(y) = I(y) - (t^2 - t^1)$, i.e., $t^1 = t^2$.

PROPOSITION 9.3. Consider a dynamical system[1] [or flow] defined on a finite-dimensional manifold X by the mapping function Φ. Let C be a compact connected subset of a hypersurface without contact with the paths of the flow.[2] Then there exists a number $T > 0$ such that Φ defines a local flow on the set $Y = \Phi([0, T], C)$.

PROOF. Choosing T so small that Y is contained in a neighborhood H of the kind mentioned in footnote 2, it is clear that every point $y \in Y$ possesses a unique representation $y = \Phi(t, y_0)$ ($y_0 \in C$, $t \in [0, T]$). Then, if we define $I(y) = [-t, T - t]$, the conditions of Definition 9.1 are obviously satisfied.

So far only interior properties of local flows have been considered, i.e., properties which do not depend on an imbedding of Y into a larger space. Now we are going to study properties which depend substantially on such an imbedding. We therefore assume Y to be a subspace of a topological space X. Closure, interior, and boundary of Y will always be understood relative to X.

DEFINITION 9.2. A local flow $\mathcal{F}' = (Y', I', \varphi')$ is called a <u>subflow</u> of the local flow $\mathcal{F} = (Y, I, \varphi)$ if $Y' \subseteq Y \subset X$, $I'(y) = \{t \in I(y) : \varphi(t, y) \in Y'\}$ for all $y \in Y'$ and φ' is the restriction of φ to the set $\Omega' = \{t, y : y \in Y', t \in I'(y)\}$.

If $\mathcal{F}'$ is a subflow of $\mathcal{F}$ we write

$$\mathcal{F}' \subseteq \mathcal{F}.$$

PROPOSITION 9.4. Consider a local k-flow $\mathcal{F} = (Y, I, \varphi)$

---

1 I.e., a continuous mapping of the product space $X \times I$ (I: real line) onto X which satisfies the conditions (b), (c), (d) of Definition 9.1 in which Y and $I(Y)$ are replaced by X and I.

2 In this context we mean by a "hypersurface without contact" a hypersurface H of X with a neighborhood N such that H does not contain more than one point of any component of the intersection of any path with N.

and denote by ∂Y the boundary of Y. If a path γ of $\mathcal{F}$ contains a point $p \in \partial Y$ not lying in Y_0 or Y_1, the entire path γ is contained in ∂Y.

PROOF. To every point $p \in Y$ we associate the set[1] $Q_p = \varphi(-t_0(p), Y_0)$ [which is a (k-1)-dimensional manifold with boundary]. Since t_0 and t_1 are continuous, given two numbers $\alpha \in (t_0(p), 0)$, $\beta \in (0, t_1(p))$, there exists a neighborhood Q_p' of p relative to Q_p such that $\varphi(t, p')$ is defined for all $t \in I = [\alpha, \beta]$ and all points $p' \in Q_p'$. If p is an interior point of Q_p' (i.e., possesses a neighborhood relative to Q_p' which is homeomorphic to a (k-1)-sphere), it is also an interior point of the set $\varphi(I, Q_p')$ (since the latter is a homeomorphic image of the product set $I \times Q_p'$) and therefore of Y. The converse is also true. Indeed, consider a sequence of points $p^n \in Y$ tending to p. Then (again using the continuity of t_0) $t_0(p^n) - t_0(p) \in I$ for sufficiently large n, while, due to the continuity of φ, the points $\varphi(t_0(p^n) - t_0(p), p^n)$ ultimately belong to Q_p'. Thus $p^n \in \varphi(I, Q_p')$. Therefore, the intersections of Y and $\varphi(I, Q_p')$ with a sufficiently small neighborhood of p coincide. This implies the equivalence of $p \in \partial Y$ and $p \in \partial Q_p$.

Now consider two points $p \in \partial Y$ and $q = \varphi(t_q, p)$ $(t_q \in I(p))$. Then $p \in \partial Q_p$ and, choosing Q_p' as before (with $t_q \in I$), $\varphi(t_q, y)$ maps Q_p' homeomorphically onto $Q_q' = \varphi(t_q, Q_p')$, so that $q \in \partial Q_q'$. This implies $q \in \partial Y$ which proves the proposition.

DEFINITION 9.3. A subflow $\mathcal{F}' = (Y', I', \varphi')$ of a local k-flow $\mathcal{F} = (Y, I, \varphi)$ is called a <u>face</u> of $\mathcal{F}$ if Y' is contained in the boundary of Y. We also say, $\mathcal{F}'$ is <u>incident</u> with $\mathcal{F}$ and use the notations[2]

$$\mathcal{F}' \subseteq \partial \mathcal{F}$$

and

$$\mathcal{F}' \subseteq \partial\mathcal{F}_1 \cap \partial\mathcal{F}_2 \quad \text{if} \quad \mathcal{F}' \subseteq \partial\mathcal{F}_1 \quad \text{and} \quad \mathcal{F}' \subseteq \partial\mathcal{F}_2 .$$

Apparently, a face of a local k-flow is a local (k-1)-flow.

PROPOSITION 9.5. Consider a local k-flow $\mathcal{F} = (Y, I, \varphi)$

---

[1] If $\varphi(-t_0(p), y)$ is defined only on a subset Y_0' of Y_0, we mean by $\varphi(-t_0(p), Y_0)$ the set $\varphi(-t_0(p), Y_0')$.

[2] The symbol $\partial\mathcal{F}$ denotes the collection of all faces of $\mathcal{F}$.

and a local (k-1)-flow $\mathcal{F}' = (Y', I', \varphi')$ which satisfy the conditions:

(1) $I'(y) = \{t \in I(y) : \varphi(t, y) \in Y'\}$ for all $y \in Y' \cap Y$.

(2) The restrictions of φ and φ' to the set $\{t, y : y \in Y \cap Y', t \in I'(y)\}$ are identical.

(3) The entrance [exit] set $Y_0'[Y_1']$ of $\mathcal{F}'$ is contained in the boundary of the entrance [exit] set $Y_0[Y_1]$ of $\mathcal{F}$.

Then

$$\mathcal{F}' \subset \partial \mathcal{F}.$$

PROOF. Assume condition (3) to be satisfied, e.g., for Y_0 and Y_0'. Then it follows by an argument analogous so that in the proof of Proposition 9.4 that

$$\gamma(Y_0') = Y' \subseteq \partial Y \subset Y \quad .$$

Conditions (1) and (2), therefore, imply that $\mathcal{F}'$ is a subflow of $\mathcal{F}$ and the proposition follows.

DEFINITION 9.4. We call two local k-flows $\mathcal{F}_1 = (Y_1, I_1, \varphi_1)$, $\mathcal{F}_2 = (Y_2, I_2, \varphi_2)$ <u>coherent</u> if $Y_1 \cap Y_2 \neq \emptyset$ and there exists a (k-1)-flow $\mathcal{F}_{12} = (Y_1 \cap Y_2, I_{12}, \varphi_{12})$ such that

$$\mathcal{F}_{12} \subseteq \partial\mathcal{F}_1 \cap \partial\mathcal{F}_2$$

and I_{12} does not reduce to a single point for any y.

Apparently, $Y_1 \cap Y_2 = \partial Y_1 \cap \partial Y_2$, $I_{12}(y) = \{t \in I_1(y) : \varphi_1(t, y) \in Y_1 \cap Y_2\} = \{t \in I_2(y) : \varphi_2(t, y) \in Y_1 \cap Y_2\}$, and $\varphi_1 = \varphi_2 = \varphi_{12}$ in the set $\{t, y : y \in Y_1 \cap Y_2, t \in I_{12}(y)\}$ [vid. Definitions 9.2 and 9.3].

DEFINITION 9.5. A local flow $\mathcal{F} = (Y, I, \varphi)$ is called the <u>union</u> of the flows $\mathcal{F}' = (Y', I', \varphi')$ and $\mathcal{F}'' = (Y'', I'', \varphi'')$, if $\mathcal{F}'$ and $\mathcal{F}''$ are both subflows of $\mathcal{F}$ and Y is the union of Y' and Y''. We use the notation

$$\mathcal{F} = \mathcal{F}' \cup \mathcal{F}'' \quad .$$

Obviously the union of local flows is associative in the sense that if one of the sums $\mathscr{F}_1 \cup (\mathscr{F}_2 \cup \mathscr{F}_3)$, $\mathscr{F}_2 \cup (\mathscr{F}_1 \cup \mathscr{F}_3)$, $\mathscr{F}_3 \cup (\mathscr{F}_1 \cup \mathscr{F}_2)$ exist, they all exist and are equal.

PROPOSITION 9.6. The necessary and sufficient condition in order that two coherent local flows $\mathscr{F}_1 = (Y_1, I_1, \varphi_1)$ and $\mathscr{F}_2 = (Y_2, I_2, \varphi_2)$ possess a union is that

$$\gamma_1(Y_1 \cap Y_2) = \gamma_2(Y_1 \cap Y_2) = Y_1 \cap Y_2. \tag{9.1}$$

Here γ_1, γ_2 denote the paths of $\mathscr{F}_1$, $\mathscr{F}_2$.

PROOF. Apparently, (9.1) implies

$$I_1 = I_2 \quad \text{on} \quad Y_1 \cap Y_2 \ . \tag{9.2}$$

If (9.1) is satisfied, we define I and φ on $Y = Y_1 \cap Y_2$ by

$$I(y) = \begin{cases} I_1(y) & \text{for} \quad y \in Y_1 \ , \\ I_2(y) & \text{for} \quad y \in Y_2 \end{cases}$$

$$\varphi(t, y) = \begin{cases} \varphi_1(t, y) & \text{for} \quad y \in Y_1,\ t \in I_1(y) \ , \\ \varphi_2(t, y) & \text{for} \quad y \in Y_2,\ t \in I_2(y) \ . \end{cases}$$

[On $Y_1 \cap Y_2$ the equality $\varphi_1 = \varphi_2$ holds due to the coherence property (vid. the remark following Definition 9.4).] Then (Y, I, φ) obviously satisfies the conditions (a), (b), (c) of Definition 9.1. Condition (d) follows easily from the following lemma:

Consider two compact sets A_1, A_2 in a topological space X and two homeomorphic mappings $\varphi_1 : A_1 \longrightarrow A_1'$, $\varphi_2 : A_2 \longrightarrow A_2'$ $(A_1', A_2' \subseteq X)$. Assume $\varphi_1 = \varphi_2$ on $A_1 \cap A_2$. Then the mapping

$$\varphi = \begin{cases} \varphi_1 & \text{on} \quad A_1 \\ \varphi_2 & \text{on} \quad A_2 \end{cases}$$

is homeomorphic if and only if

$$A_1' \cap A_2' = \varphi(A_1 \cap A_2) \ . \tag{9.3}$$

Indeed, (9.3) is necessary and sufficient in order that φ is one-one. The continuity of φ is obvious; therefore, since A_1 and A_2

are compact, φ is homeomorphic if (9.3) is satisfied.

For every t the mappings $\varphi_j(y) = \varphi_j(t, y)$ $(j = 1, 2)$ of the sets $A_j = \{y \in Y_j : t \in I_j(y)\}$ satisfy the hypotheses of the lemma and condition (9.3) due to (9.1), so that φ satisfies (d).

We now assume the existence of a union $\mathscr{F} = (Y, I, \varphi) = \mathscr{F}_1 \cup \mathscr{F}_2$. Consider an interior point y of Y_1 and suppose that $I(y) \supset I_1(y)$. Then, by Definition 9.2, one endpoint y_0 of $\gamma_1(y)$ would belong to $Y_1 \cap Y_2 \subsetneq \partial Y_1$. Consider the path γ through y_0 of the common face of $\mathscr{F}_1$ and $\mathscr{F}_2$ [vid. Definition 9.4]. Since γ is contained in $\partial Y_1 \cap \gamma_1(y)$ and does not reduce to a point, a whole arc of $\gamma_1(y)$ would lie on ∂Y_1, in contradiction to Proposition 9.4. Thus $I(y) = I_1(y)$.

Now consider a point $y \in Y_1 \cap Y_2 \subsetneq \partial Y_1$. Every neighborhood of y contains an interior point y_1 of Y_1. Since $I(y_1) = I_1(y_1)$ and the endpoints of I and I_1 depend continuously on y, we conclude that also in this case $I(y) = I_1(y)$ holds and, in the same manner, $I(y) = I_2(y)$, so that (9.2) follows.

Now consider two points $p \in Y_1 \cap Y_2$, $q = \varphi_1(t, p)$. Then, since φ_1 and φ_2 are both restrictions of φ, $\varphi_2(t, p) = q$ holds. Thus $q \in Y_1 \cap Y_2$ and (9.1) follows.

PROPOSITION 9.7. If two local flows (Y', I', φ') and (Y'', I'', φ'') are linked by the relation

$$Y_1' = Y_0'' ,$$

they possess a union (Y, I, φ), where Y, I, φ are defined by

$$Y = Y' \cup Y'' ,$$

$$I(y) = \begin{cases} [t_0'(y),\ t_1'(y) + t_1''(y_1'(y))] & \text{for } y \in Y' , \\ [t_0''(y) + t_0'(y_0''(y)),\ t_1''(y)] & \text{for } y \in Y'' , \end{cases}$$

$$\varphi(t, y) = \begin{cases} \varphi'(t, y) & \text{for } y \in Y' , \\ \varphi''(t, y) & \text{for } y \in Y'' . \end{cases}$$

The proof follows easily from the definitions of the sets Y_0, Y_1.

§5. THE QUALITATIVE PICTURE OF THE SOLUTIONS NEAR A SWITCHING SPACE

10. Local flows around sets of normal switching points. Throughout this chapter U will denote a compact connected subset of S - S'' [vid. §2.5], containing no points of switching spaces other than S.

Consider first the case $U \subset T$. Since U is compact, there exist two numbers α (< 0), β (> 0) such that the functions $x^{\pm}(t, u)$ satisfy (S) in the intervals $I^+ = [0, \beta]$ and $I^- = [\alpha, 0]$, respectively, and define local flows $\mathcal{F}^{\pm}$ on the sets $x(I^{\pm}, U)$. Moreover, $x(t, u)$ defines a local flow $\mathcal{F} = \mathcal{F}^+ \cup \mathcal{F}^-$ on $x(I, U)$ $(I = I^- \cup I^+)$ [vid. Proposition 9.7].

In the case $U \subset A$ [E] the set U is the common entrance [exit] set of two local flows defined by solutions of (S) on the sets $x^+(I^{\mp}, U)$ $[x^-(I^{\mp}, U)]$, respectively, where $I^+ = [0, \beta]$, $I^- = [\alpha, 0]$; $-\alpha, \beta > 0$ and sufficiently small.

The ambiguity occurring in connection with starting and end points can be eliminated by a modification of the phase space E^n, namely by "cutting" it along the sets A, E and replacing every point $u \in A$, E by a pair of points $u^{\mp}$, situated on either "edge" of the cut. More precisely, this procedure can be described as follows: Denote by V the set

$$V = A \cup E$$

and consider an open set $X \subset E^n$ satisfying the condition $X \cap S = V$. Then complete the set $X - V$ (with respect to the euclidean metric), and finally replace E^n by the space

$$D^* = (E^n - X) \cup \overline{(X - V)} \quad ,$$

where the closure is to be understood in the sense of the new topology, defined by the Cauchy sequences in $X - V$. Obviously, the construction of the space D^* does not depend on the choice of X. Denote by φ the continuous mapping of D^* onto E^n which leaves every point of $E^n - V$ fixed. Each point $v \in V$ possesses two images under $\psi = \varphi^{-1}$ which we denote by $\psi^{\pm}(v) = v^{\mp}$ and which are distinguished from each other by the property

$$v^{\mp} \in \bar{D}_{e^{\pm}} \quad .$$

Apparently, through every point $p \in D^*$ there exists exactly one solution of (S). This we denote by $x(t, p)$.

11. Local flows around the sets $(AL)^1$ and $(EL)^1$. If U is a subset of $(AL)^1$, it is easy to see that it is contained in one of the sets $A^+ \cap L^-$ and $A^- \cap L^+$. Indeed, in the opposite case the connectedness of U would imply the existence of a point

$$u \in U \cap \overline{A^+ \cap L^-} \cap \overline{A^- \cap L^+} ,$$

which we could, e.g., assume to belong to $A^+ \cap L^-$. This would imply $c_1^+ > 0$ [vid. Theorem 6.1]. Since, however, every neighborhood of u would contain points of $A^- \cap L^+$, on which $c_1^+ \equiv 0$, c_1^+ could not be continuous. Therefore, without loss of generality, we may assume

$$U \subset A^+ \cap L^- .$$

Relation (8.3) implies $U \subseteq \partial A \cap \partial T$. Moreover, U is contained in the intersection of the boundaries of two components A_1, T_1 of A and T:

$$U \subseteq \partial A_1 \cap \partial T_1 . \tag{11.1}$$

This follows immediately from the connectedness of U and from Theorem 7.1,[1] according to which every point $u \in U$ [$\subseteq S^1$] lies on the boundary of exactly two components of S^0.

For every point $u \in U$ the relation

$$x^-(t, u) \in D_{e^+} \qquad (0 < |t| \leq t') \tag{11.2}$$

holds for sufficiently small t' [Theorem 6.1]. If $v \in A_1$, we have

$$x^-(t, v) \in D_{e^-} \qquad (0 < t < t_1(v)) \tag{11.3}$$

for small $t_1(v)$. The continuous dependence of x^- on the initial value, together with (11.2), implies the existence of a neighborhood N of u such that

$$x^-(t', N) \subset D_{e^+} \tag{11.4}$$

---

[1] We again assume condition A to be satisfied.

holds. We now restrict v to $A_1 \cap N$. Because of (11.3) and (11.4), we can choose $t_1(v)$ in such a way that $x^-(t_1, v) \in S$ and (11.3) holds for $v \in N \cap A_1$. Since U is compact, there exist a connected compact set $\mathscr{A}$, containing U and contained in $U \cup A_1$, and a (continuous) function t_1 defined on $\mathscr{A}$ such that

$$x^-(t, v) \in \begin{cases} A_1 & \text{for } t = 0 , \\ D_e- & \text{for } 0 < t < t_1(v) , \\ T_1 & \text{for } t = t_1(v) \end{cases} \tag{11.5}$$

holds for all $v \in \mathscr{A} - U$. [It is sufficient that $\mathscr{A}$ is contained in the union of a finite number of sufficiently small open sets N_κ (satisfying, in particular, (11.4)) which together cover U.] Moreover, we can assume $\mathscr{A}$ to be a (n-1)-dimensional manifold with boundary (because A_1 is an open subset of the hypersurface S). Since $U \subseteq (AL)^1$ and $\mathscr{A} \subseteq U \cup A_1$, (11.1) implies

$$U \subseteq \partial T_1 \subseteq \bar{T}_1 \subseteq \overline{S - (A_1 \cup (AL)^1)} \subseteq \overline{S - (A_1 \cup U)} \subseteq \overline{S - \mathscr{A}} .$$

This, together with $U \subset \mathscr{A}$, yields

$$U \subseteq \partial \mathscr{A} . \tag{11.6}$$

By cutting the space along A, as described in the preceding section, the set $\mathscr{A}$ is split into two "sheets" $\mathscr{A}^\pm = \psi^\pm(\mathscr{A})$. Since $\mathscr{A}$ is compact, the mappings $\psi^\pm : \mathscr{A} \longrightarrow \mathscr{A}^\pm$ are homeomorphic. This implies the compactness of $\mathscr{A}^\pm$ and, together with (11.6), the relation

$$U \subseteq \partial \mathscr{A}^+ \cap \partial \mathscr{A}^- . \tag{11.7}$$

Then it follows immediately that

$$\mathscr{A}^+ \cap \mathscr{A}^- = U . \tag{11.8}$$

After these preparations it is easy to analyse the local flows around U. According to (11.5), the solutions of (S) define a local n-flow $\mathscr{F}^- = (Y^-, I^-, x^-)$ on the set

$$Y^- = \{x(t, v)\}_{t \in I^-(v), v \in \mathscr{A}^-} , \quad I^-(v) = [0, t_1(v)] \quad \text{for} \quad v \in \mathscr{A}^- .$$

Using the notations of §4 (and indicating by a superscript - the relatedness to $\mathscr{F}^-$), $t_1(v) = t_1^-(v)$ holds for $v \in \mathscr{A}^-$. Since $t_1(v) \longrightarrow 0$

with $v \longrightarrow U$, it follows that

$$t_0^-(y) \longrightarrow 0 \quad \text{with} \quad y \longrightarrow U \ . \tag{11.9}$$

Since $x^+(t, v) \in D_{e^+}$ holds for all $v \in U \cup A_1 \cup T_1$ and small t and $\mathscr{A} \cup Y_1^-$ is a compact subset of $U \cup A_1 \cup T_1$, there exists an interval I_0 $[0, t^*]$ $(t^* > 0)$ such that

$$x^+(I_0, \mathscr{A} \cup Y_1^-) \subset D_{e^+} \ .$$

As a consequence of (11.9), the inequality $t_0^-(v) < t^*$ holds for all $v \in Y_1^-$ if $\mathscr{A}$ is sufficiently small. Therefore, the solutions $x(t, v)$ of (S) define a local flow $\mathscr{F}^+$ with the entrance set $\mathscr{A}^+ \cup Y_1^- = Y_0^+$ and the intervals of definition

$$I^+(v) = \begin{cases} [0, t^*] & \text{for } y \in \mathscr{A}^+ \ , \\ [0, t^* + t_0^-(v)] & \text{for } v \in Y_1^- \ . \end{cases} \tag{11.10}$$

Now we split up the local flow $\mathscr{F}^+$ into two subflows $\mathscr{F}^1$ and $\mathscr{F}^2$, defined on the sets $x(I^+, \mathscr{A}^+)$ and $\bigcup_{v \in Y_1^-} x(I^+(v), v)$, respectively. Then $Y_1^- = Y_0^2$ holds and, due to Proposition 9.7, $\mathscr{F}^- \cup \mathscr{F}^2$ is a local flow which we denote by $\mathscr{F}' = (x(I_0, \mathscr{A}^-), I', x)$. Since $\mathscr{A}$ is an $(n-1)$-dimensional manifold with boundary, $\mathscr{F}^1$ and $\mathscr{F}'$ are both local n-flows. Denoting by $\mathscr{F}^{12}$ the $(n-1)$-flow defined on $x(I_0, U)$, (11.7) together with Proposition 9.5 implies

$$\mathscr{F}^{12} \subset \partial\mathscr{F}^1 \cap \partial\mathscr{F}' \ .$$

From this and (11.8) it follows that $\mathscr{F}^1$ and $\mathscr{F}'$ are coherent.

Finally, if $v \in Y_1^-$, Proposition 9.7 and (11.10) yield

$$I'(v) = [t_0^-(v), t^* + t_0^-(v)]$$

which, together with (11.9), implies

$$I'(u) = [0, t^*] = I_0 \qquad (u \in U) \ .$$

Since also $I^1(u) = I_0$, condition (9.1) of Proposition 9.6 is satisfied for the flows $\mathscr{F}^1$ and $\mathscr{F}'$, their union thus being a local flow:[1]

[1] Vid. remark following Definition 9.5.

$$\mathscr{F} = \mathscr{F}^1 \cup \mathscr{F}' = \mathscr{F}^1 \cup (\mathscr{F}^2 \cup \mathscr{F}^-) = (\mathscr{F}^1 \cup \mathscr{F}^2) \cup \mathscr{F}^-$$
$$= \mathscr{F}^+ - \mathscr{F}^- .$$

This yields the following theorem:

THEOREM 11.1. Consider a connected compact subset U of $(AL)^1$ [$= AL \cap S^1$; vid. §2, Sections 4 and 5], on which condition A is satisfied [vid. §3.7]. Assume the phase space to be cut along the set A of starting points. Then U is contained in a compact connected set Y which is an n-dimensional manifold with boundary and has the property that all solutions of (S) which contain points of Y form a single local flow $\mathscr{F}$ on Y. The entrance set of $\mathscr{F}$ consists of U and the two "edges" $\mathscr{A}^{\pm}$ of the cut along $A \cap Y$. The paths originating on $\mathscr{A}^-$ [or on $\mathscr{A}^+$ in the case $U \subset A^- \cap L^+$] traverse the switching space exactly once. The exit set is contained in D_{e^+} [D_{e^-}]. (Vid. Figures 2, 3).

The case $U \subset (EL)^1$ is exactly analogous and can be reduced to the one considered above by reversing the sign of t.

12. Local flows around the sets $(AR)^1$ and $(ER)^1$. If U is a subset of $(AR)^1$ it can be assumed without loss of generality that it is contained in $A^+ \cap R^-$. [The proof is analogous to the corresponding one in the preceding section.] Denote again by A_1, T_1 the components of A, T which are adjacent to U. Let $\mathscr{A}$ and $\mathscr{T}$ be compact connected sets satisfying the conditions

$$U \subset \mathscr{A} \subset U \cup A_1, \quad U \subset \mathscr{T} \subset U \cup T_1 .$$

We can assume $\mathscr{A}$ and $\mathscr{T}$ to be (n-1)-dimensional manifolds with boundary; $\mathscr{A}$ will again be replaced by two "sheets" $\mathscr{A}^{\pm}$.

Through every point $u \in U$ there exist exactly two solutions $x_1(t, u)$, $x_2(t, u)$ of (S). Since U is compact, the inclusions

$$(12.1) \qquad x_1(t, U) \subset D_{e^-} \qquad (-t_1 < t < t_2;\ t \neq 0)$$

and

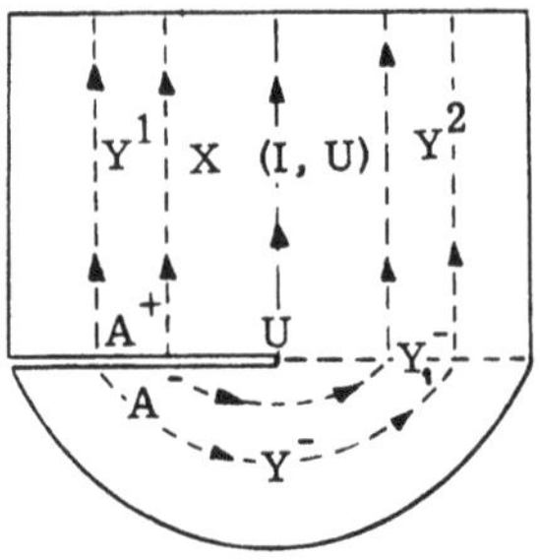

Fig. 2.

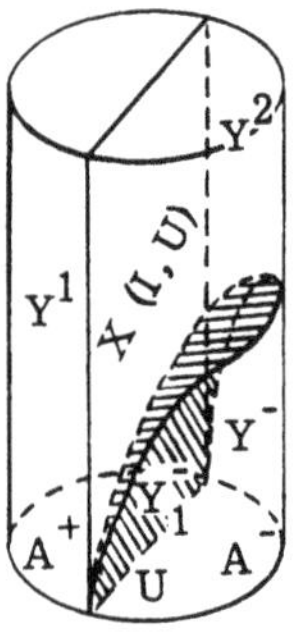

Fig. 3.

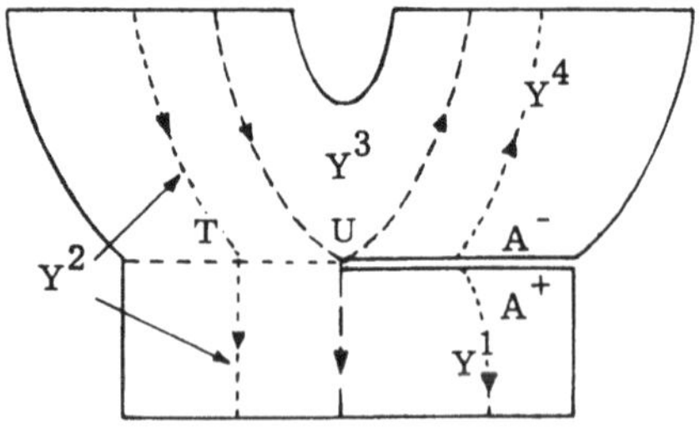

Fig. 4.

$$(12.2) \qquad x_2(t, U) \subset \begin{cases} D_{e^-} & \text{for } -t_1 < t < 0 , \\ D_{e^+} & \text{for } \quad 0 < t < t_3 \end{cases}$$

hold for sufficiently small positive numbers t_1, t_2, t_3 [vid. §2.6]. Due to the continuous dependence of $x^{\pm}$ on the initial values, it follows from (12.1) and (12.2) that the solutions of (S) define three local flows $\mathcal{F}_1, \mathcal{F}_2, \mathcal{F}_4$ on the sets $x(I_1, \mathcal{A}^+)$, $x(I_2, \mathcal{T})$, $x(I_4, \mathcal{A}^-)$, respectively, where $I_1 = [0, t_3']$, $I_2 = [-t_1', t_3']$, $I_4 = [0, t_2']$, $0 < t_j' < t_j$ $(j = 1, 2, 3)$ and $\mathcal{A}$ and $\mathcal{T}$ are sufficiently small. The local (n-1)-flows defined on the sets $x_2(I_1, U)$, $x_1([-t_1', 0]\, U)$, $x_1(I_4, U)$ we denote by $\mathcal{F}_{12}, \mathcal{F}_{23}, \mathcal{F}_{34}$. The relations (11.1) and (11.7) obviously hold also in the case under consideration. Since (11.1) implies $U \subset \partial\mathcal{T}$, Proposition 9.5 yields the incidences

$$\mathcal{F}_{12} \subseteq \partial\mathcal{F}_1 \cap \partial\mathcal{F}_2, \quad \mathcal{F}_{23} \subseteq \partial\mathcal{F}_2, \quad \mathcal{F}_{34} \subseteq \partial\mathcal{F}_4 \quad .$$

The first of these implies the coherence of the local flows $\mathcal{F}_1$ and $\mathcal{F}_2$.[1] Moreover, due to Proposition 9.7, the unions $\mathcal{F}_{23} \cup \mathcal{F}_{12}$ and $\mathcal{F}_{23} \cup \mathcal{F}_{34}$ exist.[2]

Finally, a local flow which is incident with the (n-1)-flows $\mathcal{F}_{23}$ and $\mathcal{F}_{34}$ is obtained by considering a hypersurface H without contact[3] with the solutions of (S, e^-), which contains U,[4] and a compact connected subset C of $H \cap \bar{D}_{e^-}$ the boundary of which contains U. Then, denoting by I_3 the interval $[-t_1', t_2']$, every arc $x^-(I_3, p)$, where $p \in C$, intersects C only at p. Consequently, the solutions of (S) define a local flow $\mathcal{F}_3$ on the set $x^-(I_3, C)$. Since $U \subseteq \partial C$, the incidence relations

$$\mathcal{F}_{23} \subseteq \partial\mathcal{F}_3, \quad \mathcal{F}_{34} \subseteq \partial\mathcal{F}_3$$

follow. Therefore, $\mathcal{F}_2$ and $\mathcal{F}_3$ as well as $\mathcal{F}_3$ and $\mathcal{F}_4$ are coherent.[5] We therefore have the result:

---

[1] However, the union of $\mathcal{F}_1$ and $\mathcal{F}_2$ is not a flow [vid. Proposition 9.6].

[2] This situation can be described as a "bifurcation" of the flow $\mathcal{F}_{23}$ into $\mathcal{F}_{12}$ and $\mathcal{F}_{34}$.

[3] Vid. p. , footnote 1.

[4] The set $x^+(I, U)$, e.g., has this property if I is a suitable interval around $t = 0$.

[5] Again (due to Proposition 9.6) neither of the unions is a local flow.

THEOREM 12.1. If U is a compact connected subset of $(AR)^1$ [$= AR \cap S^1$], satisfying condition A, and the space is cut along the set of starting points, splitting it into two "sheets" $\mathscr{H}^{\pm}$, there exists a compact connected set Y which is an n-dimensional manifold with boundary containing U and has the following property: The solutions of (S) which contain points of Y constitute a complex of four local flows $\mathscr{F}_1, \ldots, \mathscr{F}_4$, separated from each other by (n-1)-flows $\mathscr{F}_{12}, \mathscr{F}_{23}, \mathscr{F}_{34}$ such that $\mathscr{F}_{\mu,\mu+1}$ ($\mu = 1, 2, 3$) is incident with $\mathscr{F}_\mu$ and $\mathscr{F}_{\mu+1}$ and the two latter are coherent. Upon reaching the set U, the flow $\mathscr{F}_{23}$ bifurcates into $\mathscr{F}_{12}$ and $\mathscr{F}_{34}$ in the sense that the exit set of $\mathscr{F}_{23}$ and the entrance sets of $\mathscr{F}_{12}$ and $\mathscr{F}_{34}$ are all equal to U. The local flows $\mathscr{F}_1$ and $\mathscr{F}_4$ originate on $\mathscr{H}^+$ and $\mathscr{H}^-$, respectively. The paths of $\mathscr{F}_2$ (or, if $U \subset A^- \cap R^+$, of $\mathscr{F}_3$) and only these traverse the switching space.

Figure 4 gives a symbolic picture of the local flows and their interrelations.

The analogous result for the case $U \subset (ER)^1$ is again obtained by reversing the orientation of the paths.

BIBLIOGRAPHY

[1a] [1b] André, J. and Seibert, P., "Über stückweise lineare Differentialgleichungen, die bei Regelungsproblemen auftreten I, II," Archiv d. Math., Vol. 7 (1956), p. 148-156, 157-164.

[2] Bellman, R., Glicksberg, I., and Gross, O., "On the "bang-bang" control problem," Quart. Appl. Math., Vol. 14 (1956), p. 11-18.

[3] Boltyanskiĭ, V. G., Gamkrelidze, R. V., Pontryagin, L. S., "On the theory of optimal processes," (Russian), Dokl. Akad. Nauk SSSR (N.S.) Vol. 110 (1956), p. 7-10.

[4] Bushaw, D., "Optimal discontinuous forcing terms," these Contributions, Vol. IV, Annals of Math. Studies 41 (1958), p. 29-52.

[5] Flügge-Lotz, I., Discontinuous automatic control, Princeton, 1953.

[6] Gamkrelidze, R. V., "Theory of time-optimal processes in linear systems," (Russian), Izvest. Akad. Nauk SSSR, Ser. Mat., Vol. 22 (1958), p. 449-474.

[7] Krasovskiĭ, N. N., "On stability in the case of large initial perturbations," (Russian), Prikl. Mat. Meh., Vol. 21 (1957), p. 309-319.

[8] ——————, "On the theory of optimal control," (Russian), Avtomat. i Telemeh., Vol. 18 (1957), p. 960-970.

[9] Pontryagin, L. S., "Optimal control processes," (Russian), Uspehi Mat. Nauk, Vol. 14 (1959), p. 3-20.

[10] Solncev, Yu. K., "On stability according to Lyapounov of the equilibrium positions of a system of two differential equations in the case of discontinuous right hand sides," (Russian), Moscov. Gos. Univ. Ucenye Zapaki, Vol. 148, Mat. 4 (1951), p. 144-180.

XII. ASYMPTOTIC STABILITY IN 3-SPACE

Courtney Coleman

1. Consider the system,

$$\frac{dx}{dt} = f^{(m)}(x) + g(x, t) \quad , \tag{1}$$

where $f^{(m)}(x)$ and $g(x, t)$ satisfy the following conditions:

a) Both are n-vector functions of the n-vector x. g may depend on t. $f^{(m)}(0) = g(0, t) = 0$.

b) In some neighborhood, D, of $x = 0$ and for all t greater than some t_0, $f^{(m)}(x)$ and $g(x, t)$ are continuous in x and t and satisfy a Lipschitz condition in x.

c) $f^{(m)}(x)$ is a homogeneous function of degree m — (m an integer > 1) in D; i.e., $f^{(m)}(x) = c^m f(x)$ for any constant $c > 0$ and for all x in D.

d) $g(x, t) = o\,(\|x\|^m)$ uniformly in t in D.

Malkin [1] and Massera [2] have shown that if the trivial solution, $x = 0$, of the system of first approximation,

$$\frac{dx}{dt} = f^{(m)}(x) \quad , \tag{2}$$

is asymptotically stable, then so is the trivial solution of (1). The problem, then, is to find conditions ensuring the asymptotic stability of the trivial solution of (2). This will be done for $n = 3$ (see the theorem below) using some of the results obtained by the author [3]. For $n = 2$ such conditions can be found implicitly in a number of papers -- e.g., see Forster [4]. The conditions are simply that the origin be an attracting focus if the function $h(x_1, x_2) = x_2 f_1^{(m)} - x_1 f_2^{(m)}$ has no real zeros, and that

$$\frac{dr}{dt} < 0 \quad (r^2 = x_1^2 + x_2^2)$$

on the zeros of h if they exist.

It should be noted that for $m = 1$ and for any n the theory of characteristic roots holds. Clearly, if m is even, there can be no asymptotic stability. Hence, m is assumed odd.

2. Let $n = 3$, $m > 1$. Let $r = (x_1^2 + x_2^2 + x_3^3)^{\frac{1}{2}}$, $y = \frac{1}{r}x$. Then (2) becomes in r, y,

$$(3)\qquad \begin{aligned} &\text{a.} \quad \frac{dr}{dt} = (y, f^{(m)}(y))r^m \\ &\text{b.} \quad \frac{dy}{dt} = (f^{(m)}(y) - (y, f^{(m)}(y))y)r^{m-1}, \end{aligned}$$

where (,) indicates scalar product. Let $r^{m-1}(t)dt = d\tau$. Then (3) becomes,

$$(4)\qquad \begin{aligned} &\text{a.} \quad \frac{dr}{d\tau} = (y, f^{(m)}(y))r \\ &\text{b.} \quad \frac{dy}{d\tau} = f^{(m)}(y) - (y, f^{(m)}(y))y \quad . \end{aligned}$$

If $y(\tau)$ is a solution of (4b) and $r(\tau)$ is a solution of (4a) in which y stands for $y(\tau)$ (the solution of 4b), then $(r(\tau), y(\tau))$ is a solution of (4). Similarly for (3a, b). System (4) is more suitable for our purposes since no solution of (4) may have a finite "escape time" while a solution of (3) may. Ignoring the differing parametrizations, the trajectories in 3-space of (3) coincide with those of (4); for the only difference between the two systems is the factor r^{m-1}, which is positive everywhere except at the origin. This implies in addition that the trajectories of the two systems agree in the sense of parametric description. If all the trajectories of (4) close enough to 0 tend to 0 as limit with increasing τ and if none tend to 0 with decreasing τ, then the same is true for the trajectories of (3) and hence of (2). Since systems (2), (3), (4) are autonomous, to prove asymptotic stability it is sufficient to show that there is a neighborhood $D^* \subset D$ such that for any trajectory Γ of (4) which cuts D^* $\Omega(\Gamma)$ (the omega limit set of Γ) contains 0 only and $A(\Gamma)$ (the alpha limit set of Γ) has no points in D^*. Actually for systems (2), (3) and (4) D and D^* might as well be all of 3-space.

The following properties of system (4) and its trajectories were observed in [3].

P I. Each trajectory $\gamma : y = y(\tau, y_0, \tau_0)$ of (4b) lies on the unit sphere S^2. The Poincaré-Bendixson theory can then be applied to $\Omega(\gamma)$ and $A(\gamma)$.

P II. Each trajectory γ of (4b) determines a cone $S(\gamma)$ generated by rays through 0 and the points of γ. $S(\gamma)$ is an integral surface of (4).

P III. On $S(\gamma)$ each Γ of (4) is given by

$$\left[K \cdot \exp \int_{\tau_0}^{\tau} (y, f^{(m)}(y)) d\tau \right] \cdot y(\tau) \quad ,$$

where K is an arbitrary positive constant.

P IV. If $\lambda(p)$ is a ray on $S(\gamma)$ through 0 and and p, then all trajectories of (4) on $S(\gamma)$ cut $\lambda(p)$ at the same non-zero angle in the same sense (except in the trivial case that γ is a singular point of (4b)).

The following general properties of a system of autonomous differential equations,

$$\frac{dx}{dt} = h(x), \quad x \text{ an n-vector}, \tag{5}$$

and its trajectories will be needed. All functions are assumed analytic.

P Va. If $N(p)$ is a small enough neighborhood of the non-critical point p in the normal hyperplane to the trajectory of (5) through p, then the trajectories of (5) through the points of $N(p)$ cross $N(p)$ from one side to the other in the same sense.

P Vb. If S is a smooth surface in x-space and C a simple smooth arc on S such that all trajectories of (5) cutting C do so crossing C from one side of S to the other and not tangent to S, then there is a neighborhood $N(C)$ in S of C such that the same is true for all points of $N(C)$.

P VI. The A limit set (Ω limit set) of a bounded trajectory of (5) is closed, non-empty and connected. If a point belongs to a limit set of some trajectory, then so does the full trajectory through the point.

In the proof of the theorem that follows frequent use will be made of PI - PVI. Unless required for clarity specific reference will not be made.

THEOREM. Suppose

a) $(y, f(y)) < 0$ on all zeros of $f^{(m)}(y) - (y, f^{(m)}(y))y$

and

b) $\int_0^T (y, f^{(m)}(y))d\tau < 0$ for each periodic solution $y(\tau)$ of (4b) (T is the period of $y(\tau)$).

Then the solution $x = 0$ of (1) for $n = 3$ is asymptotically stable.

PROOF. It is sufficient to show that under assumptions a) and b) $\Omega(\Gamma)$ contains only 0 and $A(\Gamma)$ has no finite points for any trajectory Γ of (4).

In the following γ, γ^*, γ_i, $i = 1, \ldots, k$, denote trajectories of (4b) on S^2 and P, P_i, $i = 1, \ldots, k$ critical points of (4b) on S^2. $S(\gamma_i)$, $S(P_i)$, ... denote the integral surfaces determined by γ_i, P_i, ... Γ, Γ^*, Γ_p, Γ_i, $i = 1, \ldots, k$ denote trajectories of (4) on the integral surfaces $S(\gamma)$, $S(\gamma^*)$, $S(P)$, $S(\gamma_i)$, ... respectively. The Poincaré-Bendixson theory states that a limit set of a trajectory γ of (4b) on S^2 is one of the following:

(L 1a.) γ itself, where γ is a single point;

(L 2b.) γ itself, where γ is a closed curve;

(L 2a.) a critical point P, not γ;

(L 2b.) a closed curve γ^*, not γ;

(L 3) a closed graph $G : P_1 \cup \gamma_1 \cup \ldots \cup P_k \cup \gamma_k \cup P_{k+1}$, where $\Omega(\gamma_i) = \Gamma_i$, $A(\gamma_i) = \Gamma_{i+1}$, $\Gamma_{k+1} = \Gamma_1$.

The various combinations of these possibilities for $A(\gamma)$ and $\Omega(\gamma)$ must now be considered.

A. (L 1a.) for $A(\gamma)$ and hence for $\Omega(\gamma)$. Hypothesis a) ensures that for the ray trajectory Γ of (4) through γ, $\Omega(\Gamma)$ contains 0 alone and $A(\Gamma)$ is infinite. (See P III).

B. (L 1b.) for $A(\gamma)$ and hence for $\Omega(\gamma)$. Here hypothesis b) ensures that $\Omega(\Gamma)$ consists only of 0 and $A(\Gamma)$ is infinite (See P III and [3]).

C. (L 2a.) for $A(\gamma)$ and for $\Omega(\gamma)$. Whether or not $A(\gamma) = \Omega(\gamma)$, $A(\Gamma)$ is infinite and $\Omega(\Gamma)$ contains just 0. (See P III, Iv and [3]).

D. (L 2a.) for $A(\gamma)$ and (L 2b.) for $\Omega(\gamma)$. Let $A(\gamma) = P$. Then Γ_P is as in A above. By (P VI) for any Γ on $S(\gamma)$, $A(\Gamma)$ cannot contain any point of Γ_P unless it contains Γ_P itself. This is impossible by (P Va.) which also implies that $A(\Gamma)$ cannot contain 0 alone. Hence $A(\Gamma)$ has no finite points. On $S(\gamma^*)$ B. above obtains. Hence $\Omega(\Gamma)$ has no finite points, or contains 0 and at least one Γ^* on $S(\gamma^*)$, or contains 0 alone. Let p be any point of γ^*. Let

$$\int_0^T (y^*, f(y^*))d\tau = -K < 0 \quad ,$$

where γ^* is given by $y = y^*(\tau)$, $y^*(\tau + T) = y^*(\tau)$. Let σ be an arc (bounded on one end by p) of the great circle on S^2 through p normal to γ^*. Let γ be given by $y = y(\tau)$ and let d denote Euclidean distance. If $y(\tau)$ is on σ for some τ, denote by T_τ the smallest positive number such that $y(\tau + T_\tau)$ is on σ. If the length of σ is small enough, T_τ is defined for all τ for which $y(\tau)$ is on σ; and if $y(\tau_0)$ is on σ, then

$$\max_{\tau \geqq \tau_0} d(y(\tau), \gamma^*)$$

(for <u>all</u> $\tau \geqq \tau_0$) and

$$\max_{T_\tau, \tau \geqq \tau_0} |T - T_\tau| \ (y(\tau) \text{ on } \sigma)$$

are so small that

$$\int_{\tau}^{\tau+T_\tau} (y,\ f(y)d\tau \leqq -K/2 < 0 \ (y(\tau)\ \text{ on } \ \sigma) \ .$$

But this can be done for every point of γ^*. Hence, $\Omega(\Gamma)$ can contain 0 alone (recalling that Γ is given by

$$\left[K \exp \int_{\tau_0}^{\tau} (y,\ f(y))d\tau \right] \cdot y(\tau)) \ .$$

E. (L 2a.) for $A(\gamma)$ and (L 3.) for $\Omega(\Gamma)$. As in D, for any Γ on $S(\gamma)$ $A(\Gamma)$ contains no finite points. We must show that $\Omega(\Gamma)$ contains 0 alone. In the first place for any Γ_i or Γ_P on the cone $S(G)$, $A(\Gamma_i)$ and $A(\Gamma_{P_i})$ have no finite points and $\Omega(\Gamma_i)$ and $\Omega(\Gamma_{P_i})$ contain 0 alone. For, each $S(\gamma_i)$ or $S(P_i)$ is covered by cases A. or C. above (Figure 1). Hence, as in D, for any Γ on $S(\gamma)$, $\Omega(\Gamma)$ has no finite points, contains 0 and at least one Γ_i or Γ_{P_i} on $S(G)$, or else contains 0 alone. The first two alternatives must be excluded.

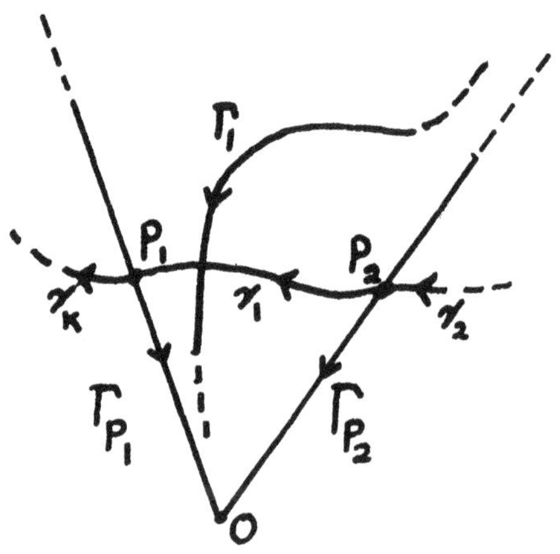

FIGURE 1

First a simple closed curve without contact C is constructed on $S(G)$ such that one of the components of the complement of C on $S(G)$ has 0 as a boundary point and is bounded. Let q_0 be any point on $S(\gamma_1)$. Suppose each Γ_1 on $S(\gamma_1)$ tends to 0 with increasing τ tangent to Γ_{P_1} and to infinity parallel to Γ_{P_2} with decreasing τ (see C. above and [3]). Let

$$y(\tau_0) = \frac{1}{\|q_0\|} \cdot q_0 \ .$$

Considering all rays as being directed toward 0, let $0 < \theta(q) < \pi$ be the angle defined by the field vector at q and the directed ray $\lambda(q)$ for each q on $S(\gamma_1)$. $\theta(q)$ is a continuous function of q as q varies along any continuous curve on $S(\gamma_1)$. Define $\varepsilon(\tau)$ as any positive continuous function of τ such that

$$\lim_{\tau \to \infty} \int_{\tau_0}^{\tau} [(y, f^{(m)}(y)) + \varepsilon(\tau)]d\tau = \ell n \frac{\|q_1\|}{\|q_0\|}$$

where q_1, is some point of Γ_{P_1}. There are many ways of defining $\varepsilon(\tau)$. For example, since by hypothesis (a) $(y, f(y))$ is negative and bounded away from 0 for τ large enough, $\varepsilon(\tau)$ could be taken as

$$\frac{1}{(\tau+1-\tau_0)^2} \cdot \ell n \frac{\|q_1\|}{\|q_0\|} - (y, f(y))$$

for τ large enough.

Define the curve $C_1^+(q_0, q_1, \tau)$ as

$$q_1 U \left\{ \|q_0\| \exp \int_{\tau_0}^{\tau} [(y, f(y)) + \varepsilon(\tau)]d\tau \right\} \cdot y(\tau)$$

for $\tau \geq \tau_0$. C_1^+ is then a smooth arc bounded by q_0 and q_1. $\frac{dr}{d\tau}$ at any point q, $q \neq q_0, \neq q_1$, of $C_1^+(q_0, q_1, \tau)$ is greater by the positive quantity $\varepsilon(\tau)$ than $\frac{dr}{d\tau}$ given by (4a). Hence the angle $\phi(q)$ defined by the directed tangent to $C_1^+(q_0, q_1, \tau)$ at q and the directed ray $\lambda(q)$ is greater than $\theta(q)$. Then $C_1^+(q_0, q_1, \tau)$ is an arc without contact. In a similar way, $C_1^-(q_0, q_2, \tau)$ for $\tau \leqq \tau_0$, q_2 on Γ_{P_2}, can be defined as a smooth curve without contact bounded by q_0 and q_2. After small changes in C_1^- and C_1^+ near q_0, if necessary, define $C_1(q_1, q_2)$ as $C_1^- U C_1^+$. $C_1(q_1, q_2)$ is then a smooth curve without contact on $S(\gamma_1)$ bounded by q_1 on Γ_{P_1} and q_2 on Γ_{P_2}. All trajectories on $S(\gamma_1)$ cross $C_1(q_1, q_2)$ into the region on $S(\gamma_1)$ bounded by 0, Γ_{P_1}, Γ_{P_2}, and $C_1(q_1, q_2)$. Note that q_1 and q_2 are arbitrary points on Γ_{P_1} and Γ_{P_2}.

In like fashion a curve without contact C_i is constructed on each $S(\gamma_i)$. Since the boundary points of each C_i may be chosen arbitrarily, the C_i's may be connected to form a simple closed curve C on $S(G)$ with the desired properties. (See Figure 2.)

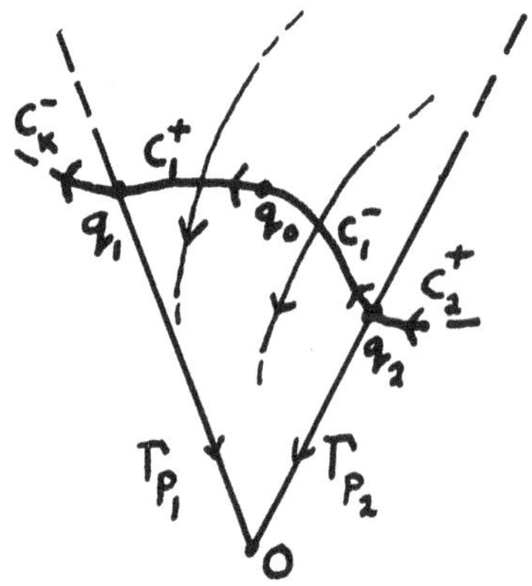

FIGURE 2

Now, let $N(C, \sigma)$ be the surface generated by the normals of length σ to $S(G)$ along C. The normals are taken in the component of the complement of $S(G) \cup O$ which contains Γ. By (P Vb) and a compactness argument, for σ small enough $N(C, \sigma)$ is a smooth (except on the boundaries) non-self-intersecting surface without contact bounded by C and the curve generated by the ends of the normals.

Let G' be a smooth simple closed curve on S^2 enclosing G and cut by γ. There is a τ_1 such that for all $\tau \geqq \tau_1$ $y(\tau)$ lies in the region on S^2 bounded by G and by G'. Let the maximal distance between G and G' be so small that the cone (not an integral surface), $S(G')$ determined by G' and O is cut by $N(C, \sigma)$ in a simple closed curve. For example, G' might be constructed by using arcs of fixed length (sufficiently small) of great circles of S^2 normal to the γ_i and then joining the subarcs around the P_i in some smooth fashion.

Let R be the region bounded by $S(G)$, $S(G')$, and O. Let R^* be the subregion of R bounded by $S(G)$, $S(G')$, O and $N(C, \sigma)$. For all

$$\tau \geqq \tau_1, \; \Gamma : \left\{ r(\tau_0) \exp \int_{\tau_0}^{\tau} (y, f(y)) d\tau \right\} \cdot y(\tau)$$

is in R. Furthermore, if Γ is in R^* for some τ, it is in R^* for all greater τ. For evidently Γ could not leave R^* via $S(G')$, $S(G)$ or O. It could not leave through $N(C, \sigma)$ because by construction of $N(C, \sigma)$ all trajectories cutting the interior of $N(C, \sigma)$ enter R^*. Hence in this case using (P VI) $\Omega(\Gamma)$ contains only O. Suppose it is not known that Γ intersects R^*. Let p be a point of Γ corresponding to some τ_2, $\tau_2 \geqq \tau_1$. Then for some k, $0 < k < 1$, $k \cdot p$ is in R^*. Then the trajectory $k \cdot \Gamma$:

$$\left\{ k \cdot r(\tau_0) \cdot \exp \int_{\tau_0}^{\tau} (y, f(y)) d\tau \right\} \cdot y(\tau)$$

must lie in R^* for all $\tau \geqq \tau_2$ and $\Omega(k \cdot \Gamma)$ contains 0 alone. By (P III) this implies that $\Omega(\Gamma)$ consists of just 0 (see Figure 3).

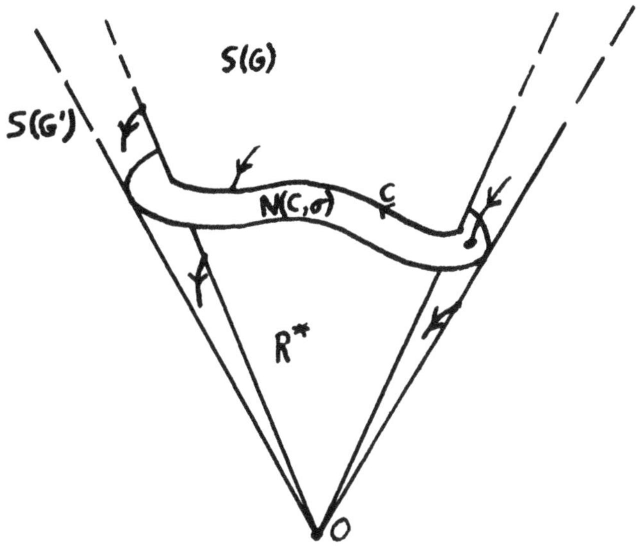

FIGURE 3

F-I. The remaining combinations of possibilities for $A(\gamma)$ and $\Omega(\gamma)$ are not discussed since in each case the argument is similar to the preceding.

Thus, in every case $A(\gamma)$ and $\Omega(\gamma)$ have the desired properties and the theorem is proved.

3. EXAMPLE

Consider the system

$$\frac{dx_1}{dt} = -\mu(x_1^3/3 - x_1x_3^2) + ax_1x_3^2 + bx_1x_2^2 + x_2x_3^2 + g_1(x, t) = f_1(x) + g_1(x,t)$$

$$(5) \quad \frac{dx_2}{dt} = -x_1x_3^2 + ax_2x_3^2 + bx_2^3 + g_2(x, t) = f_2(x) + g_2(x, t)$$

$$\frac{dx_3}{dt} = ax_3^3 + bx_2^2x_3 + g_3(x, t) = f_3(x) + g_3(x, t) \quad ,$$

where $a < 0$, $b < 0$, $\mu > 0$, $x = (x_1, x_2, x_3)$. Suppose that $g = (g_1, g_2, g_3)$ satisfies the hypotheses given in Section 1 for $m = 3$.

In r, y coordinates (5) become

$$\text{(6)}\quad \text{(a)}\quad \frac{dr}{d\tau} = r(y, f(y))$$

$$\text{(b)}\quad \frac{dy}{d\tau} = f(y) - (y, f(y))y \quad ,$$

where $f = (f_1, f_2, f_3)$ and $r^2 dt = d\tau$. It is not hard to verify that the only roots of $f(y) - (y, f(y))y = 0$ are $(\pm 1, 0, 0)$, $(0, \pm 1, 0)$, $(0, 0, \pm 1)$ and that $(y, f(y)) < 0$ on each of these roots (a and b must be negative). Projecting the hemisphere $y_3 > 0$ from the origin onto the plane $y_3 = 1$, we obtain

$$\text{(7)}\quad \frac{dz_1}{d\tau} = z_2 - \mu(z_1^3/3 - z_1)$$

$$\frac{dz_2}{d\tau} = - z_1 \quad ,$$

where

$$z_1 = y_1(1 - y_1^2 - y_2^2)^{-\frac{1}{2}}, \quad z_2 = y_2(1 - y_1^2 - y_2^2)^{-\frac{1}{2}} \quad .$$

This is just Van der Pol's equation, and there is a unique periodic solution, for μ small enough, which is near the circle: $z_1^2 + z_2^2 = 4$. Thus, there is a unique periodic solution to (6b) on the hemisphere $y_3 > 0$ (and, of course, another on the hemisphere $y_3 < 0$). This solution is near the circle $y_3 = 5^{-1/2}$ on the two sphere. But then $(y, f(y))$ is approximately $-\mu y_1^2(y_1^2/3 - 1/5) + 1/5a + by_2^2$ which is negative for $|a|$ large enough, $a < 0$. Hence,

$$\int_0^T (y, f(y))d\tau < 0$$

where T is the period of the periodic solution. Similarly for the periodic solution on the hemisphere $y_3 < 0$.

The theorem then applies, and the solution $x = 0$ of (5) is asymptotically stable.

4. REMARKS

A. The two conditions given in the theorem for asymptotic stability are usually easy to apply. The first is always easy, and the second condition can be verified without knowing either a precise expression for a periodic solution of (4b) or its exact period -- as was seen in the above example.

B. The two conditions are in a certain sense necessary for asymptotic stability. That is, if either of the inequalities is reversed, the critical point of (2) at the origin is not even stable. It is evident that if one of the two inequalities is replaced by an equality, the trivial solution of (1) may or may not be asymptotically stable, depending on the form of $g(x, t)$.

C. If $m = 1$, the linear case, the two conditions are equivalent to the condition that the characteristic roots of the matrix of coefficients of the linear terms have negative real parts. The proof of this fact is not difficult, but is a little long because of the necessity of considering the various possible canonical forms for the matrix. The proof will be omitted.

D. Zubov [5] has given conditions for asymptotic stability for $m = p/q$(p, q integers, q odd) where n is arbitrary. His conditions do not seem easy to use inasmuch as one must have a priori bounds on the solutions $x(t)$ of (1) for which $\|x(0)\| = 1$, namely $\|x(t)\| < At^{-\alpha}$, A, α positive constants.

RIAS, Baltimore

BIBLIOGRAPHY

[1] Malkin, I. G., "A theorem on stability in the first approximation," (in Russian), Doklady Akademii Nauk SSSR, 76, (1951), pp. 783-784.

[2] Massera, J. L., "Contributions to stability theory," Annals of Math., 64, (1956), pp. 182-205.

[3] Coleman, C. S., "A certain class of integral curves in 3-space," Annals of Math., 69, May, 1959.

[4] Forster, H., "Über das Verhalten der Integralkurven einer Gewöhnlichen Differentialgleichung erster Ordnung in der Umgebung eines singulären Punktes," Mathematische Zeitschrift, 43, (1937), pp. 271-320.

[5] Zubov, V. I., "An investigation of the stability problem for a system of equations with homogeneous first terms," Doklady Akademii Nauk (in Russian), 114, (1957), pp. 942-944.

XIII. EXISTENCE AND UNIQUENESS OF THE PERIODIC SOLUTION OF AN EQUATION FOR AUTONOMOUS OSCILLATIONS

Rui Pacheco de Figueiredo

I. PRELIMINARY REMARKS

Consider the differential equation

(1) $$\ddot{x} + R(\dot{x}) + x = 0$$

where $R(\dot{x})$ is a real-valued function having a piecewise continuous derivative** and from now on the dot on a variable denotes differentiation with respect to t. Both (1) and its alternate form

(2) $$\ddot{y} + r(y)\dot{y} + y = 0$$

where $y = \dot{x}$ and $r(y) = R'(y)$, have been widely used to represent the behavior of physical systems undergoing self-sustained oscillatory motion.

In this paper we propose a general set of conditions for the existence and uniqueness of the periodic solution of (1).

* The results of this paper are from a part of the author's doctorate thesis at Harvard University and were presented at the annual meeting of the American Mathematical Society in Philadelphia on January 20, 1959. It is a pleasure to thank Professor P. Le Corbeiller of Harvard for his very valuable advice throughout this reasearch, and Dr. F. Huckemann, now at the Mathematical Institute of the University of Giessen, Giessen-Lahn, Germany, for his careful reading of, and very useful remarks on the original manuscript of this thesis. This work was supported in part by a grant from the Junta de Investigações do Ultramar, Portugal.

** For such a function, it can be readily shown (cf. the author's thesis) that the two-dimensional system equivalent to (1): $\dot{x} = y$, $\dot{y} = -R(y) - x$ possesses locally a unique solution everywhere on the (x, y) plane. This and the continuity of the right-sides of this system assure (see reference [1]) that the Poincaré-Bendixson theorem is applicable to this system under the above assumption.

Noting that $r(y)$ $(= R'(\dot{x}))$ represents the damping in the physical system, we may describe our existence conditions simply as follows: (1) has at least one periodic solution if, for small $|\dot{x}|$, the damping in the system is negative and provided that, for large $|\dot{x}|$, the damping is positive for positive $\dot{x}$ and greater than a negative minimum for negative $\dot{x}$ (or vice-versa). We further show that the periodic solution is unique (except for translation in t) if $R(\dot{x})$ satisfies some additional restrictions.

Among the previous contributions on this subject, notable are those of Liénard [2] and Levinson and Smith* [3]. In the present conditions, we remove the restriction of these authors that, for large $\dot{x}$, the damping in the system be positive for both positive and negative $\dot{x}$, and in particular we do not require $R(\dot{x})$ to be odd.

It is convenient to replace (1) by the equivalent system

$$\dot{x} = y, \quad \dot{y} = - R(y) - x \tag{3a, b}$$

which is a special case of the two-dimensional autonomous system

$$\dot{x} = X(x, y), \quad \dot{y} = Y(x, y) \tag{4a, b}$$

where X and Y are real continuous functions defined on the real (x, y) plane and such that (4) is assured of possessing locally a unique solution everywhere on this plane. Some of the well known definitions and results of the Poincaré-Bendixson theory [1] relative to (4) will be now briefly recalled, the terms defined being assumed to refer to (4) unless otherwise stated.

A point of the (x, y) plane is said to be <u>critical</u> if both the right sides of (4) vanish there and it is said to be <u>regular</u> otherwise. If $x(t)$ and $y(t)$ constitute a solution of (4) then for a given t they define the x- and y-coordinates of a <u>representative</u> point $P(t)$ on the (x, y) plane for this solution. The set $\{x(t), y(t)\}$ for all real t, that is the curve on the (x, y) plane which is the locus of all the representative points for this solution, is called an <u>orbit</u>. Uniqueness of the solutions of (4) implies that no two orbits can intersect. A <u>closed orbit</u> (that is an orbit which is represented by a closed curve in the (x, y) plane) made of regular points only is <u>periodic</u> (i.e., it is such that only a finite change in t is required for a representative point to describe a complete circuit on it), and vice-versa. In particular a

---

* Levinson and Smith actually considered the more general equation $\ddot{y} + r(y, \dot{y})\dot{y} + g(y) = 0$. Their results may be readily particularized to (2).

periodic orbit is said to be a stable (unstable) limit cycle if it is approached asymptotically by its neighboring orbits as $t \longrightarrow \infty$ $(-\infty)$. If these remarks are particularized to the system (3), it is clear that, for the purpose of this paper, it is sufficient to determine the conditions for the existence and uniqueness of a periodic orbit of (3).

In order to establish the existence conditions we shall need the conclusions of the following section and also the following general result [4].

LEMMA 1 (Poincaré-Bendixson). If an orbit of (4) remains for $t \longrightarrow \infty$ in a finite region K of the (x, y) plane and if K is free of critical points of (4), then either this orbit is periodic or it approaches asymptotically a limit cycle as $t \longrightarrow \infty$.

2. AN AUXILIARY AUTONOMOUS SYSTEM OF TWO FOCI OR OF A FOCUS AND A NODE

Consider the piecewise linear autonomous system

$$\dot{x} = y, \quad \dot{y} = -R_1(y) - x \tag{5a, b}$$

where

$$R_1(y) = \begin{cases} -\rho_1 y & (y \leq 0) \quad \text{(6a)} \\ \rho_2 y - \gamma & (y > 0) \quad \text{(6b)} \end{cases}.$$

Here, ρ_1, ρ_2 and γ are positive constants and $2 > \rho_1 < \rho_2$.

The patterns of the orbits of this system for $\rho_2 <, =$ and > 2 are illustrated respectively in Figures 1(a), (b) and (c), where we make use of the well-known results for the (x, y) plane solutions of a linear second order differential equation in t with constant coefficients.

Since $0 < \rho_1 < 2$, the origin 0 of the (x, y) plane is an unstable focus for the parts of orbits of (5) lying in the half-plane $y \leq 0$. Hence the arc of an orbit of (5) beginning at an arbitrary point A on the positive x-axis, when followed in the clockwise (increasing t) direction, although getting farther from the origin, must necessarily turn and meet the negative x-axis at some point B. Then from the well-known results alluded in the preceding paragraph, the ratio of the distances OB and OA is

$$\frac{OB}{OA} = e^{\frac{\pi}{2}\frac{\rho_1}{\omega_1}} \tag{7}$$

where

$$\omega_1 = \sqrt{1 - \frac{1}{4}\rho_1^2} \tag{8}$$

From the conditions on ρ_2 it follows that for the parts of orbits of (5) located in the half-plane $y > 0$, the point $(0, \gamma)$, labeled F in Figures 1(a), (b) and (c), is a stable focus or node depending on whether $\rho_2 <$ or ≥ 2. Let us assume first that $\rho_2 < 2$ and refer to Figure 1(a). As we proceed clockwise on the orbit mentioned

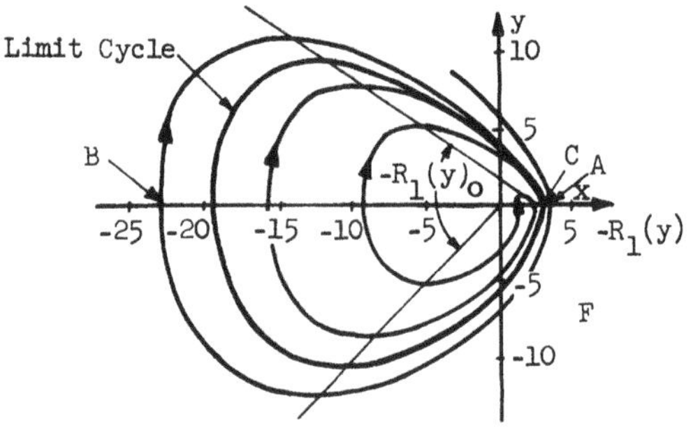

(a) $\rho_1 = 1$, $\rho_2 = \frac{3}{2}$, $\gamma = \frac{5}{2}$

FIGURE 1(a)

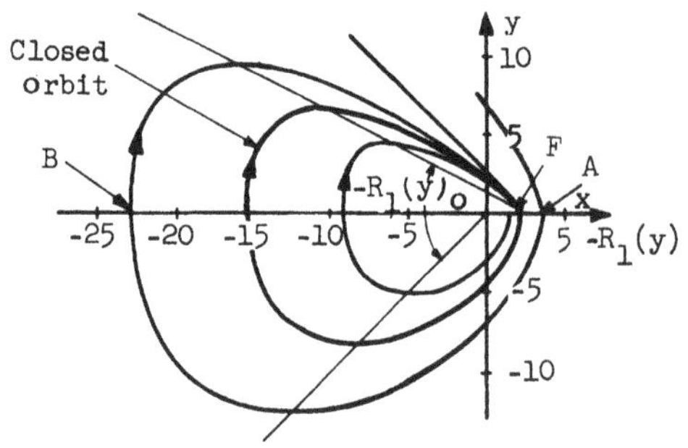

(b) $\rho_1 = 1$, $\rho_2 = 2$, $\gamma = \frac{5}{2}$

FIGURE 1(b)

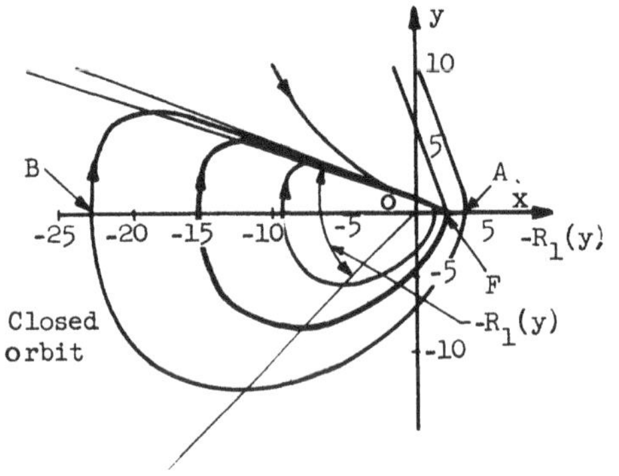

(c) $\rho_1 = 1$, $\rho_2 = 3$, $\gamma = \frac{5}{2}$

FIGURE 1(c)

above byond B, we shall meet the positive x-axis at some point C to the right of F. Then again from the well-known results alluded above we deduce the relationship between the distances OC and OB

$$\frac{OC - \gamma}{OB + \gamma} = e^{-\frac{\pi}{2}\frac{\rho_2}{\omega_2}} \tag{9}$$

where

(10) $$\omega_2 = \sqrt{1 - \frac{1}{4}\rho_2^2}$$

From (7) and (9) there is

$$\frac{OC}{OA} = e^{-\frac{\pi}{2}\left(\frac{\rho_2}{\omega_2} - \frac{\rho_1}{\omega_1}\right)} + \gamma\left(e^{-\frac{\pi}{2}\frac{\rho_2}{\omega_2}} + 1\right) e^{\frac{\pi}{2}\frac{\rho_1}{\omega_1}} \frac{1}{OA}$$

which shows that as OA increases from 0 to ∞, the ratio OC/OA decreases monotonically from ∞ to the value

$$e^{-\frac{\pi}{2}\left(\frac{\rho_2}{\omega_2} - \frac{\rho_1}{\omega_1}\right)}$$

which is positive and less than unity. Hence there is one and only one position of A, say A', at which A and C coincide, this occurring when the ratio (11) is unity, that is, when

(12) $$OA = OA' = \frac{\gamma\left(e^{-\frac{\pi}{2}\frac{\rho_2}{\omega_2}} + 1\right)}{e^{-\frac{\pi}{2}\frac{\rho_1}{\omega_1}} - e^{-\frac{\pi}{2}\frac{\rho_2}{\omega_2}}}$$

In virtue of (11), it is clear that the resulting closed orbit $A'BA'$ is a limit cycle of (5) to which all the other orbits of the system tend as $t \longrightarrow \infty$.

We next assume that $\rho_2 \geq 2$ (see Figures 1(b) and (c)). Then, whatever the position of the point B on the negative x-axis, the orbit through AB always tends to F when followed clockwise from B, since now F acts as a stable node for the arc BF. Thus as A varies its position on the positive x-axis from 0 to ∞, there is only one position of A corresponding to a closed orbit of the system, namely when A coincides with F. All other orbits of the system tend to F (and hence approach the closed orbit FBF) as $t \longrightarrow \infty$.

The above conclusions may be summarized as follows:

LEMMA 2. Under the conditions stated, the system (5) has in the (x, y) plane one and only one closed orbit which is approached by all the other orbits of the system as $t \longrightarrow \infty$.

3. ON THE EXISTENCE OF A PERIODIC SOLUTION

We now state the following result for the original equation (1):

THEOREM 1. The equation (1) has at least one periodic solution of

(a) $R(0) = 0$

(b) $R'(0)$ exists and is negative

and provided there is a $y_o > 0$ such that

(c) $R'(\dot{x}) > 0$, min $(\dot{x} \geq y_o)$

(d) $2 > -\min R'(\dot{x}) < R'(-\dot{x})$, $(\dot{x} \leq -y_o)$

except for the values of $\dot{x}$ at which $R'(\dot{x})$ undergoes simple discontinuities.

REMARK. Below it will be necessary to compare the slopes of the elements of orbits, through a point P in the (x, y) plane, of two systems Σ_1 and Σ_2 of the form

$$\Sigma_1 : \dot{x} = y, \quad \dot{y} = -F_1(y) - x \tag{13}$$

$$\Sigma_2 : \dot{x} = y, \quad \dot{y} = -F_2(y) - x \tag{14}$$

where F_1 and F_2 are real, continuous (piecewise differentiable) functions. For this purpose we let z denote the distance from the origin of the (x, y) plane to a point on an orbit of Σ_1 and Σ_2, i.e.,

$$z = \sqrt{x^2 + y^2} \quad ,$$

which after differentiation with respect to t and combination with (13) or (14) gives

$$\dot{z} = -\frac{1}{z} F_i(y)y, \quad i = 1, 2 \tag{15}$$

This shows that at P, if $y > 0$ and $F_1 < F_2$, $\dot{z}$ is larger along the element of orbit of Σ_1 than along that of Σ_2, in other words, when taken in the direction of increasing t, the first of these elements intersects the second one outward or away from the origin. The same is the case if $y < 0$ and $F_1 > F_2$, the opposite being true if $y > 0$ and $F_1 > F_2$ or $y < 0$ and $F_1 < F_2$.

PROOF. Consider the two-dimensional system (3) equivalent to (1).

Since $R(y)$ is continuous on $-y_0 \leq y \leq y_0$, $|R(y)|$ is bounded by a positive constant M there.

It will be convenient to introduce the variable

$$u = x + M \quad . \tag{16}$$

The conditions (c) and (d) of the theorem allow us to define the positive constants ρ_2 and ρ_1 satisfying

$$\rho_2 = \min R'(y), \qquad (y \geqq y_0) \tag{17}$$

$$2 > \rho_1 < \rho_2 \tag{18}$$

$$\rho_1 \geq -\min R'(y), \qquad (y \leq -y_0) \quad . \tag{19}$$

Finally, let

$$\gamma = \rho_2 y_0 + 2M \quad . \tag{20}$$

ρ_1, ρ_2 and γ thus defined satisfy the conditions assumed for these constants in (6) and hence by Lemma 2, the system

$$\dot{u} = y, \qquad \dot{y} = -R_1(y) - u \tag{21}$$

(22a)
(22b)
$$R_1(y) = \begin{cases} -\rho_1 y, & (y \leq 0) \\ \rho_2 y - \gamma, & (y > 0) \end{cases}$$

has a unique orbit in the (u, y) plane and therefore in the (x, y) plane. We label this orbit Γ_2.

Now consider the annular region K (see Figure 2) in the (x, y) plane bounded outside by Γ_2 and inside by the circle Γ_1 defined by

$$x^2 + y^2 = \delta \tag{23}$$

where δ is a sufficiently small positive constant such that $R'(y) < 0$ in the interior of and on Γ_1.

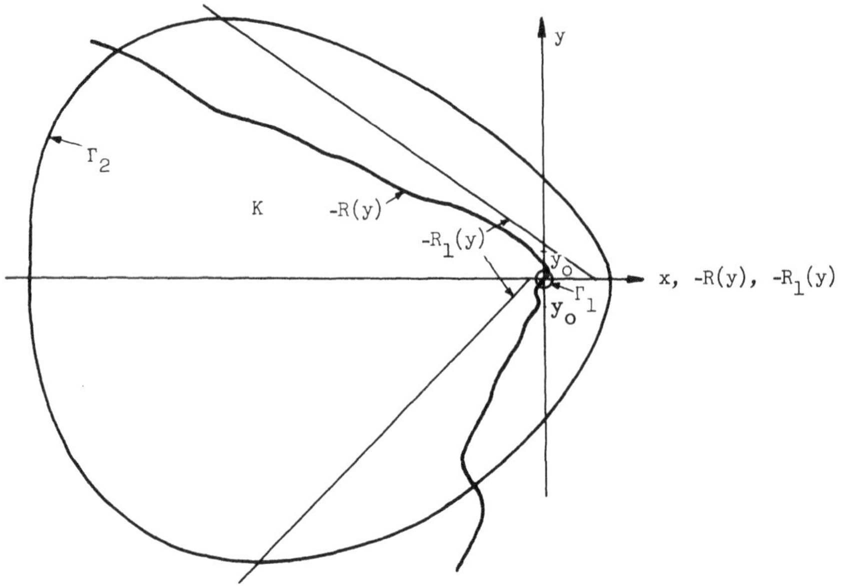

FIGURE 2

Let (21) and (22), after combination with (16) and (20), be re-written as

(24a, b) $$\dot{x} = y, \qquad \dot{y} = -F(y) - x$$

(25a), (25b) $$F(y) = \begin{cases} -\rho_1 y + M, & (y \leq 0) \\ \rho_2(y - y_o) - M, & (y > 0) \end{cases} .$$

Then since $|R(y)| \leq M$ for $0 \leq |y| \leq y_o$ and according to (17) and (18) we have

(26a) $$\text{For } y < 0: \; R(y) \leq -\rho_1 y + M = F(y)$$

(26b) $$\text{For } y > 0: \; R(y) \geq \rho_2(y - y_o) - M = F(y)$$

where the rightmost terms in (26a) and (26b) arise from the definitions (25a) and (25b). It follows, from (26a) and (26b) (when considered together with (3) and (24)) and by virtue of the remark just preceding this proof, that the elements of orbits of (3) on Γ_2, taken in the direction of increasing t, remain in K.

Also, according to the conditions (a) and (b) of the theorem,

$R(y) < 0$ for $y > 0$ and $R(y) > 0$ for $y < 0$. Moreover (23) is a solution of the system $\dot{x} = y$, $\dot{y} = -x$. Hence, by the remark just referred to, the elements of orbits of (3) on Γ_1, when taken in the direction of increasing t, intersect Γ_1 outward from the origin and hence into K.

Thus an arc of orbit of (3) starting at a point on K cannot leave this region for increasing t. Moreover, the origin of the (x, y) plane, which is the only critical point of (3), is external to K. Hence by Lemma 1, K contains at least one periodic orbit of (3). This concludes the proof.

4. UNIQUENESS OF THE PERIODIC SOLUTION

We now make the following proposition:

THEOREM 2. Assume that the conditions of Theorem 1 are satisfied for the equation (1) and let there exist a $y_1 > 0$ such that*

(a) $R(y_1) = R(0) = 0$

(b) $\dot{x}\, R(\dot{x}) < 0, \quad (0 < |\dot{x}| < y_1)$

(c) $R(\dot{x}) > 0, \quad (\dot{x} > y_1)$

(d) $R'(\dot{x}) \geq \frac{1}{\dot{x}} R(\dot{x}), \quad (\dot{x} < 0,\ \dot{x} > y_1)$

except at the values of $\dot{x}$ where $R'(\dot{x})$ undergoes simple discontinuities. Then (3) has a periodic solution which is unique except for the translations in t.

REMARKS. As before, instead of (1), the equivalent system (3) will be considered.

Figure 3 illustrates the direction, for increasing t, of the elements of orbits of (3) under the conditions of this theorem. In this and the following figures, $-R(y)$ is also shown plotted along the axis of the abscissae. It follows from (3) that, for increasing t, the elements of orbits of (3) above the x-axis are directed rightward and those below it leftward; the ones at the left of $-R(y)$ are directed upward and those at its right, downward. The curve $-R(y)$ itself is intersected

* By evident changes in the proof it can be shown that the result of this theorem holds if the conditions (a), (c) and (d) are replaced respectively by (a) $R(-y_1) = R(0) = 0$, (c) $R(\dot{x}) > 0$, $(\dot{x} < -y_1)$, (d) $R'(\dot{x}) \geq \frac{1}{\dot{x}} R(\dot{x})$, $(\dot{x} < -y_1,\ \dot{x} > 0)$.

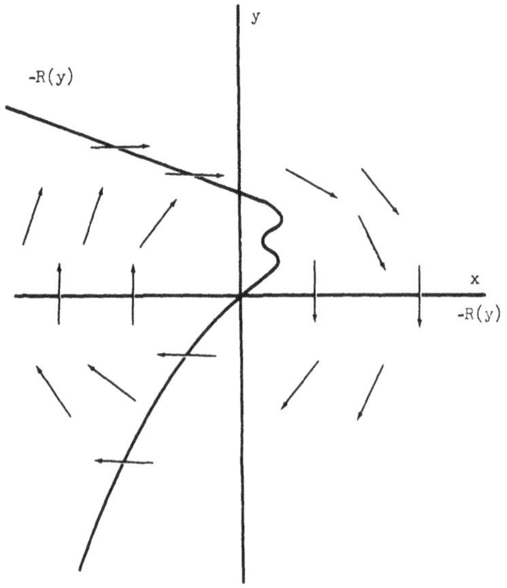

FIGURE 3

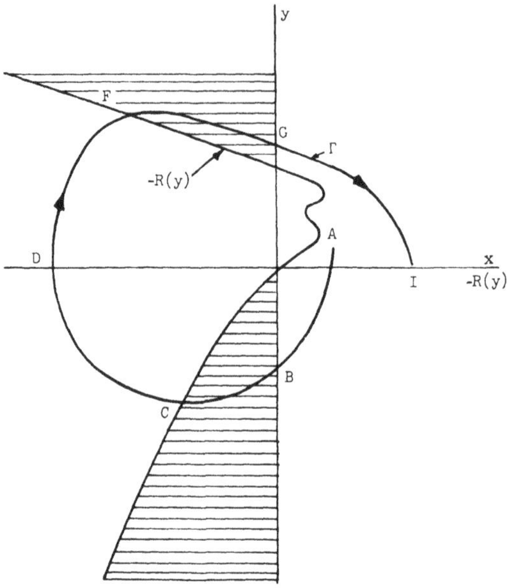

FIGURE 4

by every orbit horizontally and the x-axis is crossed vertically.

Let Γ in Figure 4 represent an arc of orbit of (3) that describes a complete clockwise turn around the origin, leaving the positive x-axis at A and meeting this axis again for the first time at some point I. Then nowhere on Γ is an alement of Γ along a radial direction. In fact, this follows directly from the discussion in the preceding paragraph, for that part of Γ that does not lie in the region between the curve $-R(y)$ and the y-axis (indicated by the hatched area in the Figure). To show that the same holds for the arcs FG and BC of Γ that lie in this region, consider first FG. In the y-interval on which FG is defined there is a $N > 0$ such that $R(y)/y \leq N$. Since an orbit through F of the linear system $\dot{x} = y$, $\dot{y} = -Ny - x$, when followed clockwise from F, does not approach a radial direction in this y-interval, the arc FG cannot certainly approach a radial direction (since $R(y) \leq Ny$). Similar argument holds for BC, but in this case one follows BC counterclockwise from C to B.

If $P(x_P, y_P)$ and $Q(x_Q, y_Q)$ are two points on an orbit of (3), we shall often need to relate their distances OP and OQ from the origin of the (x, y) plane. For this purpose (3) is reduced to a single first-order equation in x and y and then integrated to give

$$OQ^2 - OP^2 = (x_Q^2 + y_Q^2) - (x_P^2 + y_P^2) = -2\int_P^Q R(y)\,dx \tag{27}$$

where the rightmost term is a line integral computed on the orbit from P to Q.

PROOF. Since the conditions of Theorem 1 are satisfied, (3) has at least one periodic orbit. It remains to be proved that this periodic orbit is unique under the present set of restrictions.

Condition (b) of the theorem clearly implies* that the periodic orbit must be partly above and below the band $|y| \leqq y_1$.

Consider the arc of an orbit of (3) ACDFI (see Figure 5) which

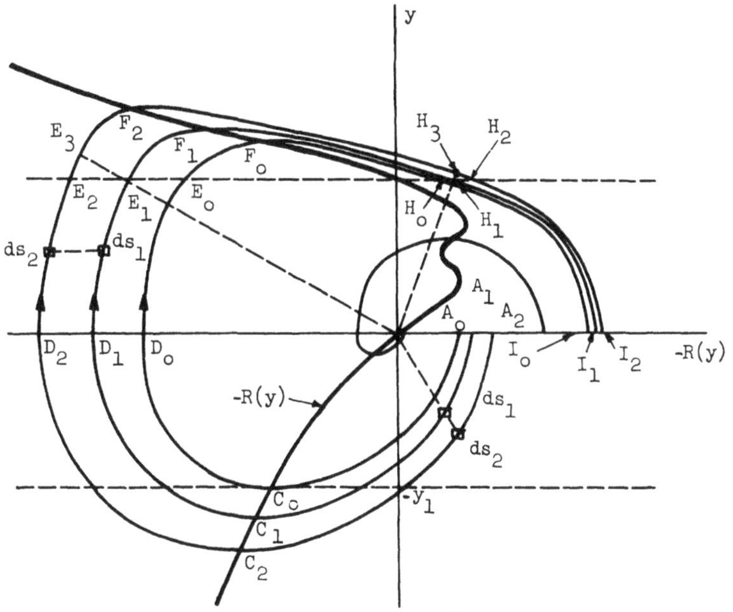

FIGURE 5

starts at a point A on the positive x-axis and continues clockwise from A intersecting: the curve $-R(y)$ for $y < 0$ at C, the negative x-axis at D, the curve $-R(y)$ for $y > 0$ at F, and again the positive x-axis at I. Clearly, C and F are respectively the minimum and maximum of y for this arc.

If C and F (the y-coordinates of which are denoted by y_C

---

* This follows from application of (27) to the arcs of orbits of (3) lying in the band $|y| \leq y_1$, taking into account the condition (b) of the theorem.

and y_F) are inside or at the edges of the band $|y| \leq y_1$, A is certainly at the left* of I. Let $A_oC_oD_oF_oI_o$ be the arc ACDFI such that $\min\{-y_{Co}, y_{Fo}\} = y_1$. The proof of the theorem now proceeds via the following lemma.

LEMMA 3. As the position of the point A moves rightward on the positive x-axis from A_o, the ratio of the distances OI/OA decreases monotonically.

PROOF OF LEMMA 3. Syppose A_1 and A_2 are two points on the positive x-axis such that A_2 is to the right of A_1 and A_1 is to the right of (or coincides with) A_o. The corresponding arcs of orbits $A_1C_1D_1F_1I_1$ and $A_2C_2D_2F_2I_2$ will be denoted by Γ_1 and Γ_2 and the quantities associated with these arcs will be indicated respectively by the subscripts 1 and 2. The lemma is proved by showing that

$$\frac{OI_2}{OA_2} \leq \frac{OI_1}{OA_1} \quad . \tag{28}$$

Let the left- and right-side intercepts of $y = y_1$ with Γ_1 be denoted respectively by E_1 and H_1, and those with Γ_2 respectively by E_2 and H_2. Assume that OE_1 and OH_1 when produced in the directions from O to E_1 and H_1 intersect Γ_2 respectively at E_3 and H_3.

Now the change in the radial distance of a representative point moving clockwise on Γ_1 will be compared to that on Γ_2. For this purpose, each of the parts into which Γ_1 is divided by the points D_1, E_1 and H_1 will be considered together with the corresponding part on Γ_2.

Consider first the arcs $A_1C_1D_1$ and $A_2C_2D_2$.

Introducing the polar coordinates z (already defined) and θ (which is the angle measured in the clockwise direction from the positive x-axis), we write

$$x = z \cos \theta \tag{29a}$$

$$y = - z \sin \theta \tag{29b}$$

with which (3) may be expressed as the single first order equation

$$\frac{dz}{d\theta} = \frac{z R \sin \theta}{z + R \cos \theta} \tag{30}$$

* See footnote on preceding page.

where $R = R(-z \sin \theta)$.

(30) may be conveniently rewritten in the form

$$\frac{d}{d\theta}(\log z) = \frac{R \sin \theta}{z + R \cos \theta} \tag{31}$$

Consider now two differential elements ds_1 and ds_2 belonging respectively to $A_1C_1D_1$ and $A_2C_2D_2$ and lying along the same radial direction, as indicated in Figure 5. For both these elements the terms in θ in the right side of (31) are the same.

Partial differentiation of (31) with respect to z gives

$$\frac{\partial}{\partial z}\frac{d}{d\theta}(\log z) = \frac{\left(\frac{\partial R}{\partial z} z - R\right) \sin \theta}{(z + R \cos \theta)^2} \tag{32}$$

The denominator in the right side of (32) does not vanish at any point on $A_1C_1D_1$ and $A_2C_2D_2$. For if it did, then (30) would imply that the element of $A_1C_1D_1$ or $A_2C_2D_2$, there, is along a radial direction, and, as remarked earlier, this is not possible.

Also, for the elements ds_1 and ds_2

$$\sin \theta > 0 \tag{33}$$

and, according to (29b),

$$\frac{\partial R}{\partial z} z - R = R'(y)y - R \leq 0 \tag{34}$$

the right-most member of (34) resulting from the condition (d) of the theorem.*

(33) and (34) together with (32) imply that

$$\frac{\partial}{\partial z}\left[\frac{d}{d\theta}(\log z)\right] \leq 0 \tag{35}$$

and since ds_2 is at a greater radial distance than ds_1, (35) shows that the right side of (31) is less (or equal) for ds_2 than (and) for ds_1, or symbolically:

---

* Note that since $y < 0$, multiplication by y changes the sign of the inequality given in the condition (d) of the theorem.

$$\left\{ \frac{R \sin \theta}{z + R \cos \theta} \right\}_{ds_2} \leq \left\{ \frac{R \sin \theta}{z + R \cos \theta} \right\}_{ds_1} \tag{36}$$

Integrating (31) along $A_1C_1D_1$ and $A_2C_2D_2$, making, in this integration, the differential elements of these arcs along the same radial direction correspond, and using (36), we obtain

$$\int_{OA_2}^{OD_2} d \log z = \int_{0\,A_2C_2D_2}^{\pi} \frac{R \sin \theta}{z + R \cos \theta} d\theta \leq \int_{0\,A_1C_1D_1}^{\pi} \frac{R \sin \theta}{z + R \cos \theta} d\theta =$$

$$= \int_{OA_1}^{OD_1} d \log z \tag{37}$$

and hence

$$\frac{OD_2}{OA_2} \leq \frac{OD_1}{OA_1} \tag{38}$$

Next we consider the arcs D_1E_1 and D_2E_2 using a modification of Liénard's [2] approach.

Now y is taken as the independent variable (in the line integration to be performed) and a differential element ds_1 on D_1E_1 is made to correspond to the element ds_2 of D_2E_2 having the same y-coordinate, as shown in the figure.

Let (x', y') and (x'', y'') denote the coordinates of ds_1 and ds_2. Since $x'' < x' < 0$ (see Figure 5)), $y' = y'' = y > 0$, and according to the condition (b) of the theorem, $R(y') = R(y'') = R(y) < 0$, we deduce from (3) (by reducing (3) first to a single order equation in x and y and expressing dx in terms of the other variables):

$$dx' > dx'' \geq 0 \tag{39}$$

and

$$R(y)\, dx' \leq R(y)\, dx'' \leq 0 \quad . \tag{40}$$

Then integrating (40) along D_1E_1 and D_2E_2 in the way indicated in (27), we have:

$$OE_1^2 - OD_1^2 = -2 \int_{D_1E_1} R(y)\, dx' \geq -2 \int_{D_2E_2} R(y)\, dx'' = \tag{41}$$

$$= OE_2^2 - OD_2^2 \geq 0$$

or simply

$$OE_1^2 - OD_1^2 \geq OE_2^2 - OD_2^2 \geq 0 \tag{42}$$

which may be rearranged as

$$OD_1^2 \left(\frac{OE_1^2}{OD_1^2} - 1 \right) \geq OD_2^2 \left(\frac{OE_2^2}{OD_2^2} - 1 \right) \quad . \tag{43}$$

Since, according to (42), both sides of (43) are non-negative and because $OD_2 \geq OD_1$ we derive from (43)

$$\frac{OE_1^2}{OD_1^2} - 1 \geq \frac{OD_2^2}{OD_1^2} \left(\frac{OE_2^2}{OD_2^2} - 1 \right) \geq \frac{OE_2^2}{OD_2^2} - 1 \tag{44}$$

and therefore

$$\frac{OE_2}{OD_2} \leq \frac{OE_1}{OD_1} \quad . \tag{45}$$

For the arcs E_2E_3 and H_3H_2 of Γ_2, $R(y) > 0$ (since $y > y_1$). Hence applying (27), separately, to each of these arcs, we get

$$\frac{OE_3}{OE_2} < 1 \tag{46}$$

$$\frac{OH_2}{OH_3} < 1 \quad . \tag{47}$$

The arcs $E_1F_1H_1$ and $E_3F_2H_3$ are treated in the same way as $A_1C_1D_1$ and $A_2C_2D_2$. For those arcs (33) is negative and (34) is positive. Hence, the right side of (32) is again negative and therefore, as in (38), there is

(48) $$\frac{OH_3}{OE_3} \leq \frac{OH_1}{OE_1} \quad .$$

Finally, dealing with the arcs H_1I_1 and H_2I_2 in much the same way as with D_1E_1 and D_2E_2, we deduce

(49) $$\frac{OI_2}{OH_2} \leq \frac{OI_1}{OH_1} \quad .$$

Multiplying together separately the left and right sides of (38), (45), (46), (47), (48) and (49), the result (28) is established.

CONCLUSION OF THE PROOF OF THEOREM 2. It follows from the above lemma that if the points A and I coincide, they can do so only once. Since (3) has at least one periodic orbit (since it is assumed to satisfy the conditions of Theorem 1), A and I must coincide at least once. In other words, (3) possesses one and only one periodic orbit and this orbit corresponds to the periodic solution of (1). This completes the proof of the theorem.

BIBLIOGRAPHY

[1] Coddington, E. A., and Levinson, N., "Theory of Ordinary Differential Equations," McGraw-Hill, (1955), p. 389.

[2] Liénard, A., "Étude des Oscillations Entretenues," Rev. Gen. d'Eléctricité, 23, (1928), pp. 901-912.

[3] Levinson, N., and Smith, O. K., "A General Equation for Relaxation Oscillations," Duke Math. Jnal., 9, (1941), pp. 382-403.

[4] Bendixson, I., "Sur les Courbes Définies par des Équations Différentielles," Acta Mathematica, 24, (1901), pp. 1-88.

APPENDIX

E. Pinney

Errata to "Nonlinear Differential Equations Systems", Contributions to the Theory of Nonlinear Oscillations, vol. III, pp. 31-56. Princeton University Press, 1956.

I. In formulas (2.7), (3.3), (5.1), (8.4), (9.13), (10.1), (10.7), (10.8) replace

$$\frac{\mu_j!}{D^{(\mu_j)}(z_j)} G_1^{(q)}(z_j) \qquad \text{by} \qquad G_{1j}^{(q)}(z_j) \quad .$$

II. In (2.7) replace the upper limit of the second summation sign by $\mu_j - 1$.

III. Replace (2.9) by

$$(2.9) \qquad G_{1j}(z) = -(z - z_j)^{\mu_j} D_1(z)/[D(z)\bar{D}(z)]$$

IV. Replace (2.11) and the material above (2.12) by

$$(2.11) \quad y_1(t) = -\frac{1}{2\pi i} \int_C \sum_{\alpha=1}^{m} \frac{D_{1\alpha}(z)}{D(z)} [e^{zt} b_\alpha + \int_0^t e^{z(t-\tau)} u_\alpha(\tau) d\tau] \, dz \quad ,$$

where C is a contour enclosing the zeros of $D(z)$ but not those of $\bar{D}(z)$. Then

$$\frac{dy_1}{dt} - \sum_{\beta=1}^{m} A_{1\beta} y_\beta = \frac{1}{2\pi i} \int_C \sum_{\alpha=1}^{m} \sum_{\beta=1}^{m} \frac{D_{\beta\alpha}(z)}{D(z)} (A_{1\beta} - \delta_{1\beta} z) [e^{zt} b_\alpha$$

$$+ \int_0^t e^{z(t-\tau)} u_\alpha(\tau) d\tau] dz - \frac{1}{2\pi i} \sum_{\alpha=1}^{m} u_\alpha(t) \int_C [D_{1\alpha}(z)/D(z)] \, dz \quad .$$

By (2.4) the first term on the right vanishes because its integrand has no poles inside C. Therefore

V. Replace equation (2.13) by

$$S_{1\alpha} = -\frac{1}{2\pi i}\int_C [D_{1\alpha}(z)/D(z)]\, dz \quad . \tag{2.13}$$

VI. Replace the material below equation (2.14) and above equation (2.16) by, "To evaluate $S_{1\alpha}$ note that as $z \longrightarrow \infty$,"

VII. Change formula number (2.16) to (2.15).

VIII. Line 11, p. 36 should read, "degree $m - 1$ or $\leq m - 2$ according as ... ."

IX. Replace "inserting (2.16) into (2.15)," by "inserting (2.15) into (2.13),".

X. Omit lines 15-18, p. 36.

XI. Replace line 19, p. 36 by "Solving (2.5) for $D_{1\alpha}(z)$ and inserting in (2.11),"

XII. Replace lines 4 and 5, p. 37 by

$$y_1(t) = -\frac{1}{2\pi i}\int_C \frac{D_1(z)}{D(z)\bar{D}(z)}\, [e^{zt}B(z) + \int_0^t e^{z(t-\tau)}U(z,\tau)d\tau]\, dz \quad .$$

The integrand has poles at the characteristic roots only, so by the theory of residues and (2.9),

$$y_1(t) = \sum_{S_j} \frac{1}{(\mu_1-1)!}\frac{\partial^{\mu_j-1}}{\partial z_j^{\mu_j-1}}\Big\{G_{1j}(z_j)[e^{z_jt}B(z_j) + \int_0^t e^{z_j(t-\tau)}U(z_j, \tau)d\tau]\Big\} \quad .$$

www.ingramcontent.com/pod-product-compliance
Ingram Content Group UK Ltd.
Pitfield, Milton Keynes, MK11 3LW, UK
UKHW030659161224
452542UK00003B/10

9 780691 079332